EARTHQUAKE DISPLACEMENT FIELDS AND
THE ROTATION OF THE EARTH

ASTROPHYSICS AND SPACE SCIENCE LIBRARY

A SERIES OF BOOKS ON THE RECENT DEVELOPMENTS

OF SPACE SCIENCE AND OF GENERAL GEOPHYSICS AND ASTROPHYSICS

PUBLISHED IN CONNECTION WITH THE JOURNAL

SPACE SCIENCE REVIEWS

VOLUME 20

EARTHQUAKE DISPLACEMENT FIELDS AND THE ROTATION OF THE EARTH

A NATO ADVANCED STUDY INSTITUTE

CONFERENCE ORGANIZED BY THE DEPARTMENT OF
GEOPHYSICS, UNIVERSITY OF WESTERN ONTARIO,
LONDON, CANADA, 22 JUNE–28 JUNE 1969

Organizing Committee

Dr. A. E. Beck, *Director of Institute*

Dr. L. Mansinha, *Secretary of Institute*

Dr. D. E. Smylie / Mr. R. W. Tanner / Dr. W. H. Wehlau

Edited by

L. MANSINHA, D. E. SMYLIE AND A. E. BECK

SPRINGER-SCIENCE+BUSINESS MEDIA, B.V.

Library of Congress Catalog Card Number: 72-118130

ISBN 978-94-010-3310-7   ISBN 978-94-010-3308-4 (eBook)
DOI 10.1007/978-94-010-3308-4

FOREWORD

The seeds of this conference were sown with the publication by Press, in 1965, of a paper in which he suggested that the displacement field due to a major earthquake may extend over much greater distances than had been thought possible before. Later on, Mansinha and Smylie pointed out that if Press was correct then, since the redistribution of significant quantities of mass was involved, the inertia tensor of the earth would be altered and thus cause the earth to wobble; this revived the idea that earthquakes might be the long sought source for maintaining the Chandler Wobble. They argued that since earthquakes are sudden events it should be worthwhile trying to determine if there was any correlation between sudden changes in the Chandler term of the pole path and major earthquakes. Furthermore, since displacements occur both before and after an earthquake it might be possible to obtain a few days warning of a major earthquake by making instantaneous observations of the pole path.

Analysis of the data indicated some correlation but, as often happens in science in general and in geophysics in particular, the results were not conclusive because of imperfect theory and the need for more accurate determinations of the pole position. It soon became clear that a meeting between geophysicists and astronomers involved in this type of work would be of mutual benefit. It was not a major objective to solve controversial problems; rather it was hoped that the geophysicists would gain a better appreciation of the problems encountered by the astronomers in obtaining accurate data, and that the astronomers would become sufficiently interested in the objectives of the geophysicists to be stimulated to employ improved instruments or new methods to increase the accuracy of their data. Hence the NATO Advanced Study Institute was organized to bring about detailed discussion in the general areas of the theory and practice pertaining to earthquakes, strain fields, astronomical observations to determine pole positions, and the rotation of the earth.

Because of the controversial nature of many aspects of the topics covered by the Institute, considerable allowance was made for discussions, these were sometimes long, sometimes short, usually polite, occasionally heated but nearly always informative and stimulating. In reporting parts of the discussion after some of the papers, much of this atmosphere has regrettably been lost but we hope that sufficient comes through to give at least some idea of the sense of participation. The editors have been highly selective in the portions of the discussions that are reproduced but it is hoped that these selections are reasonably logical. For instance, some segments of the discussions were made redundant by authors who produced manuscripts that were revised in the light of the question and answer periods; basically, we have tried to retain those sections of the discussion which would add to the papers, or emphasize aspects which the editors felt might be obscure to some groups of readers, or underline areas of continuing differences of opinion, and to convey some of the atmosphere.

Acknowledgments are made to the Science Committee of the North Atlantic Council for the major financial contribution required to organize the Institute; to the National Research Council of Canada and the University of Western Ontario for generous grants to cover additional expenses, particularly those relating to participants from non NATO countries; and to the numerous organizations who by many means, helped their staff attend the Institute.

TABLE OF CONTENTS

DEFORMATION FIELDS: OBSERVATION

PRECISE MEASUREMENT OF THE EARTH'S ROTATION AND POLAR MOTION BY NEW METHODS

LIST OF REGISTERED PARTICIPANTS

Adotevi-Akue, G. M., Oregon State University, Corvallis, Ore., U.S.A.
Alley, C. O., University of Maryland, College Park, Md., U.S.A.
Beck, A. E., University of Western Ontario, London, Canada.
Behari, S., University of Western Ontario, London, Canada.
Bender, P. L., Joint Institute for Laboratory Astrophysics, Boulder, Colo., U.S.A.
Berg, J. W., National Academy of Sciences, Washington, D.C., U.S.A.
Berger, J., Institute of Geophysics, U. of California, La Jolla, Calif., U.S.A.
Binh, N. T., University of Western Ontario, London, Canada.
Borra, H., University of Western Ontario, London, Canada.
Bostrom, R. C., University of Washington, Seattle, Wash., U.S.A.
Broten, N. W., National Research Council, Ottawa, Canada.
Busse, F. H., Max-Planck-Institut für Physik, Munich ,W. Germany.
Cannon, W. H., University of British Columbia, Vancouver, Canada.
Carmichael, C. M., University of Western Ontario, London, Canada.
Carson, J. M., University of Western Ontario, London, Canada.
Chinnery, M. A., Brown University, Providence, R. I., U.S.A.
Couch, R., Oregon State University, Corvallis, Ore., U.S.A.
Crossley, D. J., University of British Columbia, Vancouver, Canada.
De la Cruz, S., University of Toronto, Toronto, Canada.
Deutsch, E. R., Memorial University of Newfoundland, St. John's, Canada.
Dubey, A. C., University of Toronto, Toronto, Canada,
Elias, D. P., Scientific Group for Space Research, Athens, Greece.
Ellis, R. M., University of British Columbia, Vancouver, Canada.
Frangos, A. C., Scientific Group for Space Research, Athens, Greece.
Frost, A. D., University of New Hampshire, Durham, N.H., U.S.A.
Fuchs, K., Geophysikalisches Institut, Karlsruhe, W. Germany.
Gangi, A. F., Texas A & M University, College Station, Tex., U.S.A.
Garg, O. P., University of Saskatchewan, Saskatoon, Canada.
Garland, D. G., University of Toronto, Toronto, Canada.
Gemperle, M., Oregon State University, Corvallis, Ore., U.S.A.
Guinot, B., Bureau International de l'Heure, Paris, France.
Gurkan, T., University of Western Ontario, London, Canada.
Harding-Pederson, G., Memorial University, St. John's, Canada.
Hall, R. G., U.S. Naval Observatory, Washington, D.C., U.S.A.
Hamza, V. M., University of Western Ontario, London, Canada.

Hastie, L. M., University of Toronto, Toronto, Canada.
Haubrich, R. A., Institute of Geophysics, U. of California, La Jolla, Calif., U.S.A.
Hayatsu, A., University of Western Ontario, London, Canada.
Hodgson, J. H., Dominion Observatory, Ottawa, Canada.
Hofmann, R. B., California Dept. of Water Resources, Sacramento, Calif., U.S.A.
Hunter, J. A., University of Western Ontario, London, Canada.
Jady, R. J., University of Exeter, Exeter, England.
Jensen, O. G., University of British Columbia, Vancouver, Canada.
Jobidon, G., University of Western Ontario, London, Canada.
Johnson, L., Institute of Geophysics, U. of California, La Jolla, Calif., U.S.A.
Johnson, S. H., Oregon State University, Corvallis, Ore., U.S.A.
Jones, H. E., Geodetic Survey, Ottawa, Canada.
Kakuta, C., Memorial University of Newfoundland, St. John's, Canada.
Kaula, W. M., Institute of Geophysics, U. of California, Los Angeles, Calif., U.S.A.
Kellner, H. A., European Space Operations Center of ESRO, Darmstadt, W. Germany.
Killeen, P. G., University of Western Ontario, London, Canada.
Koch, F. H., Dominion Observatory, Priddis, Alberta, Canada.
Knight, C., Massachusetts Institute of Technology, Cambridge, Mass., U.S.A.
Lasch, D. K., Texas A & M University, College Station, Tex., U.S.A.
Levy, G., University of Western Ontario, London, Canada.
Lewis, T., University of Western Ontario, London, Canada.
Locke, J. L., National Research Council, Ottawa, Canada.
Major, M. W., Colorado School of Mines, Golden, Colo., U.S.A.
Majumdar, S. C., University of Western Ontario, London, Canada.
Mansinha, L., University of Western Ontario, London, Canada.
Markowitz, Wm., Marquette University, Milwaukee, Wisc., U.S.A.
Marlborough, J. M., University of Western Ontario, London, Canada.
Marlow, R. B., University of Western Ontario, London, Canada.
McCamy, K., IBM Thomas J. Watson Research Centre, New York, N.Y., U.S.A.
McCarthy, D. D., U.S. Naval Observatory, Charlottesville, Va., U.S.A.
Mercer, R., Aerospace Corporation, Los Angeles, Calif., U.S.A.
Mereu, R. F., University of Western Ontario, London, Canada.
Moorhead, J. M., University of Western Ontario, London, Canada.
Myerson, R. J., Princeton University, Princeton, N.J., U.S.A.
Nason, R. D., ESSA Earthquake Mechanism Laboratory, San Francisco, Calif., U.S.A.
O'Hora, N. P., Royal Greenwich Observatory, Hailsham, Sussex, England.
Palmer, H. C., University of Western Ontario, London, Canada.
Patel, J. P., University of Western Ontario, London, Canada.
Paul, M. P., University of Florida, Gainesville, Fla., U.S.A.
Pulpan, H., University of Alaska, College, Alas., U.S.A.
Reiter, L., University of Michigan, Ann Arbor, Mich., U.S.A.
Robinson, J. L., Northern Alberta Institute of Technology, Calgary, Canada.

Rochester, M. G., Memorial University of Newfoundland, St. John's, Canada.
Rodgers, D. A., Brown University, Providence, R. I., U.S.A.
Romig, P. R., Colorado School of Mines, Golden, Colo., U.S.A.
Rossiter, R. J., University of Western Ontario, London, Canada.
Rossiter, J. R., University of Toronto, Toronto, Canada.
Roy, R., University of Western Ontario, London, Canada.
Rudman, A. J., Indiana University, Bloomington, Ind., U.S.A.
Runcorn, S. K., University of Newcastle, Newcastle, England.
Russel, R. D., University of British Columbia, Vancouver, Canada.
Schloessin, H. H., University of Western Ontario, London, Canada.
Singh, S. J., Kurukshetra University, Kurukshetra, India.
Smylie, D. E., University of British Columbia, Vancouver, Canada.
Stacey, F. D., University of Queensland, Brisbane, Australia.
Strange, W. E., Geonautics Inc., Falls Church, Va., U.S.A
Strankay, M , University of Western Ontario, London, Canada
Steiner, H. A., USAF Science Advisory Board, U.S.A.F., Washington, D.C., U.S.A.
Swetnick, M. J., NASA Headquarters, Washington, D.C., U.S.A.
Tanner, R. W., Dominion Observatory, Ottawa, Canada.
Thapar, M. R., University of Western Ontario, London, Canada.
Trask, D., Jet Propulsion Laboratory, Pasadena, Calif., U.S.A.
Turner, D., University of Western Ontario, London, Canada.
Ulrych, T. J., University of British Columbia, Vancouver, Canada.
Vali, V., Boeing Scientific Research Laboratory, Seattle, Wash., U.S.A.
Van Flandern, T. C., U.S. Naval Observatory, Washington, D.C., U.S.A.
Vonbun, F., NASA-Goddard Space Flight Center, Greenbelt, Md., U.S.A.
Walcott, R. I., Dept. of Energy Mines & Resources, Ottawa, Canada.
Wells, F. J., Brown University, Providence, R. I., U.S.A.
Wehlau, A., University of Western Ontario, London, Canada.
White, R. E., University of Western Ontario, London, Canada.
Whitten, C. A., U.S. Department of Commerce, Rockville, Md., U.S.A.
Winkler, G. M., U.S. Naval Observatory, Washington, D.C., U.S.A.
Yen, J. L., University of Toronto, Toronto, Canada.
Yumi, S., International Latitude Observatory, Mizusawa-shi, Japan.

REVIEW

POLAR WOBBLE AND DRIFT: A BRIEF HISTORY

M. G. ROCHESTER

Dept. of Physics, Memorial University of Newfoundland, St. John's, Newfoundland, Canada

Abstract. The history of observations of periodic and secular motion of the Earth's rotation pole relative to a geographic frame of reference, and of theoretical attempts to account for the dynamical characteristics of the several modes of polar motion, is summarized through 1968.

1. Introduction

The history of this subject is tarnished with a certain lack of precision, both in the physical concepts and in the terminology employed, which has not been entirely dispelled by the appearance, nearly ten years ago, of the now-classic monograph by Munk and MacDonald (1960). It is perhaps worthwhile to make some points in this connection.

First of all, the changes in rotation which are directly observed are not those of the Earth, which is a deformable body with oceans and a liquid core, but rather are changes in the angular velocity of what Munk and MacDonald have called a geographic frame of reference. Such a coordinate frame is as nearly fixed as can be with respect to the particular set of observatories being used to measure these changes. Only to the extent that we can regard the crust and mantle as rigid can we accurately refer to the rotation of this frame as 'rotation of the solid Earth'. In these days of sea-floor spreading, continental drift, and convection in the mantle the distinction is no longer quite trivial. Having recognized this we can for brevity henceforth refer to the rotation of 'the Earth'.

The second distinction to be made is between motions of the Earth's axis of rotation with respect to two quite different frames of reference:

(a) Changes in the orientation of the rotation axis in space, i.e. with respect to a frame of reference prescribed by the ecliptic, very nearly. Such motions include precession, nutation, secular change in the obliquity, and sway.

(b) Geographic motion of the pole, i.e. with respect to the geographic frame discussed above. Such motion may be periodic (in which case it is called wobble) or secular (sometimes referred to as drift of the pole). The term 'polar wandering' is reserved to describe the large amplitude geographic motion which has taken place on the geologic time scale.

The literature continues to reveal a looseness of terminology by referring to the Chandler wobble as 'free precession' or 'free nutation'. For the sake of clarity the terms precession and nutation should be restricted to the forced motions of the rotation axis in space, driven by the gravitational torques exerted by the Moon and Sun (and to a far smaller extent by the other planets) on the Earth's equatorial bulge due to its inclination to the ecliptic (obliquity). Sway is the (very small) departure of the rotation axis from the total angular momentum vector of the Earth, which is itself

L. Mansinha et al. (eds.), Earthquake Displacement Fields and the Rotation of the Earth, 3–13.

invariable in space, during polar motion in the absence of external torque. Nutation and sway in obliquity are measured by the change in declination of a star, whereas wobble and polar drift are measured by variation in latitude (Figure 1).

The third point to emphasize is that all these phenomena depend crucially on the fact of the Earth's dynamical ellipticity, i.e. the difference between the principal moments of inertia about its axis of figure and about an axis through the equatorial bulge. Unfortunately even the current literature provides examples of the neglect of the vital restrictions which the equatorial bulge imposes on polar motion.

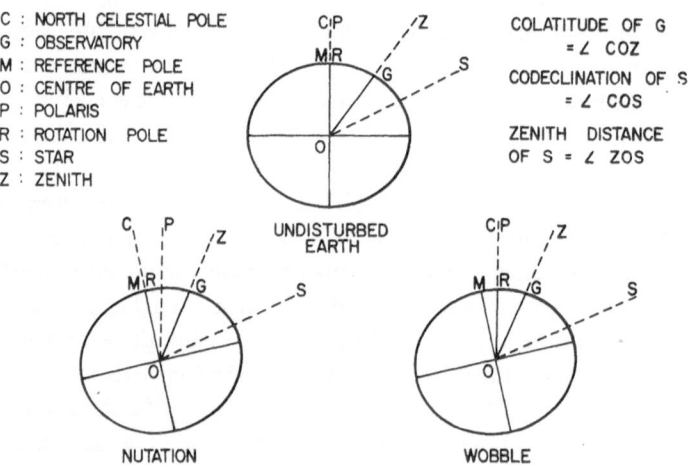

Fig. 1.   Nutation vs. Wobble. (After Munk and MacDonald 1960, Fig. 2.1.)

2. Dynamics of Polar Motion

The rotation of the deformable Earth about its centre of mass is described by the equation of motion

$$\dot{\vec{H}} = \vec{L},$$ (1)

where $\vec{L}$ is the external torque on the system being called 'the Earth', $\vec{H}$ is its total angular momentum, and $(\dot{\square})$ denotes differentiation of $(\square)$ with respect to time. In terms of the instantaneous angular velocity $\vec{\omega}$ of the frame of reference,

$$\vec{H} = \mathbb{I} \cdot \vec{\omega} + \vec{h},$$ (2)

where $\mathbb{I}$ is the inertia tensor of 'the Earth' and $\vec{h}$ is its angular momentum relative to the reference frame. For all but the ample excursions involved in polar wandering, the dynamics of polar motion on the deformable Earth is described in a geographic frame of reference by a set of linearized equations derived from (1)–(2) (Munk and Mac-Donald, 1960, p. 38):

$$\dot{\tilde{m}} = i\sigma_r(\tilde{m} - \tilde{\phi})$$ (3)

where the excitation function $\tilde{\phi}$ is defined by

$$\Omega^2(C - A)\,\tilde{\phi} = \Omega^2\tilde{c} - i\Omega\dot{\tilde{c}} + \Omega\tilde{h} - i\dot{\tilde{h}} + i\tilde{L} \qquad (4)$$

The other symbols used are defined as follows:

$\hat{e}_j$ = unit vector in the x_j-direction in the geographic frame ($j=1, 2, 3$). Conventionally, the geographic axes are prescribed so that $\hat{e}_3$ points nearly along the axis of figure to the reference pole, and $\hat{e}_1$ and $\hat{e}_2$ point to the intersections of the equator with the meridians $0°$ and $90°$ E. (Astronomers, however, use a left-handed coordinate system with $\hat{e}_2$ reversed to describe the motion of the pole.)

$$(\square)_j = (\overrightarrow{\square})\cdot\hat{e}_j = j\text{'th component of the vector } (\overrightarrow{\square})$$

$$(\tilde{\square}) = (\square)_1 + i(\square)_2 = \text{complex equatorial component of } (\overrightarrow{\square})$$

Ω = reference value of ω_3 = mean sidereal rate of spin about the axis of figure = 7.29×10^{-5} rad sec^{-1}

$m_j = \omega_j/\Omega$ = component of displacement of the instantaneous pole from the reference pole, in angular measure

A, C = principal moments of inertia about an equatorial axis, the axis of figure respectively

$$\sigma_r = \left(\frac{C - A}{A}\right)\Omega = \text{Euler frequency}$$

$$\tilde{c} = c_{13} + ic_{23}$$

$$c_{j3} = \hat{e}_j\cdot\mathbb{I}\cdot\hat{e}_3 - C\delta_{j3}.$$

For a rigid Earth (3) reduces to the Euler equations on which the theory of spinning tops and gyroscopes is founded. If there is no torque then

$$\dot{\tilde{m}} = i\sigma_r\tilde{m},$$

which implies the possibility of a periodic motion of the rotation pole on a circle of radius $|\tilde{m}|$ about the reference pole, traced out in the same sense as the Earth's axial spin, with a period

$$T_r = \frac{2\pi}{\sigma_r} = \frac{A}{C - A}\text{ sidereal days}.$$

To an observer fixed on the Earth the rotation axis sweeps out a circular cone of semi-apex angle $\tan^{-1}|\tilde{m}| \simeq |\tilde{m}|$ about the axis of figure $\hat{e}_3$. To an observer at rest in space the Earth itself spins and wobbles about its axis of rotation while the latter remains nearly fixed in direction (not quite, because of sway with amplitude $\simeq (C-A)/A$ times the wobble amplitude $|\tilde{m}|$). Since $(C-A)/A \simeq \frac{1}{305}$ for the Earth, a period of roughly 10 months (the Eulerian period) is indicated for this torque-free wobble.

3. The Chandler Period

Evidence for a variation of latitude with 10-month period was sought by a number of astronomers from time to time during the latter part of the 19th century, but observational errors prohibited a conclusive interpretation of such data as were obtained. In 1891 two events occurred to change the situation:

(1) Förster presented Lord Kelvin with the first decisive evidence for latitude variation, obtained from a year of very accurate synchronous observations made at Berlin and Honolulu (Munk and MacDonald, 1960).

(2) Chandler (1891) announced that a study of all the earlier latitude observations available to him revealed two periodic components in the polar motion – one annual, and the other with a period of about 14 months rather than the expected 10.

In 1892 Newcomb showed that the 14-month Chandler motion was simply the Eulerian free wobble with its period lengthened by the Earth's departure from rigidity. A non-rigid Earth deforms under the centrifugal force due to its own rotation about an axis inclined to the original axis of figure, in such a way as to shift the matter in the equatorial bulge toward the instantaneous equator. This decreases the magnitude of the *effective* dynamical ellipticity and hence increases the period of wobble. The effect can be described by writing

$$\tilde{c} = \frac{k}{k_f} (C - A) \, \tilde{m} \tag{5}$$

where k/k_f denotes the fraction of the equatorial bulge which instantaneously adjusts to the new axis of rotation. k and k_f are Love numbers describing the response of the actual Earth, and of a completely fluid Earth of the same shape and density distribution, to a disturbing force derived from a potential which is a second-degree spherical harmonic (as is the extra centrifugal potential due to wobble).

The natural frequency of wobble for the deformable Earth is found by substituting (5) into (3)–(4): it is

$$\sigma_0 = \left(1 - \frac{k}{k_f}\right) \sigma_r. \tag{6}$$

For $\sigma_0 \simeq 0.84$ cycles per year (cpy), $k \simeq 0.29$. This value of k embraces the effects of the liquid core, which shortens the period of free wobble by about 30 days (Hough, 1895), the oceans, which lengthen it by some 40 days (Larmor, 1915), and the elasticity of the mantle as inferred from seismology, which further lengthens the period by about 120 days (Jeffreys and Vicente, 1957). There is no difficulty in accounting for the net increase in period of the free wobble.

4. Observations of Latitude Variation

The interest in latitude variation which the events of 1891 aroused among astronomers led to the establishment in 1899 of the International Latitude Service (ILS), consisting

of five stations specifically charged with the observation of latitude, spaced out along
a single parallel ($\simeq 39°8'$ N) so as to make use of the same stars, thus eliminating the
effects of errors in catalogues of star positions and proper motions. The longitudes of
the five ILS stations presently operating are shown in Figure 2. Precise determinations
of latitude are therefore now available for a 70 year time span. In 1962 the ILS was
succeeded by the International Polar Motion Service (IPMS), whose Central Bureau
at Mizusawa collects and reduces data from the five ILS observatories. Unfortunately
there is as yet no second chain of observatories spaced out along another parallel of
latitude, though there are now a number of pairs of stations able to observe the same
stars at zenith.

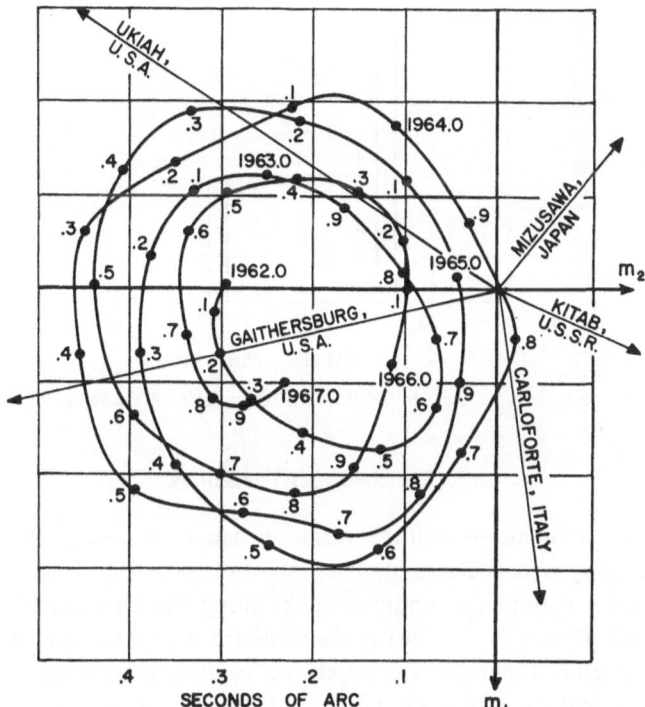

Fig. 2. The motion of the pole 1962.0 – 1967.0 (IPMS stations). (After Mueller, 1969, Fig. 4.9.)

The Bureau International de l'Heure (BIH) at Paris publishes latitude measurements obtained from several score of instruments, both photographic zenith tubes
(PZT) and Danjon astrolabes, scattered over the globe. The disadvantage that these
measurements are made from observations of a large number of different sets of stars
is offset by the statistical advantage of data from a large number of stations.

A typical polar path is shown in Figure 2. Both the IPMS and the BIH now refer
polar motion to the Conventional International Origin (CIO), which was the mean
pole position during 1900–1905.

There are a variety of ways of handling the raw data. Rudnick (1956) carried out a spectral analysis of a 54.4 year record of observations, computing the power in units $(0\rlap{.}''01)^2$ contributed by the harmonics between the 40th and 62nd, for both forward and retrograde motion of the pole. The annual line and Chandler peak are clearly distinguished (Figure 3). The rms amplitude at the Chandler frequency is $\simeq 0\rlap{.}''14$. Rudnick used data collected up to 1945. Now that a significantly longer record of ILS observations is available it would be worthwhile to recompute the power spectrum.

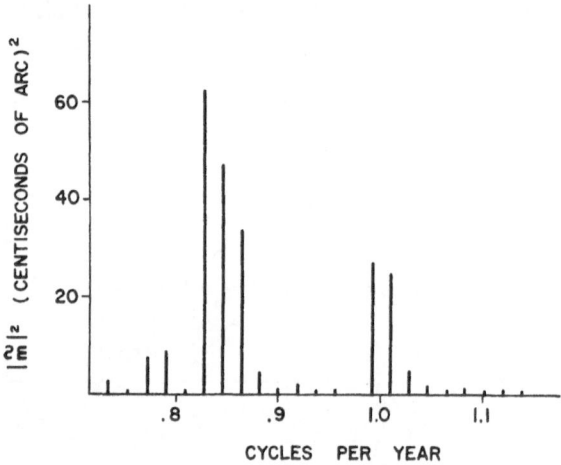

Fig. 3.   Fourier spectrum of latitude variation. (From Rudnick, Table 1.)

5. The Seasonal Variations

The seasonal components of latitude variation can be removed from the data by subtracting monthly means. The annual motion of the pole is quite well represented by an ellipse with its major axis, oddly enough, along the Greenwich meridian! The amplitude is $\simeq 0\rlap{.}''09$, and is very nearly the same from year to year. As long ago as 1901 Spitaler concluded that this was largely due to the seasonal changes in the inertia tensor of the atmosphere, whose effect is amplified by the annual period's proximity to resonance (i.e. the Chandler period). Subsequent studies, the most recent of which was carried out by Munk and Hassan (1961), have confirmed this result. Significant contributions to the annual wobble are made also by snow load, seasonal changes in groundwater distribution, and shifts in ocean mass induced by wind stress. While the latter are difficult to estimate there appears to be no problem in adequately accounting for the observed excitation.

Likewise a semiannual wobble, with a tenth the amplitude of the annual component, presents no difficulty. While in principle the wobble spectrum should show lines at all the lunar and solar tidal frequencies, due to the asymmetry of continents and oceans, all but the Chandler, annual and semiannual components appear to be well below noise level.

6. Diurnal Wobble

The presence of the liquid core not only affects the Chandler frequency through the Love number k, but also provides an extra degree of freedom in wobble. The core and mantle can be regarded as a system of coupled oscillators having two normal modes, one the Chandler wobble and the other with a period differing from the sidereal day by a fraction of the order of the ellipticity of the core-mantle boundary, i.e. by a few hundred seconds. This diurnal mode was predicted independently by Sludskii and by Hough in 1895. The most recent theoretical discussions are by Jeffreys and Vicente (1957) and Molodenskii (1961), based on seismologically-inferred models of the density and elasticity distributions in the Earth's interior. The diurnal mode of free wobble is characterized by much greater relative motion between the core and mantle than takes place in the Chandler mode.

Popov (1963), Thomas (1964), and Kulagin and Kovbasyuk (1965) claim to have detected latitude variation with the period $23^h56^m54^s$ deduced for the diurnal wobble from one of the Earth models considered by Molodenskii. Vicente and Jeffreys (1964) discuss the question. The calculated amplitudes ($0\rlap{.}''006$ to $0\rlap{.}''020$) are hardly above noise level and may be seriously affected by the period assumed, which is very close to one of the main diurnal tides (K_1).

7. Damping of the Chandler Wobble

The irregularity of the polar motion, even with the seasonal components removed, makes an exact determination of the Chandler period difficult. The most recent analysis of the unsmoothed ILS data (1899–1966) by Jeffreys (1968) yields 434 sidereal days, with a probable error of ± 2 days. The literature still from time to time refers to the concept of a Chandler period which varies with time by as much as 10%. Solely from the point of view of data analysis, an instantaneous Chandler period which varies with time is completely equivalent to broadening of the Chandler peak in a frequency analysis such as that carried out by Rudnick. There is however a strong physical argument against the concept of such a large variation in σ_0, since it would require equally large changes in k in only a few years, changes for which there is no imaginable mechanism.

One therefore looks for another explanation of the broadening of the spectral peak centred at the Chandler frequency. The most obvious suggestion is that it is due to damping. Indeed Rudnick and others have assumed that the wobble spectrum could be represented by regarding the Earth as a damped linear oscillator tuned to the Chandler frequency and irregularly excited. This is equivalent to replacing σ_0 by $\sigma_0 + i/\tau$ in the equation of motion, where τ is the relaxation time associated with the damping. On this basis values of τ ranging anywhere from 2 years (Walker and Young, 1957) to 90 years (Panchenko, 1959) have been obtained. Jeffreys (1968) finds $\tau \simeq$ 23 years, with the range allowed by the standard error between 14 and 70 years. The distribution is skewed and Jeffreys is inclined to suppose that $\tau > 30$ years.

The 'figure of merit' of the Earth oscillator, Q, is defined by the fractional energy loss per cycle due to damping:

$$\frac{1}{Q} = \frac{1}{2\pi E} \oint_{\text{cycle}} - \dot{E} \, dt \simeq \frac{2}{\sigma_0 \tau} \tag{7}$$

where $E \simeq (C-A)\,\Omega^2 |\tilde{m}|^2$ is the total energy of wobble. A damping time of 11 years (Rudnick) corresponds to a Q of 30; for $\tau \simeq 30$ years, $Q \simeq 80$.

The mean dissipation rate, of the order $10^{15}/Q$ erg sec^{-1}, is minute compared to that involved in tidal friction, for example, but the location of the energy sink for the Chandler wobble damping has not yet been conclusively identified. Friction at the ocean floor, imperfections of elasticity in the mantle, and viscous or electromagnetic dissipation at the core-mantle boundary have each at one time or another been suggested as candidates, and most have been rejected and revived at least once in the history of the problem. Even if all the rotational inertia of the core is available to take up wobble angular momentum from the mantle, purely electromagnetic damping would take at least 10^5 years (Rochester and Smylie, 1965). Calculations by Munk and MacDonald (1960, pp. 170–171) indicate that the core's viscosity would need to exceed 10^9 cm^2 sec^{-1} to provide the necessary damping. While such a value may not be absolutely ruled out (the viscosity of the core is perhaps its most enigmatic property) it is regarded by most geophysicists as extremely, if not impossibly, high (Rochester, 1970, this volume p. 136). Jeffreys dismissed ocean bottom friction as inadequate but this mechanism has been reinstated as a possible contender by Munk and MacDonald (1960, p. 172) and deserves a deeper study than it has yet been given.

Whether or not anelastic dissipation in the body of the mantle is adequate to absorb all the Chandler wobble energy depends critically on the Q of wobble. Values for Q of several hundred are deduced from the free oscillations of the solid Earth, and Q is generally considered to be nearly independent of frequency. If the wobble Q is of the order 100 or less, it becomes difficult to ascribe the damping entirely to the mantle. This provides an additional argument for referring the damping back to the oceans (Lagus and Anderson, 1968).

A different interpretation of the broadening of the Chandler peak in the power spectrum has been proposed by Colombo and Shapiro (1968). If the mantle itself is regarded as two separate oscillators coupled through a region of low viscosity between 400 and 1000 km depth, the broad peak associated with a single Chandler frequency would be split into two sharper peaks in close proximity. This would remove the problem of an anomalously low Q if the oceans cannot damp the wobble. However, the details of the dynamics envisaged by Colombo and Shapiro have not yet been published, and one wonders about the effect of such a structure on the forced nutations. It is perhaps worth noting that a reported bifurcation of the Chandler frequency obtained from Yashkov's (1965) analysis of latitude observations has been examined and dismissed by Fedorov and Yatskiv (1965).

8. Excitation of the Chandler Wobble

The record of observations of the Chandler wobble is already longer than the most likely damping time, hence the wobble must somehow be regenerated. The irregularity of the pole path suggests that the excitations themselves are irregular. In principle, of course, any excitation which is even slightly aperiodic will generate some response at the natural frequency of an oscillator.

The obvious place to look is in the departure of the changes in the atmospheric inertia tensor from purely seasonal periodicity. For a long time geophysicists were satisfied that sufficient power to sustain the Chandler wobble was contained in the sidebands of the annual line in the frequency spectrum of the air mass distribution over the Earth's surface. Then Munk and Hassan (1961), using 75 years of atmospheric pressure data, showed that atmospheric fluctuations at the Chandler frequency failed by a factor of 10 to 100 to provide the necessary excitation. They argued that changes in the ocean load on the solid Earth are even less effective in generating the Chandler wobble, since they contribute less than the atmosphere to the annual wobble. Sekiguchi (1966) disagrees, and attributes excitation of the Chandler wobble primarily to the oceans, but his argument is difficult to follow.

As the geomagnetic secular variation fields penetrate the electrically-conducting lower mantle from the core, angular momentum is transferred across the core-mantle boundary by means of electromagnetic torques. Rochester and Smylie (1965) found that this mechanism was inadequate by a factor of 10^3 to excite the observed amplitude of wobble. However a proposal permitting the wobble energy to be ultimately derived from the precessional stirring of the core, through electromagnetic core-mantle coupling of a non-linear kind, is outlined at this Institute by Stacey (1970, this volume p. 176).

In 1960 Munk and MacDonald (1960, pp. 163–164) presented a simple argument based on crustal block movement to rule out, by a factor of 10^2 to 10^4, generation of wobble by changes in the inertia tensor accompanying earthquakes. After the 1964 Alaska earthquake Press (1965) pointed out that the displacement fields associated with major earthquakes were much more extensive than had been realized before. Mansinha and Smylie saw that this would involve much greater changes in the products of inertia than Munk and MacDonald had envisaged, and in their preliminary study (1967) concluded that the cumulative effects of large earthquakes could sustain the observed amplitude of the Chandler wobble against damping. That only 10^{-4} of the energy annually released by earthquakes is required to maintain the wobble at first suggests that a significant correlation between the two phenomena is unlikely to be found. Nevertheless any change in the pole path associated with a major seismic event should be sensitive to the geographical location of the epicentre and the nature of the fault movement. According to Mansinha and Smylie (1968) the BIH data (1957–1968) do show a significant correlation between breaks in the non-seasonal pole path and the times of occurrence of large earthquakes. This Institute's *raison d'être* is largely to be attributed to the revival of earthquakes as a possible source of detectable polar motion, but a glance at the programme of papers in this volume shows that the hypothesis is not yet firmly re-established.

9. Secular Motion of the Pole

Although a drift of the mean pole of epoch, at a rate of some 0″.003/year toward New-foundland, was claimed by Wanach as long ago as 1917, the reality of a general trend of this kind was questioned until relatively recently. This was due in part to different interpretations of the data in terms of horizontal displacements of individual stations, or to attributing apparent drift of the pole to asymmetries and irregularities in the observing programme of the ILS. Since the late 1950s the reality of a secular motion of the pole has come to be acknowledged, particularly after reworking the data adopting the CIO as the reference pole. However, there is still uncertainty about the rate and direction of drift, owing to differences both in the data used and in the treatment of it. Yumi and Wako (1968) obtain an average rate of drift since 1900 of 0″.002/year in the direction 78 °W, while Jeffreys (1968) on the other hand finds no significant trend till the 1940s and since then a movement of the pole by nearly 0″.2 in the general direction of Greenland. The cause of this progressive drift could be changes in sea level due to glacial melting (Munk and MacDonald, 1960, pp. 233–236) or earthquakes (if Mansinha and Smylie are correct).

Markowitz (1960) has suggested that there is, superimposed on the secular drift (for which he obtains a rate of 0″.003/year in the direction 60°W), a libration of the pole with period 24 years and amplitude 0″.02 along the meridian 58°E–122°W. Until now the evidence for this remarkable feature, which might be christened the *Markowitz wobble*, has been purely empirical, but Busse (1970, this volume p. 88) has suggested that it may be due to the presence of the Earth's solid inner core, wobbling inside its liquid matrix and communicating its motion to the mantle through inertial coupling.

Secular motion of the pole, and continental drift, require several decades to be detected by present astronomical techniques. Their far-reaching implications spur enormous interest in the new devices now being developed, which promise to enable much more rapid detection, some of which are described in other papers in this volume.

References

Busse, F. H.: 1970, 'The Dynamical Coupling between the Inner Core and Mantle of the Earth and the 24-Year Libration of the Pole', this volume, p. 88.
Chandler, S.: 1891, 'On the Variation of Latitude', *Astron. J.* **11**, 65 and 83.
Colombo, G. and Shapiro, I. I.: 1968, 'Theoretical Model for the Chandler Wobble', *Nature* **217**, 156.
Fedorov, E. P. and Yatskiv, Ya. S.: 1965, 'The Cause of the Apparent 'Bifurcation' of the Free Nutation Period', *Soviet Astron.* **8**, 608.
Hough, S. S.: 1895, 'The Oscillations of a Rotating Ellipsoidal Shell Containing Fluid', *Phil. Trans. Roy. Soc. London*, **A186**, 469.
Jeffreys, H.: 1968, 'The Variation of Latitude', *Monthly Notices Roy. Astron. Soc.* **141**, 255.
Jeffreys, H. and Vicente, R. O.: 1957, 'The Theory of Nutation and the Variation of Latitude', *Monthly Notices Roy. Astron. Soc.* **117**, 142.

Kulagin, S. G. and Kovbasyuk, L. D.: 1965, 'Diurnal Free Nutation from Observations at the Gor'kii Station', *Soviet Astron.* **8**, 603.

Lagus, P. L. and Anderson, D. L.: 1968, 'Tidal Dissipation in the Earth and Planets', *Phys. Earth Planetary Interiors* **1**, 505.

Larmor, J.: 1915, 'The Influence of the Oceanic Waters on the Law of Variation of Latitudes', *Proc. Lond. Math. Soc., Ser. 2*, **14**, 440.

Mansinha, L. and Smylie, D. E.: 1967, 'Effect of Earthquakes on the Chandler Wobble and the Secular Pole Shift', *J. Geophys. Res.* **72**, 4731.

Mansinha, L. and Smylie, D. E.: 1968, 'Earthquakes and the Earth's Wobble', *Science* **161**, 1127.

Markowitz, W.: 1960, 'Latitude and Longitude, and the Secular Motion of the Pole', in *Methods and Techniques in Geophysics* (ed. by S. K. Runcorn), Vol. I, Publishers John Wiley and Sons – Interscience, p. 325.

Molodenskii, M. S.: 1961, 'The Theory of Nutation and Diurnal Earth Tides', *Comm. Obs. Roy. Belgique* **188**, 25.

Mueller, I. I.: 1969, *Spherical and Practical Astronomy as Applied to Geodesy*, Frederick Ungar, New York.

Munk, W. H. and Hassan, E. S. M.: 1961, 'Atmospheric Excitation of the Earth's Wobble', *Geophys. J.* **4**, 339.

Munk, W. H. and MacDonald, G. J. F.: 1960, *The Rotation of the Earth*, Cambridge University Press.

Newcomb, S.: 1892, 'Remarks on Mr. Chandler's Law of Variation of Terrestrial Latitudes', *Astron. J.* **12**, 49.

Panchenko, N. I.: 1959, 'Discussion', *Astron. J.* **64**, 94.

Popov, N. A.: 1963, 'Nutational Motion of the Earth's Axis', *Nature* **198**, 1153.

Rochester, M. G.: 1970, 'Core-mantle Interactions: Geophysical and Astronomical Consequences', this volume, p. 136.

Rochester, M. G. and Smylie, D. E.: 1965, 'Geomagnetic Core-Mantle Coupling and the Chandler Wobble', *Geophys. J.* **10**, 289.

Rudnick, P.: 1956, 'The Spectrum of the Variation in Latitude', *Trans. Am. Geophys. Union* **37**, 137.

Sekiguchi, N.: 1966, 'On the Damping Coefficient of the Polar Motion', *Publ. Astron. Soc. Japan* **18**, 116.

Stacey, F. D.: 1970, A Re-Examination of Core-Mantle Coupling as the Cause of the Wobble, this volume p. 176.

Thomas, D. V.: 1964, 'Evidence for a Nearly Diurnal Term in the Nutation of the Earth's Axis', *Nature* **201**, 481.

Vicente, R. O. and Jeffreys, H.: 1964, 'Nearly Diurnal Nutation of the Earth', *Nature* **204**, 120.

Walker, A. M. and Young, A.: 1957, 'Further Results on the Analysis of the Variation of Latitude', *Monthly Notices Roy. Astron. Soc.* **117**, 119.

Yashkov, V. Ya.: 1965, 'Spectrum of the Motion of the Earth's Poles', *Soviet Astron.* **8**, 605.

Yumi, S. and Wako, Y.: 1968, 'On the Secular Motion of the Mean Pole', in *Continental Drift, Secular Motion of the Pole, and Rotation of the Earth* (ed. by W. Markowitz and B. Guinot), D. Reidel Publishing Company, Dordrecht, Holland, p. 33.

ELASTICITY THEORY OF DISLOCATION

EARTHQUAKE DISPLACEMENT FIELDS

MICHAEL A. CHINNERY

Dept. of Geological Sciences, Brown University, Providence, R.I., U.S.A.

Abstract. The standard model for the calculation of the displacement fields of earthquakes is briefly reviewed. The source dimensions of a set of 27 strike-slip faults are used to obtain empirical relationships between fault dimensions and earthquake magnitude. It is shown that only large earthquakes contribute significantly to the displacement fields at large distances from a fault zone. The mechanics of fault slippage, and some of the problems of the elastic rebound theory are discussed. It is shown that earthquakes on some portions of the San Andreas system appear to account for only a small part of the total movement on the zone indicated by geodetic measurements. Furthermore, these earthquakes are very inefficient in relieving the regional strain indicated by displacement of triangulation stations. It is concluded that aseismic creep in the upper portions of the crust must be an important contribution to the fault slippage in some regions. Also, the regional strain can only be relieved by movements in the lower crust and upper mantle. It is suggested that these movements almost certainly precede surface breakage. They could be detected by geodetic observations at 30–50 km from the fault, and might give valuable premonitory indications of destructive earthquakes.

1. Introduction

The displacement fields that accompany surface faulting are of interest for several reasons. Kasahara (1957), Chinnery (1961), and Chinnery and Petrak (1968) showed how the displacement of triangulation stations near a fault can be used to estimate the depth to which fault movement extends. In this way it was shown that, even in the case of large earthquakes, the observed offset at the surface seldom extends any deeper than 10 or 15 km, and in some cases only penetrates a few km.

Displacements at larger distances were studied by Press (1965) and Chinnery (1965). Press showed that appreciable displacements occur at distances of thousands of kilometers from a large earthquake, and that measurable strain changes are expected at teleseismic distances. He showed observational evidence that such permanent strain changes do indeed occur. Chinnery pointed out that the vertical displacements associated with strike-slip faulting are comparable in size to the horizontal movements at large distances, and that they may have an important bearing on large scale tectonics.

For the purposes of this symposium, we are interested in the mass transport associated with faulting, and its effect on the moments of inertia of the earth. Mansinha and Smylie (1967) have shown, using Press's results, that the redistribution of mass accompanying a large shallow earthquake may be sufficient to perturb the axis of figure of the earth, and excite the Chandler Wobble.

The purpose of the present paper is to examine the existing calculations of displacement fields in the light of recent ideas about earthquake mechanism and the nature of the slip in fault zones. We give first a brief review of the model presently used, then a discussion of the change in displacement field with magnitude, and then some more speculative ideas about the contribution of aseismic creep to the overall movement in fault zones and to the teleseismic displacements.

L. Mansinha et al. (eds.), Earthquake Displacement Fields and the Rotation of the Earth, 17–38.
All Rights Reserved. Copyright © 1970 by D. Reidel Publishing Company, Dordrecht-Holland

2. Elastic Dislocation Theory

The theory which lies at the basis of the calculation of the displacement fields associated with a fault movement is the elasticity theory of dislocations, which has been
described by Steketee (1958a, b). He showed that a dislocation surface, which is a
surface of discontinuity in displacement, when placed into an elastic medium, may
be viewed as a distribution of 'nuclei of strain' (Love, 1944). These 'nuclei of strain'
are fundamental solutions of the Navier equations of linear elasticity, and are analogous in many ways to the magnetic dipole in potential theory. However, instead of
being a vector quantity like a magnetic dipole, they are described by a symmetrical
second rank tensor, and exist in six basic forms.

If a nucleus of strain is introduced into an infinite elastic medium, it will cause
displacements throughout the medium. These will have the form of a third rank
tensor, which is usually written ω_{ij}^k. Here the i and j specify the type and orientation
of the nucleus of strain, and the k suffix refers to the three components of displacement at a point in the medium. If the movement on the dislocation surface corresponds
to a shear fault, then those nuclei with $i \neq j$ are involved; if we are dealing with the
opening or closing of a crack, then those nuclei with $i = j$ are concerned. The analytical
expressions for the ω_{ij}^k have been given by Maruyama (1964).

If a dislocation surface Σ is created within an elastic solid bounded by a surface S,
which was in equilibrium in a strained state, and if the surface forces on S are unchanged by the introduction of Σ, then the displacements u_k of the elastic solid are
given by Steketee (1958b) to be

$$u_k = \frac{1}{8\pi\mu} \left\{ \iint_\Sigma \Delta u_i \omega_{ij}^k v_j \, d\Sigma - \iint_S u_i \omega_{ij}^k v_j \, dS \right\}, \tag{1}$$

where μ is the rigidity, Δu_i is the offset on Σ, and v_j are the direction cosines of the
normals to $d\Sigma$ and dS. The u_i in the second term are the displacements of the surface
S produced by the introduction of Σ into the medium.

If S recedes to infinity, while Σ remains finite, then the displacements u_i on S
become zero, and the second term vanishes. The remaining term is the expression
usually quoted. ω_{ij}^k are the displacements at an observation point in the medium
due to a single nucleus, Δu_i is the 'magnitude' of the nucleus, and the integral is
carried out over the dislocation surface Σ. The next step in applying this theory to the
earth is to include the effect of a plane stress-free boundary. There are two ways of
doing this. One could use the above Equation (1) and include the plane boundary
as part of the surface S, but this is algebraically complex, and apparently has never
been done. A much easier way is to view the problem as an analogy of the Green's
function method of potential theory and modify the ω_{ij}^k to include the boundary
conditions. Thus if we know the displacement field τ_{ij}^k of a nucleus of strain in a
semi-infinite elastic medium, we may use this in Equation (1) instead of ω_{ij}^k. The
τ_{ij}^k can be evaluated from the results of Mindlin (1936) and have been given in analyti-

cal form by Mindlin and Cheng (1950), and Maruyama (1964). We shall not repeat these expressions here.

The displacements caused by the introduction of a dislocation surface Σ into a semi-infinite elastic medium, where the remaining parts of the surface S are at infinity, may therefore be written

$$u_k = \frac{1}{8\pi\mu} \iint_\Sigma \Delta u_i \tau_{ij}^k v_j \, d\Sigma. \tag{2}$$

A simple application of this formula would be to the evaluation of the displacements associated with a vertical strike-slip or dip-slip fault, where movement occurs over a rectangular shape. The integral can be evaluated analytically if the offset Δu_i is constant over the face of the fault. This has been done for the near field of strike-slip faults by Chinnery (1961), for a variety of fault models by Maruyama (1964), and for the far field by Press (1965), who considered both dip-slip and strike-slip faults.

Press's results were of considerable interest, because the displacements at teleseismic distances turned out to be quite large. If, for example, an offset of 5 m occurs on a strike-slip fault of length 200 km and depth 10 km (corresponding roughly to an earthquake of magnitude 7.5–8), at 1500 km from the fault, the horizontal displacement turns out to be about 1.5 mm and vertical displacement is about 0.5 mm. Clearly a large amount of movement of material is associated with such an earthquake, and this suggests the possibility of detecting earthquakes with instruments which will respond to permanent changes in displacement and strain. Press has called this field 'zero frequency seismology'. He showed records of strain instruments that did indeed appear to show a permanent strain change at very large distances from earthquakes. It is worth noticing that his results indicate that strike-slip faults produce larger teleseismic displacements than dip-slip faults.

3. Displacement Fields and Earthquake Magnitude

In order to be able to calculate the displacement fields produced by an earthquake, we need to know the dimensions of the associated fault movement, which will, of course, vary with magnitude. The notation that we shall use for these dimensions is as follows: L = total length of the fault, W = width or depth of faulting, D = displacement or offset.

We have no theory presently available that will permit us to relate the magnitude of an earthquake to its source dimensions. This must be done empirically by taking the known parameters for a set of earthquakes, and trying to establish some empirical relationships. We shall do this for a set of 27 strike-slip faults which have recently been considered by Chinnery (1969). This set of data includes virtually all examples of faulting associated with earthquakes where estimates of the parameters are available. Dip-slip fault data are not included because they appear to be somewhat less reliable, and indeed there may be a fundamental difference between the mechanisms of strike-slip and dip-slip faults. In the remainder of this paper we shall be concerned primarily with strike-slip fault movement.

Figure 1 shows the relationship between $\log L$ and the magnitude M. In this, and succeeding diagrams, no distinction is made between the surface wave magnitude M_s and the local Richter magnitude M_L. There is obviously some question here as to whether the relationship between $\log L$ and M is single valued or not. Since there is often some difficulty in measuring the length L, and the surface breakage may not be representative of conditions at depth, it is hard to make a positive statement about this. For the present, we shall assume that there is a relationship between these two quantities, and the solid curve shown on Figure 1 is probably reasonably close to such a relationship, if it exists. Figure 2 shows the data for $\log W$ plotted against magnitude. The number of earthquakes for which an estimate of W is available is small, and in many cases is probably unreliable. However, the solid line shown on this graph gives some kind of an estimate for the variation of W with M.

The third parameter is the displacement D. $\log D$ is shown plotted against M in Figure 3. The resulting graph is an interesting one, which appears to be linear. The scatter of data points at low magnitude may not be real, since small earthquakes frequently occur as sudden movements during a period of accelerated creep, and

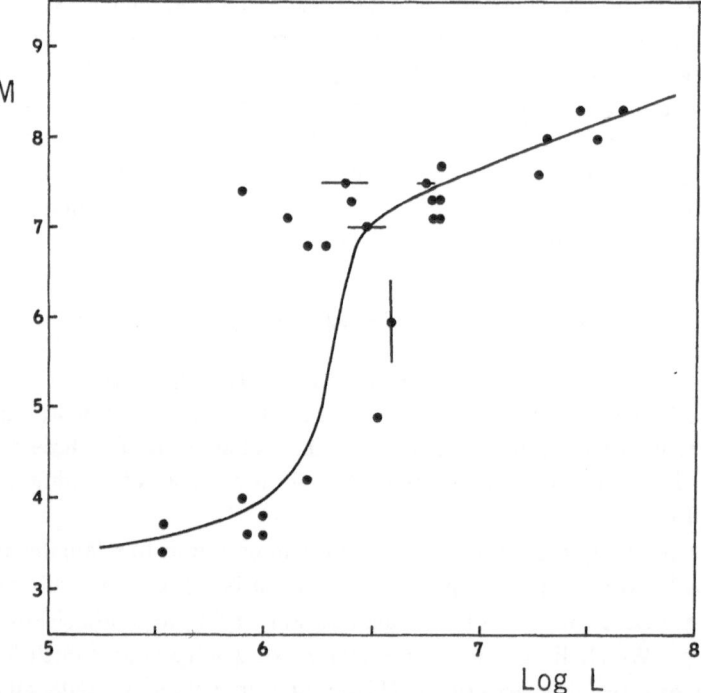

Fig. 1. $\log L$ plotted against magnitude M, for the fault data given by Chinnery (1969). L is the fault length in cm, and logs are to base 10. The solid line shows an approximate empirical relation between $\log L$ and M. In this and succeeding similar diagrams, the bars attached to the data points indicate the range of estimates to be found in the literature.

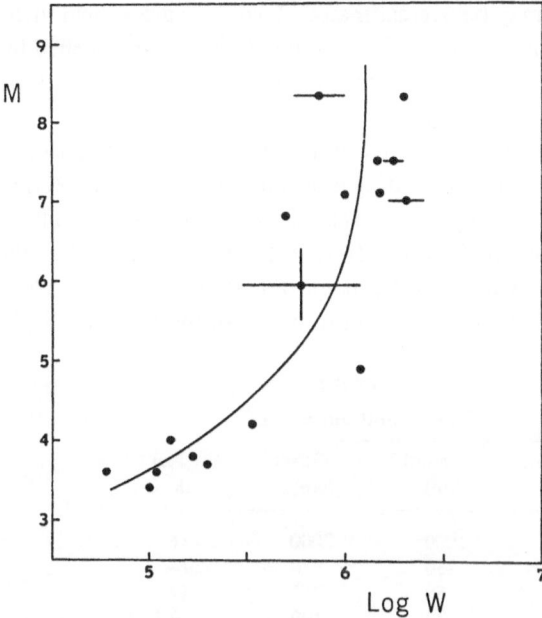

Fig. 2. Log W plotted against magnitude M, for the fault data given by Chinnery (1969). W is the fault depth in cm, and logs are to base 10. The solid line shows an approximate empirical relation between log W and M.

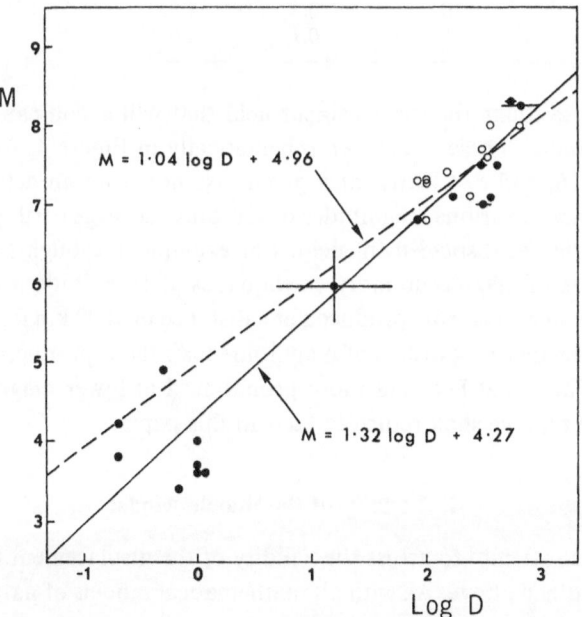

Fig. 3. Log D plotted against magnitude M, for the fault data given by Chinnery (1969). D is the total fault offset in cm, and logs are to base 10. The solid line is a least squares fit to all the data points, and the dashed line is a fit to points with $M > 6$.

some of the data may be overestimated. Two lines are shown through the data points. The solid line is a least squares fit to all the data, and the dashed line is a fit to the data at large magnitudes ($M > 6$). Presumably the true relationship, if one exists, lies between these two extremes.

Using Figures 1, 2, and 3, we are now in a position to estimate the parameters of earthquakes due to surface faulting as a function of magnitude, and these are shown in Table I. It should be emphasized that these are only mean values, and apply only to near-surface strike-slip faulting. It is interesting to note that the ratio of depth to length reaches a maximum in the magnitude range 6–6.5. It is clear from the results of Press (1965) that increasing this ratio increases the far field displacement for a given fault offset.

TABLE I

Typical fault dimensions

Magnitude	Length (km)	Offset (cm)	Depth (km)
8.5	1000	2000	16
8	250	630	14
7.5	80	250	12
7	30	100	9.3
6.5	24	40	8.0
6	22	20	6.3
5.5	21	5	5.0
5	19	2.2	3.7
4.5	16	1.0	2.6
4	10	0.3	1.7
3.5	3	0.1	1.0

We may now evaluate the displacement field that will accompany earthquakes of different magnitudes. These are shown schematically in Figure 4, for earthquakes of magnitude 8, 7, 6, and 5. Clearly, at a given distance from an active fault zone on which earthquakes of various magnitudes occur, only the large earthquakes contribute significantly to the far displacement field. For example, although there are approximately 2.5 orders of magnitude more earthquakes at magnitude 5 than there are at magnitude 8, the displacements produced at a distance of 1000 km from a magnitude 5 earthquake are smaller by 4 orders of magnitude than those produced by a magnitude 8 earthquake. This trend becomes more pronounced at lower magnitudes. This is a significant point that we shall return to later in this paper.

4. Validity of the Simple Model

It is worthwhile to attempt to assess the validity of the displacement field calculations that we have outlined above. As with all mathematical models of natural phenomena, a number of simplifying assumptions are necessary in order to obtain an analytical solution. We shall discuss some of these assumptions, with particular interest in their possible influence on the far displacement field.

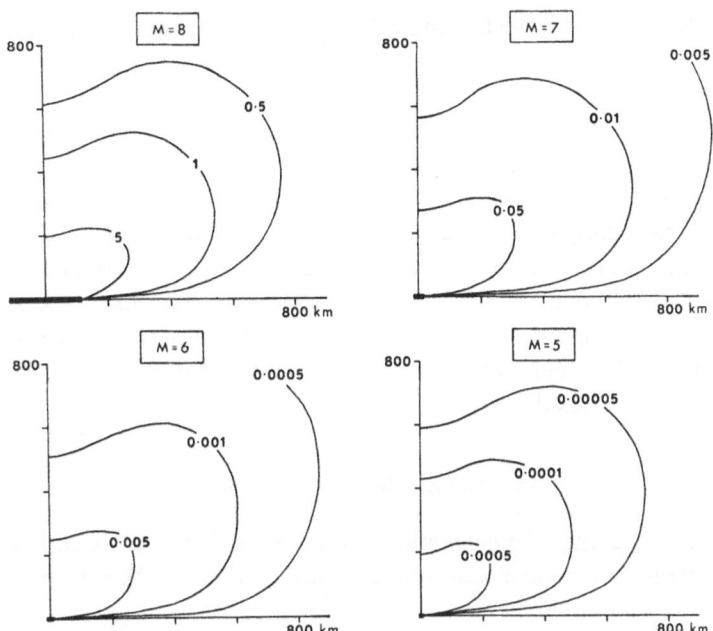

Fig. 4. The displacement fields produced by earthquakes of magnitudes 8, 7, 6 and 5, calculated using the dimensions listed in Table I. The contours show the total horizontal displacement in cm.

The model that we have described represents the earth as a uniform elastic half-space. The assumption that the earth has the characteristics of a linearly elastic solid is clearly an important one. Seismology indicates that the elastic approximation is probably good for time scales up to and including an hour. The calculations of McConnell (1968) suggest that the crust and perhaps the whole lithosphere may behave approximately elastically on time scales of 1000 years or more. We shall assume that this assumption is valid over time intervals of weeks or months, since there is little information to the contrary.

On the other hand, the assumption that the earth is a uniform half-space is clearly incorrect. Since we are concerned with the displacements caused at distances of hundreds or thousands of kilometers from a fault, the layered structure of the earth must be included, and the effects of curvature of the earth's surface are likely to be far from negligible.

The problem of calculating the effect of a dislocation surface inserted into a layered elastic half-space is a difficult but not impossible one. The basic problem which requires solution is that of a nucleus of strain in a layered structure, and this has been formulated by Braslau and Lieber (1968), and Ben-Menahem and Singh (1968). The problem, however, is extremely cumbersome, and the effects of a realistic earth structure on the calculation of displacement fields is not as yet clear. The problem of curvature has been studied by Ben-Menahem *et al.* (1969), and this work is described elsewhere in the next paper. It appears possible that the solution for the layered case could be obtained by extending existing theories of the propagation of seismic waves in layered structures to a zero frequency limit. However, if this has been done, it apparently has not been published.

In addition to these improvements in the mathematical theory, there are a number of rather basic questions that must be asked about the physics of the situation. Firstly, is it really reasonable to assume that the surface S in Equation (1) is at infinity? That is, are the boundary conditions adequate for the real situation? A second and equally important problem is concerned with the movement of the material associated with a fault zone. Are earthquakes the only way in which this movement occurs?

These two questions are very much concerned with the nature of the whole mechanism of movement in fault zones, and in the sections that follow we plan to discuss this subject in some detail.

5. The Elastic Rebound Theory

The problems of earthquake generation and fault movement are clearly indicated by the San Andreas fault zone in California. Figure 5, which is taken from Allen (1968),

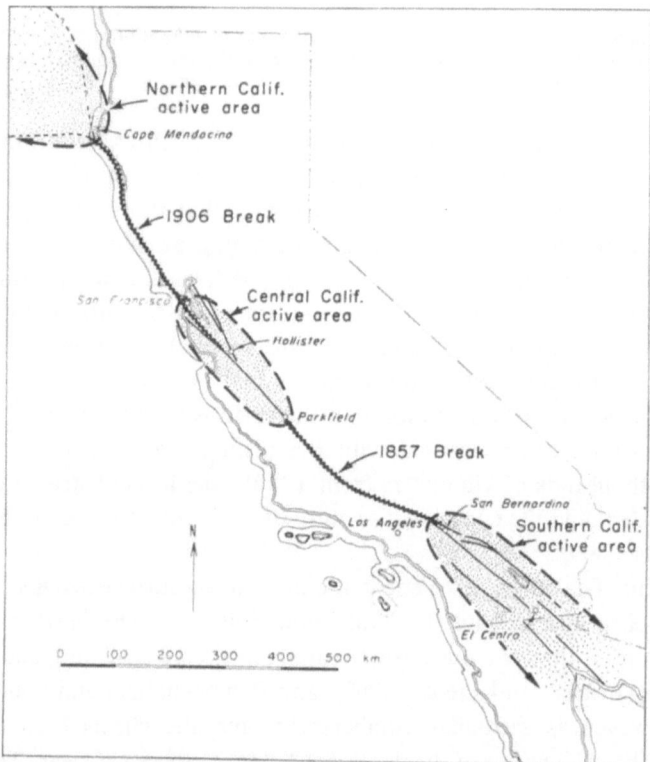

Fig. 5.   Areas of contrasting seismic behavior along the San Andreas fault zone in California (after Allen, 1968).

shows that this region can be divided into two seismic provinces. In the active areas, motion on the fault zone is clearly associated with smaller earthquakes (magnitude up to about 6.5), and aseismic creep is frequently observed in the surface layers. In the remaining regions, however, virtually no creep or seismic activity has been observed since the occurrence of the last major earthquake (magnitude about 8).

The traditional view of the mechanics of this system was that the faults resulted from a North-South compression and an East-West tension. More recently, the transform fault concept suggests that the two sides of the fault are being carried along relative to one another by fluid motions in the asthenosphere. If this is so, then the fault is the result of forces applied at the underside of the lithosphere, at horizontal distances from the fault that may not be large. Clearly this suggests that our assumption that the surface S is at infinity may not be correct. Attempts to include this possibility into the theoretical model are envisioned but as yet have not been carried out. Furthermore, if the San Andreas fault is a transform fault, it is likely that the strain in the fault zone is accumulating relatively uniformly (in some statistical sense) throughout the length of the system (Allen, 1968). It is unlikely, for example, that the two sides of the fault in northern California are moving past each other at half the rate of the two sides in southern California. Certainly, it would appear that this situation could not persist for long.

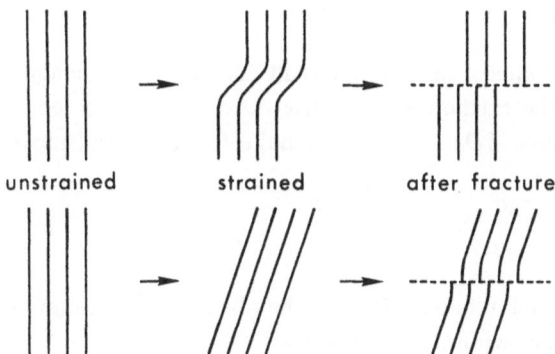

Fig. 6.   Two versions of Reid's elastic rebound theory (compare Richter, 1958).

As a result of his observations of the deformation of the earth's surface during the 1906 San Francisco earthquake, Reid (1910) produced his 'elastic rebound theory'. This theory is illustrated in Figure 6, and, following Richter (1958), we show two versions of this idea. The upper diagram shows what is essentially Reid's formulation of the theory. Hypothetical lines which originally were parallel to one another and perpendicular to the fault line are successively distorted as strain accumulates, until in the region of the fault the strength of the earth's crust is exceeded, faulting occurs, and the accumulated stress is relieved. The lower diagram was given by Richter (1958), and at that time this was the mechanism preferred by him.

These sets of diagrams clearly may be related to the two conflicting tectonic theories. Reid's picture is what we would expect if the transform fault idea is applicable to the San Andreas fault zone. On the other hand, Richter's diagram corresponds to the application of a large scale tectonic stress, perhaps by North-South compression of the entire region. In the first case, all the strain is relieved by the earthquake, while in the second, strain is relieved only near the fault, and one must therefore expect the formation of a set of parallel faults.

Let us assume for the present that the San Andreas fault is indeed a transform fault, and that Reid's diagrams are those which are applicable. The question that now arises is the scale of the region within which the deformation takes place prior to the earthquake. This in turn leads us to ask two important questions:

(a) Does the movement on the fault keep up with the applied tectonic offset? Presumably somehow it must.

(b) How does the width of the zone of strain accumulation and release compare with the zone affected by earthquakes?

The next two sections will consider these questions in turn.

6. Fault Movement and Tectonic Offset

The relationship between the fault movements accompanying earthquakes and the overall slip of a fault zone has been considered recently by Brune (1968). His argument goes as follows:

Suppose an earthquake occurs in a fault zone of overall length L_0 and depth W_0. Suppose, during the earthquake, an offset D occurs over a length L and depth W, then the contribution δD_0 of the earthquake to the overall movement on the fault zone is

$$\delta D_0 = \frac{LW}{L_0 W_0} D. \tag{3}$$

The total slip on the fault zone, D_0, can then be found by summing the contributions from all earthquakes within a given period:

$$D_0 = \frac{\Sigma(LDW)}{L_0 W_0}. \tag{4}$$

Brune (1968) estimated $\Sigma(LDW)$ by measuring the seismic moment μLDW of earthquakes from their seismic radiation. Another way in which this could be done would be to establish a relation between LDW and magnitude M, and use this to assign a value of LDW to each earthquake.

Figure 7 shows a graph of $\log LDW$ against magnitude M for the same set of data used in Figures 1, 2, and 3. A very good linear relationship is indicated, and a least squares fit to the data points gives the following relation:

$$M = 0.79 \log LDW - 4.74. \tag{5}$$

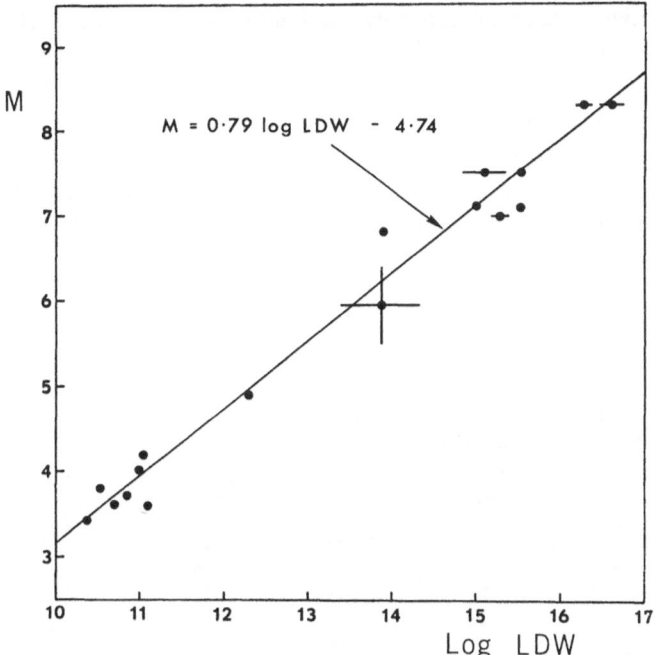

Fig. 7. Log LDW plotted against magnitude M, for the fault data given by Chinnery (1969). L is the fault length, D is the offset, and W is the fault depth. All quantities are in cm, and logs are to base 10. The solid line is a least squares fit to the data points.

Using this, we may estimate the quantity LDW for earthquakes of magnitudes greater than about 3.

In order to form the sum mentioned in Equation (4), we need to find a region where the number of earthquakes of different magnitudes is known fairly accurately. Brune (1968) used information in the Imperial Valley obtained from a study of the seismicity of southern California by Allen *et al.* (1965). This information includes the number of earthquakes occurring at each magnitude greater than 3 for the period 1934–63. These numbers are tabulated for each half magnitude interval in Table II. Also shown in this table are the values of LDW obtained from Equation (5), and the total value of $\Sigma (LDW)$ for this particular time interval. Brune (1968) chose to continue the range of magnitude considered down to magnitude zero by using a magnitude-frequency relation, but the contribution from earthquakes of low magnitude appears to be small and they are not included here.

If, following Brune, we take the total length of the fault zone in the Imperial Valley to be 120 km, and the depth of movement to be 20 km, the total slip of the fault zone during the period 1934–63 turns out to be about 74.5 cm. This corresponds to an annual rate of about 2.6 cm/year. Brune (1968) obtained 3.2 cm/year, which is very close to this estimate. However, Whitten (1956), as we shall see in the next section, has observed a large scale geodetic displacement in this area of at least 8 cm/year for the period 1941–54.

The discrepancy between these two estimates is considerable. If we have really accounted for all of the movement on the fault zone by summing over all observed earthquakes, we must conclude, as suggested by Brune, that within this comparatively short period approximately 5 m of displacement have accumulated. If this were relieved all at once, it would result in an extremely large earthquake, of magnitude about 8. However, earthquakes of this size have never been recorded in the Imperial Valley region, so we must question the calculation outlined in Table II.

TABLE II

Imperial Valley Earthquakes 1934–63

Magnitude M	Number of Earthquakes N	$\log LDW$	$N \cdot LDW \, (\times 10^{12})$
7.1	1	15.00	1000
6.75	1	14.53	340
6.25	1	13.90	80
5.75	9	13.27	171
5.25	18	12.65	81
4.75	55	12.00	55
4.25	131	11.38	31
3.75	354	10.74	19
3.25	885	10.10	12

$$\Sigma(LDW) = 1789 \times 10^{12}$$

Is there any way in which we can reduce the discrepancy between these two estimates? There appear to be three possibilities:

(a) We could increase D_0 by decreasing the product $L_0 W_0$. It is not obvious that the length L_0 can be changed significantly. Brune (1968) has suggested reducing the depth W_0 to about 7 km. This would increase D_0 by a factor of about 3, and produce a nice agreement between the two estimates. The reason why this solution is not acceptable is discussed in the next section.

(b) It is possible that the number of events at low magnitude has been underestimated. If these numbers could be increased, it is possible that a significant contribution could arise from events of low magnitude. However, this is again not an acceptable solution, for reasons outlined in the next section.

(c) The only other possibility is that aseismic creep is a significant and perhaps dominant characteristic of the motion in this area. This in turn suggests that the displacement field observed at points away from the fault zone may be largely influenced by aseismic processes, and that displacement fields calculated using only earthquake data may be underestimates.

7. Strain Accumulation and Strain Release

The second problem that we have posed has received much less attention in the past. We must account for the strain release not only in the immediate vicinity of the fault, but also at some distance from it. In other words, we must consider the relative sizes of the zones of strain accumulation and strain release.

To illustrate this problem, Figure 8 shows the observed displacements of the ground that accompanied the San Francisco earthquake of 1906 (this diagram is taken from Chinnery, 1961). Significant displacements extended only to 10 or 15 km from the fault, and the zone of strain release was very narrow, in spite of the length of the fault, which was in excess of 400 km. As we have already mentioned, if the lower diagram in Figure 6 is applicable, we would expect to find numerous faults parallel to this section of the San Andreas, but these are not found.

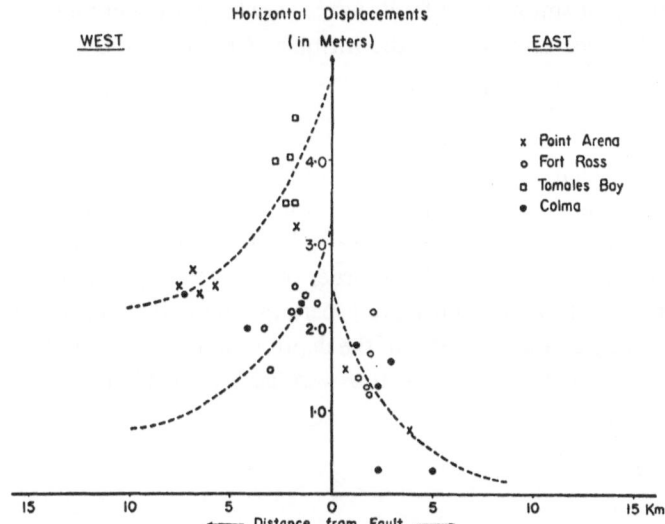

Fig. 8. Displacement of triangulation stations observed during the 1906 San Francisco earthquake (after Chinnery, 1961).

If the elastic rebound theory as illustrated by the upper portion of Figure 6 is valid for this case, we must conclude that the region of strain accumulation is also extremely narrow. This is hard to believe, whether the tectonic stress is produced by currents in the asthenosphere or by a large scale compression. The only way in which it might occur is if the crust in the immediate vicinity of the fault has a very low rigidity compared to the surroundings. This was the conclusion suggested by Byerly and DeNoyer (1958), who postulated that the rigidity μ varied with distance x from the fault according to the following formula

$$\mu = \mu_0(1 + \alpha^2 x^2),\tag{6}$$

where μ_0 is the rigidity at the fault and α is a constant. μ_0 and α were estimated from the observed displacements, and Equation (6) then gave:

rigidity at the fault $= 3 \times 10^{10}$ cgs
rigidity at 10 km from fault $= 3 \times 10^{11}$ cgs
rigidity at 30 km from fault $= 3 \times 10^{12}$ cgs.

To the present author's knowledge there is no seismological or geological information that would support this large variation in rigidity.

The only other possibility is that the zone of strain accumulation is much larger than the zone of strain release, and that at some distance from the fault strain is relieved as the result of aseismic fault movements that may have preceded or followed the earthquake. Let us consider this possibility in more detail.

The fall-off of horizontal displacement with distance from a fault such as the San Andreas fault can be calculated using the dislocation theory outlined earlier in this paper. Choosing, for simplicity, a fault that is very long, we find that the displacement falls off with distance according to the formula (Chinnery, 1961)

$$u(x) = \frac{D}{\pi} \tan^{-1} \frac{W}{x}, \tag{7}$$

where $u =$ displacement at distance x from fault, $D =$ total offset on fault, $W =$ depth of fault. This formula is shown graphically for faults of different depth in Figure 9, which shows the distortion of a line drawn perpendicular to the fault on the ground just before faulting commenced. Clearly, in order to relieve a significant amount of strain at (say) 50 km from the fault, a deep fault, perhaps 50 or 100 km deep, is needed. This diagram shows why all the estimates of the depth of faulting during the San Francisco earthquake were so low. Some estimates were as low as 2 km (Chinnery, 1961), or 3 km (Knopoff, 1958).

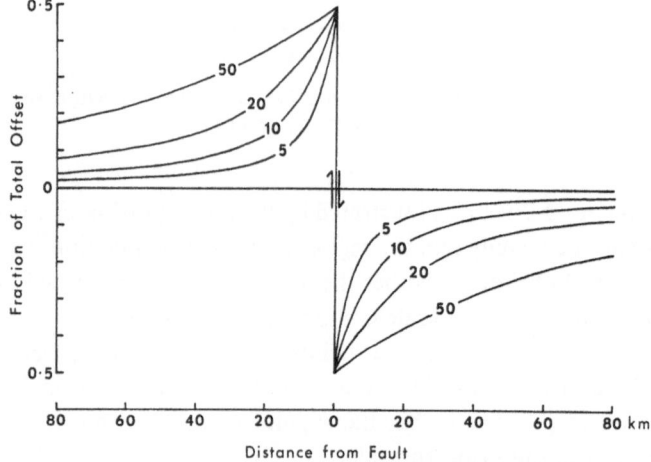

Fig. 9. Distortion of a line drawn on the ground surface just before an earthquake by fault movements that extend to the indicated depths (in km).

At this point we need some information about the zone of strain accumulation. Very few examples of large scale geodetic surveys of the San Andreas system seem to have been published in the literature. The best information that the present author has been able to find is included in a paper by Whitten (1956), which included a large scale survey of the Imperial Valley. Figure 10 shows the displacements that he observed on one line in part of the Imperial Valley during the period 1941–54. The relative movement of points A and B on this diagram, each 60 km from the active fault line, corresponds to 8.2 cm/year (this is an often quoted-number). The angular strain implied by this information is similar to that found on other segments of the San Andreas fault (Whitten, 1956), and is about 1 sec of arc in 10 years. This corresponds to a strain accumulation of about 0.25×10^{-6}/year, or a stress accumulation of about 0.1 bars/year.

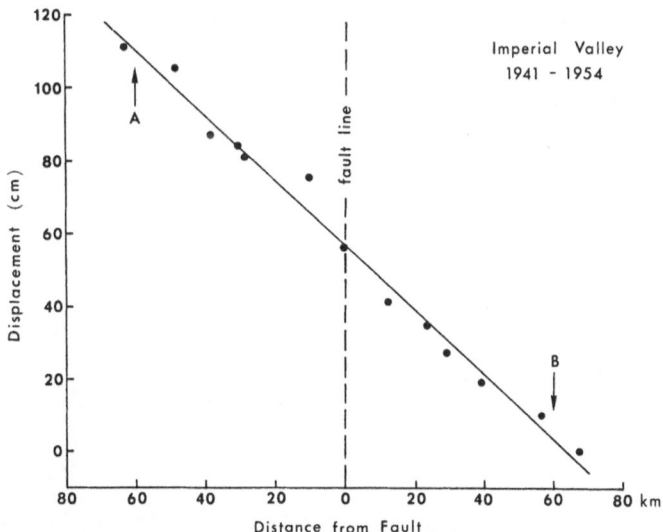

Fig. 10. Displacements of triangulation stations in the Imperial Valley during the period 1941–54. This set of stations forms the southernmost line of the network described by Whitten (1956).

The linearity of the data shown in Figure 10 is remarkable. Notice, however, that the boundaries of the zone of strain accumulation are not apparent, and the tectonic offset of about 8 cm/year must be a minimum estimate. It is clearly of great importance that geodetic surveys extend further from the San Andreas fault in an attempt to define the width of the zone of strain accumulation.

We can compare Whitten's data with the strain release that has occurred in the Imperial Valley as a result of earthquakes. Figure 11, taken from Allen et al. (1965), shows the occurrence of earthquakes of magnitude 6 and greater in the southern California region during the period 1912–63. The active segment of the fault in the Imperial Valley region is clear, and very roughly one earthquake with a magnitude between 6 and 7 has occurred on each section of the fault during the 50 year period.

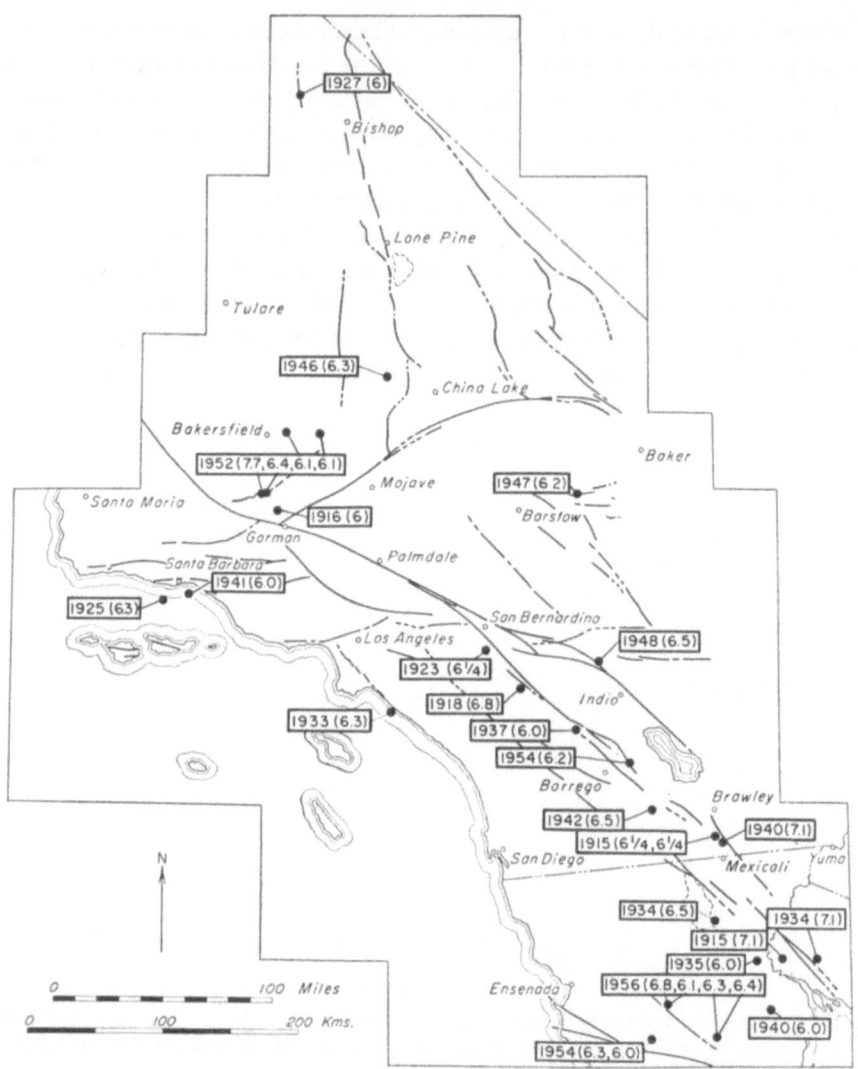

Fig. 11. Earthquakes of magnitude 6.0 and greater in the southern California region during the period 1912–63 (after Allen *et al.*, 1965).

Let us suppose that the present rate of accumulation of strain in the Imperial Valley has continued for a period of 52 years. This would result in displacements which were four times larger than those shown in Figure 10, and a line drawn perpendicular to the fault in 1912 would be distorted to the straight line shown in Figure 12 by 1964. Let us now suppose that at the end of this period an earthquake occurs with an offset of 50 cm at the fault, and a depth of about 7 km (following the suggestion of Brune, 1968). This would correspond roughly to an earthquake with a magnitude of about 6.6. The relief of strain which would result is shown in Figure 12. Even if the entire accumulated offset is relieved by a single large earthquake, Figure 12 shows that the relief of the strain at points away from the fault would be very small if the movement only extends to a depth of 7 km.

Figure 12 summarizes the problem that we are outlining in this paper. A magnitude 6.5 earthquake in the Imperial Valley, which is about the largest size to be expected,

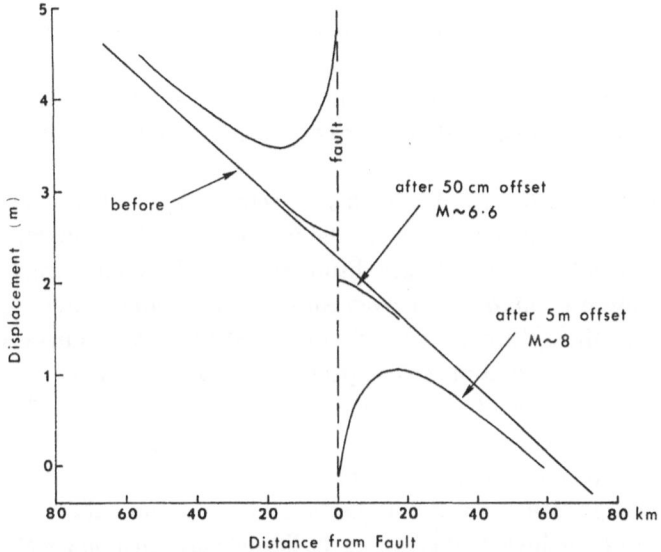

Fig. 12. If the strain accumulation shown in Figure 10 were to continue for 52 years, the line of stations would lie close to the straight line labelled 'before'. This diagram shows the relief of the accumulated strain that would accompany earthquakes with offsets of 50 cm and 5 m, where these movements extend to a depth of 7 km.

produces a negligible change in the overall pattern of accumulated strain. As we have seen in Figure 4, the presence of smaller earthquakes on the fault zone will not help the situation, and though aseismic creep in the upper layers of the crust may produce more offset at the fault, it will not produce relief of strain where it is needed, some tens of kilometers away from the fault zone. This can only be accomplished by some kind of aseismic movement on much deeper portions of the fault plane.

8. Fault Movements at Depth

In a recent paper, Scholz *et al.* (1969) discussed in detail the nature of the slippage on the San Andreas fault. They defined three depth ranges with different slip mechanisms:

> 0–4 km: stable, mostly aseismic sliding.
> 4–12 km: stick-slip motion together with earthquakes.
> below 12 km: stable sliding or plastic friction.

They discussed the third of these layers very little. In their conclusions, however, they make the following comment: "The bottom is no problem; the region below 12 km may have been sliding smoothly prior to the earthquake and, rather than have been loaded by it, was strain relieved."

There can surely be no doubt that movement does indeed occur at levels deeper than 12 km in the earth's crust, and perhaps all the way down to the base of the lithosphere, if the transform fault model is valid. If this is so, the mass transport associated with this movement at depth will be much larger than that accompanying seismic or aseismic movement near the top of the crust. This movement at depth is likely to have a larger influence on the moment of inertia of the earth, and will permit the release of strain at horizontal distances of greater than 30 or 40 km from the fault line.

We must first make the point that this movement at depth cannot occur at the same instant as an earthquake occurs in the upper layers. If it did, we would expect to find displacements at much larger distances from the fault during earthquakes. Presumably, therefore, this deeper movement must occur either before or after a main shock. In view of the apparent observation of Smylie and Mansinha (1968) that disturbances of the earth's rotation pole appear to precede earthquakes by 5 or 10 days, it is tempting to suggest that this movement may occur at that time. This immediately raises the question of the rapidity of the movement at depth: does it occur in rapid episodes, or by a very slow continuous plastic flow?

We can at least partially answer this question. Let us suppose that movement occurs from a depth of 12 km down to the base of the lithosphere, on a time scale such that the lithosphere will behave approximately as an elastic solid. We can calculate the distortion of the ground surface which would accompany such a movement, using the dislocation theory model. If the top of the slipped zone is at depth w and the bottom is at W, then the displacements as a function of distance from the surface trace of the fault have the following form:

$$u(x) = \frac{D}{\pi} \tan^{-1} \frac{x(W - w)}{x^2 + wW}. \tag{8}$$

The predicted displacements are plotted for $w = 12$ km and $W = 50$ and 100 km in Figure 13. It is clear that large displacements of the ground surface are to be expected, and these should be easily detectable by standard geodetic networks.

If we now refer back to the geodetic observations in the Imperial Valley shown in Figure 10, the linearity of the data strongly suggests that no movement occurred at depth in this region during the period 1941–54. Such a linear plot must imply that the whole lithosphere was strained very uniformly during this period. This one observation is probably enough to enable us to dismiss the possibility that transform fault movements in the lithosphere (below a depth of about 12 km) occur continuously. Let us assume, at least provisionally, that they are episodic in character.

We can also argue that it is unlikely that these episodes follow surface faulting. The Imperial Valley geodetic observations were made in the period immediately following the large ($M=7.1$) earthquake of 1940, and there is no evidence of the distortions shown in Figure 13. There is also no evidence of such distortions following earthquakes in other parts of the San Andreas system.

We must conclude, therefore, that the movement at depth occurs prior to a surface movement. Presumably strain will accumulate throughout the lithosphere until the deeper layers begin to move relative to one another, perhaps by plastic flow in a fairly narrow zone beneath the surface trace of the fault. This, in turn, will strain the surface layers, which are much stronger, and which will eventually break either as an earthquake or in the form of creep.

If the sequence of events that we have outlined really occurs, we can modify the standard diagrams of the elastic rebound theory given in Figure 6 to the slightly more complex scheme shown diagrammatically in Figure 14. This is a kind of 'double elastic rebound'. We must also admit the possibility that even this picture may be

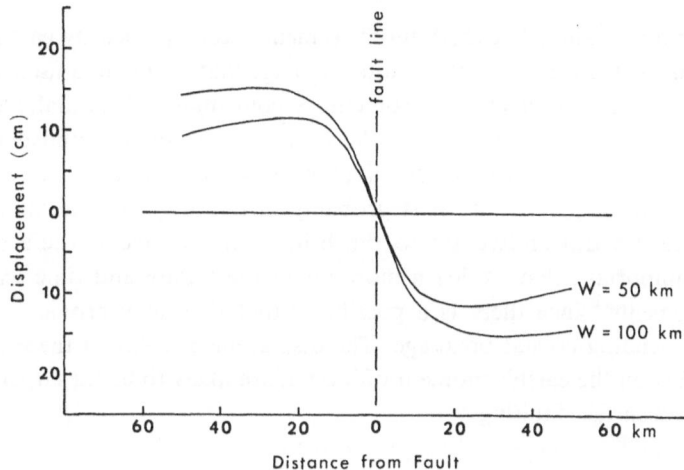

Fig. 13. The distortion of a line on the ground surface perpendicular to a fault trace by movement that extends from a depth of 12 km down to the depth W.

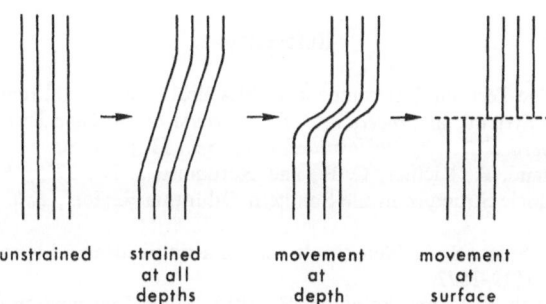

Fig. 14. A possible extension of the elastic rebound theory to include movement of material at depth.

drastically oversimplified, and that there may be many episodes of movement in the deeper layers before the strain at the surface is sufficient to cause breakage.

The major problem that remains is to establish the time scale of this sequence of events. If, for example, the movement at depth occurred just before the main shock, then one would expect geodetic stations near the fault to register the difference between the second and fourth stages of Figure 14. Thus, although only the surface layers may have moved during the earthquake, the observed displacements should indicate a very much deeper fault. This does not appear to be found. Nevertheless, the movement at depth must occur sometime, and when it does it should be easily detectable by stations placed 30–50 km from the fault zone. Even the faint possibility that this movement may sometimes precede the earthquake by a few days should be sufficient to encourage very careful monitoring of such stations.

9. Conclusions

There seems little doubt that the fault movements accompanied by earthquakes can represent only a small fraction of the total slippage that occurs in a fault zone. In the surface layers, aseismic creep may sometimes contribute substantially to the total offset. More important, however, are the deeper layers of the lithosphere, from a depth of about 12 km down to the base, at about 100 km. A seismic relative movement must occur in these layers, and it is probably in this way that the motions of the asthenosphere are transmitted up to the brittle surface layers where earthquakes occur. It is important that we learn more about the nature and time scale of these deeper movements, since there is a possibility that they may provide premonitory signals of impending crustal breakage. The displacement fields of these movements, and their effect on the earth's moment of inertia, are likely to be far larger than those associated with surface faulting.

Acknowledgement

This research was supported by the National Science Foundation Grant GA-708.

References

Allen, C. R.: 1968, 'The Tectonic Environments of Seismically Active and Inactive Areas along the San Andreas Fault System', in *Proceedings of Conference on Geologic Problems of San Andreas Fault System, Stanford University Publications, Geol. Sci.*, **11**, p. 70–80.
Allen, C. R., St. Amand, P., Richter, C. F., and Nordquist, J. N.: 1965, 'Relationship between Seismicity and Geologic Structure in the Southern California Region', *Bull. Seism. Soc. Am.* **55**, 753–797.
Ben-Menahem, A. and Singh, S. J.: 1968, 'Multipolar Elastic Fields in a Layered Half-Space', *Bull. Seism. Soc. Am.* **58**, 1519–1572.
Ben-Menahem, A., Singh, S. J., and Solomon, F.: 1969, 'Static Deformation of a Spherical Earth Model by Internal Dislocations', *Bull. Seism. Soc. Am.* **59**, 813–854.
Braslau, D. and Lieber, P.: 1968, 'Three Dimensional Fields due to a Volterra Dislocation Imbedded

in a Layered Half-Space: Analytical Representation of a Seismic Mechanism', *Bull. Seism. Soc. Am.* **58**, 613–628.

Brune, J. N.: 1968, 'Seismic Moment, Seismicity, and Rate of Slip along Major Fault Zones', *J. Geophys. Res.* **73**, 777–784.

Byerly, P. and DeNoyer, J.: 1958, 'Energy in Earthquakes as Computed from Geodetic Observations', in *Contributions in Geophysics in Honor of Beno Gutenberg*, Pergamon Press, New York, pp. 17–35.

Chinnery, M. A.: 1961, 'The Deformation of the Ground around Surface Faults', *Bull. Seism. Soc. Am.* **51**, 355–372.

Chinnery, M. A.: 1965, 'The Vertical Displacements Associated with Transcurrent Faulting', *J. Geophys. Res.* **70**, 4627–4632.

Chinnery, M. A.: 1969, 'Earthquake Magnitude and Source Parameters', *Bull. Seism. Soc. Am.* **59**, 1969–1982.

Chinnery, M. A. and Petrak, J. A.: 1968, 'The Dislocation Fault Model with a Variable Discontinuity', *Tectonophysics* **5**, 513–529.

Kasahara, K.: 1957, 'The Nature of Seismic Origins as Inferred from Seismological and Geodetic Observations (1)', *Bull. Earthquake Res. Inst. Tokyo Univ.* **35**, 473–532.

Knopoff, L.: 1958, 'Energy Release in Earthquakes', *Geophys. J.* **1**, 44–52.

Love, A. E. H.: 1944, *A Treatise on the Mathematical Theory of Elasticity*, 4th ed., Dover Publications, New York.

Mansinha, L. and Smylie, D. E.: 1967, 'Effect of Earthquakes on the Chandler Wobble and the Secular Polar Shift', *J. Geophys. Res.* **72**, 4731–4743.

Maruyama, T.: 1964, 'Statical Elastic Dislocations in an Infinite and Semi-Infinite Medium', *Bull. Earthquake Res. Inst. Tokyo Univ.* **42**, 289–368.

McConnell, R. K., Jr.: 1968, 'Viscosity of the Mantle from Relaxation Time Spectra of Isostatic Adjustment', *J. Geophys. Res.* **73**, 7089–7105.

Mindlin, R. D.: 1936, 'Force at a Point in the Interior of a Semi-Infinite Solid', *Physics* **7**, 195–202.

Mindlin, R. D. and Cheng, D. H.: 1950, 'Nuclei of Strain in the Semi-Infinite Solid', *J. Appl. Phys.* **21**, 926–930.

Press, F.: 1965, 'Displacements, Strains, and Tilts at Teleseismic Distances', *J. Geophys. Res.* **70**, 2395–2412.

Reid, H. F.: 1910, The Mechanics of the Earthquake', in *The California Earthquake of April 18, 1906*, Report of the State Investigation Commission, vol. 2, Carnegie Institution of Washington.

Richter, C. F.: 1958, *Elementary Seismology*, W. H. Freeman and Co., San Francisco.

Scholz, C. H., Wyss, M., and Smith, S. W.: 1969, Seismic and Aseismic Slip on the San Andreas Fault', *J. Geophys. Res.* **74**, 2049–2069.

Smylie, D. E. and Mansinha, L.: 1968, 'Earthquakes and the Observed Motion of the Rotation Pole', *J. Geophys. Res.* **73**, 7661–7673.

Steketee, J. A.: 1958a, 'On Volterra's Dislocations in a Semi-Infinite Elastic Medium', *Can. J. Phys.* **36**, 192–205.

Steketee, J. A.: 1958b, 'Some Geophysical Applications of the Elasticity Theory of Dislocations', *Can. J. Phys.* **36**, 1168–1198.

Whitten, C. A.: 1956, 'Crustal Movement in California and Nevada', *Trans. Am. Geophys. Union* **37**, 393–398.

Discussion

Nason: With reference to the last part of your paper, I want to point out that the fault zone in the San Francisco area is perhaps 10 or 20 km wide while that in the Imperial Valley area is perhaps 100 km wide, and that it is reasonable that there is a difference in strain accumulation in the two areas. Also, a survey between the thirties and the early sixties, in the Monterey area, and another one farther South in the Parkfield area, have shown that there is significant right lateral strain accumulating in blocks close to the fault but not in blocks 10 or 20 km away from the fault.

Chinnery: I agree, but it bothers me somewhat that relative motion between the two sides of the fault in southern California occurs at the rate of 8 cm/yr while in northern California it is only 3 cm/yr.

Nason: Perhaps one of the problems is that the period of strain accumulation for an earthquake is probably 100 years or even more, yet we are basing our surveys on only 10–20 years observations.

Whitten: I believe the numbers that have been quoted were taken from our 1956 paper. In 1960 a paper was published in which the same data were analysed in a form of strain determination with a term inserted which would determine any slippage which might occur in the fault zone; I think it might be possible to redraw your Figure 10 in such a way that it would indicate the possibility of some slippage.

The Imperial Valley was resurveyed in 1967 and if we compare this survey with those of 1941 and 1954, that is the three surveys at 13 year intervals, we find that the data are in reasonably good agreement. I think the figure of 8 cm that you have been quoting is based on the 1941 and 1954 surveys and is a cumulative effect from Yuma to San Diego, which represents the total width of Southern California.

In the most recent analysis that we have made of the data, instead of holding a reference point some 50 or 60 km to the East, we held reference points reasonably close to the fault but we selected three triangulation points that fell just to the East of the fault system. When we treated the 1941, the 1954 and 1967 surveys and put them into a comprehensive adjustment, letting time be one of the dimensions, we found that in addition to an indication of continuous creep along the fault, the strain was accumulating as far away as 30 or 40 km from the fault, which is somewhat greater than the distances you show.

Nason: Did you use, in the strike-slip case, a fault model with a uniform displacement over the whole surface or a displacement graded from high in the center to zero at the end?

Chinnery: When you are very close to the fault you must allow the displacement to vary from point to point both horizontally and vertically, but the further you get away from the fault the more accurate is the approximation method which I use. This is why I purposely used only far field displacements.

DEFORMATION OF A SPHERICAL EARTH MODEL
BY FINITE DISLOCATIONS

ARI BEN-MENAHEM

Dept. of Applied Mathematics, The Weizmann Institute of Science, Rehovot, Israel

and

SARVA JIT SINGH*

Dept. of Mathematics, Kurukshetra University, Kurukshetra, India

Abstract. Numerical results are obtained for the static surface displacements, strains and tilts of a homogeneous, isotropic, non-gravitating, elastic sphere due to finite dip-slip and strike-slip faults. The theory is applied to strain observations from the Alaska earthquake of March 28, 1964. It is shown that the theoretical values are within an order of magnitude of the observed values and are of the correct sign.

The method can be extended to derive the displacements everywhere within the sphere. These are required for the determination of the changes in the components of the inertia tensor in the earth due to major earthquakes which lead to a spherical theory for the calculation of the contribution of earthquakes to the excitation of the Chandler wobble and the secular polar shift.

During the preceding few years, elasticity theory of dislocations has been developed and applied by several investigators, e.g. Steketee (1958a, b), Chinnery (1961, 1963), Maruyama (1964), Press (1965), Savage and Hastie (1966). Mansinha and Smylie (1967), computed the changes in the products of inertia of the earth due to rearrangement of masses associated with major earthquakes. As a mathematical model, they used vertical, rectangular, strike-slip and dip-slip faults in a half-space. They then calculated the contribution to the excitation of the Chandler wobble and the secular polar shift from the above changes in the products of inertia.

However, there is no justification for using a half space model in problems with intrinsic spherical geometry. Ben-Menahem and Singh (1968), obtained explicit expressions for the deformation of a uniform non-gravitating sphere due to internal dislocations of arbitrary orientation and depth. These results constitute the theoretical nucleus of a fundamental study by Ben-Menahem *et al.* (1969). Therein displacements and strains are calculated everywhere on a spherical earth model and the numerical results are displayed in various forms. Some important results of this study are summarized below.

(1) The field components are practically insensitive to changes in the Poisson ratio in the range $0.25 \leqslant \sigma \leqslant 0.33$.

(2) By using the results of this paper it is possible to calculate the displacements, strains, stresses and tilts at any point on the surface of a sphere induced by a tangential dislocation of arbitrary depth and dip and slip angles.

(3) The ratio of the displacement at the free surface of a sphere to the corresponding

---

* During a major part of the investigation, this author was a Weizmann Institute fellow.

L. Mansinha et al. (eds.), Earthquake Displacement Fields and the Rotation of the Earth, 39–42.
All Rights Reserved. Copyright © 1970 by D. Reidel Publishing Company, Dordrecht-Holland

quantities on the basis of a half-space model is computed. It is found that the absolute value of this ratio varies from zero to about four and that in some cases the algebraic sign of the ratio is negative over most part of the spherical surface. The latter effect cannot be due to a disparity in the sign convention because as the point of observation approaches the epicenter, the ratio tends to the value $+1$. Therefore, it appears that calculations based on a half-space approximation of the earth such as produced by Mansinha and Smylie (1967) may sometimes be misleading.

(4) Wideman and Major (1967), claimed that their observations yield a strain-step amplitude dependence upon distance like $\Delta^{-3/2}$. Our calculations do not show any such simple relationship.

(5) Assuming that an earthquake of top magnitude $(M \sim 8.7)$ can be simulated by taking a dislocation of 30 m and a fault area of 10^5 km^2, we conclude that a maximum strain of the order of 10^{-7} should result at a distance of $\Delta = 15°$ and of the order 10^{-10} at a distance of $\Delta = 150°$. Thus modern instruments are capable of detecting strains from major events even in the lower hemisphere.

(6) With the same source dimensions and same dislocation, the far field due to a vertical dip-slip source is much weaker than that due to a vertical strike-slip source, especially for shallow foci. Further, while the far field due to a strike-slip fault is almost independent of the focal depth, that due to a dip-slip fault is very sensitive to changes in the focal depth.

(7) It should be possible to study the earthquake mechanism from strain-step measurements if these could be made at various stations around an event.

(8) On the basis of a homogeneous spherical earth model it is concluded that the stresses resulting from one earthquake are not likely to trigger another earthquake in the far field.

(9) Singh and Ben-Menahem (1969), produced contour maps for displacement and strain components for a vertical strike-slip and vertical dip-slip faults. Detailed calculations reveal that for nearly vertical and nearly horizontal faults, the field due to a dip-slip source shows a strong dependence on the dip angle. A small variation in the latter can change both the sign and order of magnitude of the calculated field component.

A comparison is made between the strain steps recorded at Kipapa, Isabella and the Green Observatory from the Alaska earthquake of March 28, 1964, and the theoretical values. We took the area of the fault equal to 12×10^4 km^2, the dislocation equal to 10 m, the focal depth as 64 km and the azimuth of the strike of the fault as 230°. It is found that by taking the orientation of the instruments and the azimuth of the recording stations into account the calculated values are within an order of magnitude of the observed values and are of the correct sign.

Ben-Menahem *et al.* (1970), numerically integrated the results for a point source over the fault area to obtain the field due to a finite source in a sphere. Detailed contour maps are produced for a vertical strike-slip and a vertical dip-slip source. The computation of the double integrals involves a successive use of Simpson's rule, first in one coordinate, then in the second. Since it is possible to evaluate one of the integrals analytically, this is used to check the numerical integration. To obtain meaningful approximation of the integrals near the fault, a method is devised in which no computation is made in the very neighbourhood of the fault.

Most of the far field results obtained for a finite dislocation are essentially similar to those obtained for a localized dislocation. It is only in the neighbourhood of the fault that one might get radically different results.

The number and location of the nodal lines on the surface of a sphere depends on the type of source, its dimensions and the field component under consideration. For a finite source in a half-space Chinnery (1961), finds a single nodal line for the vertical component of the displacement due to a vertical strike-slip fault. He also reports a linear relationship between the vertical extent of the source, its fault length and the distance of the nodal line from the fault-line. No such simple relationship exists in the case of a sphere.

Thus far we have attempted to calculate the field at the free surface of a sphere. However, to compute the changes in the components of the inertia tensor in the earth due to an earthquake, one must know the displacements everywhere within the sphere. Some progress has been made in this direction. Once the displacement field is known throughout the sphere, it will be possible to calculate the contribution of earthquakes to the excitation of the Chandler wobble and the secular polar shift. It is hoped that this will throw some more light on the long debated question whether earthquakes excite the Chandler wobble (Ben-Menahem and Israel, 1970).

Acknowledgements

This Research has been sponsored by the Air Force Cambridge Research Laboratories under contract AF61(052)-954 through the European Office of Aerospace Research, OAR, USAF, as part of the Advanced Research Projects Agency's Project VELA-UNIFORM.

References

Ben-Menahem, A. and Singh, S. J.: 1968, 'Eigenvector Expansions of Green's Dyads with Applications to Geophysical Theory', *Geophys. J. Roy. Astron. Soc.* **16**, 417–452.

Ben-Menahem, A. and Israel, M.: 1970, 'Effects of Major Seismic Events on the Rotation of the Earth', *Geophys. J. Roy. Astron. Soc.* **19** 367–393.

Ben-Menahem, A., Singh, S. J., and Solomon, F.: 1969, 'Static Deformation of a Spherical Earth Model by Internal Dislocations', *Bull. Seismol. Soc. Am.* **59**, 813–853.

Ben-Menahem, A., Singh, S. J., and Solomon, F.: 1970, 'Deformation of a Homogeneous Earth Model by Finite Dislocations', *Rev. Geophys.* (in press).

Chinnery, M. A.: 1961, 'The Deformation of the Ground around Surface Faults', *Bull. Seismol. Soc. Am.* **51**, 355–372.

Chinnery, M. A.: 1963, 'The Stress Changes that Accompany Strike-Slip Faulting', *Bull. Seismol. Soc. Am.* **53**, 921–932.

Mansinha, L. and Smylie, D. E.: 1967, 'Effect of Earthquakes on the Chandler Wobble and the Secular Polar Shift', *J. Geophys. Res.* **72**, 4731–4743.

Maruyama, T.: 1964, 'Statical Elastic Dislocations in an Infinite and Semi Infinite Medium', *Bull. Earth Res. Inst.* **42**, 289–368.

Press, F.: 1965, 'Displacements, Strains and Tilts at Teleseismic Distances', *J. Geophys. Res.* **70**, 2395–2412.

Savage, J. C. and Hastie, L. M.: 1966, 'Surface Deformation Associated with Dip-Slip Faulting', *J. Geophys. Res.* **71**, 4897–4904.

Singh, S. J. and Ben-Menahem, A.: 1969, 'Displacement and Strain Field due to Faulting in a Sphere', *Phys. Earth Planet. Int.* **2**, 77–87.

Steketee, J. A.: 1958a, 'On Volterra's Dislocations in a Semi-Infinite Medium', *Can. J. Phys.* **36**, 192–205.

Steketee, J. A.: 1958b, 'Some Geophysical Applications of the Elasticity Theory of Dislocations', *Can. J. Phys.* **36**, 1168–1198.

Wideman, C. J. and Major, M. W.: 1967, 'Strain Steps Associated with Earthquakes', *Bull. Seismol. Soc. Am.* **57**, 1429–1444.

Discussion

Haubrich: Have you calculated the changes in the inertia tensor?

Singh: The expressions for displacements everywhere within a sphere have been developed but the actual computation has not been carried out. In the next few months changes in the inertia tensor will be calculated.

Haubrich: What will be the differences if you used a layered spherical earth model?

Singh: I do not know, but in the case of a layered half space, the difference in the displacements is not very significant.

Anonymous questioner: Have you thought about using the same basic theoretical approach taking the self gravitational field of the earth into account?

Singh: Yes, we have tried that approach but so far we have not been successful.

Mansinha: What is the behaviour of the displacement field in the lower half of the hemisphere? Does it increase or does it decrease towards the antipode?

Singh: Some people were expecting it to increase but we haven't found it.

Mansinha: When you compute that displacement inside the sphere do you find any depth at which there is an increase in displacement or in the strain, say like a resonance effect?

Singh: So far we haven't computed the displacements at depth.

Anonymous questioner: Do you agree with Press's estimate of the strain at Hawaii due to the Alaskan earthquake?

Singh: Yes.

PRESENT DAY MEASUREMENT AND ANALYSIS OF ROTATION AND POLAR MOTION

POLAR MOTION IN RECENT YEARS

S. YUMI

International Latitude Observatory, Mizusawa-shi, Japan

1. Introduction

The work of the International Polar Motion Service which aims at the study of polar motion has progressed smoothly from its beginning at 1962.0 with the collaboration of more than 50 stations in 24 countries – Algeria, Argentina, Australia, Belgium, Brazil, Canada, Chile, China, Czechoslovakia, Equador, Finland, France, Germany, Italy, Japan, Poland, Rumania, South Africa, Spain, Switzerland, UK, USA, USSR and Yugoslavia.

Preliminary summaries of the results of latitude observations are published in the *Monthly Notes of the IPMS* as a rapid service and the detailed description of the results are in the *Annual Report of the IPMS**.

Though the coordinates of the pole are to be derived using all the available data all over the globe and it is also considered to be most reasonable to derive the polar coordinates in conformity to those which were already given by the ILS, there seems to exist several problems to be solved when the results of the independent stations are combined with those of the 5 ILS stations, such as the determination of the mean latitude referred to the Conventional International Origin[†], treatment of the z-term and the comparison among the respective star catalogues adopted by the respective stations. For this reason, the coordinates of the pole have been calculated only from the results of the 5 ILS stations.

In this note, the coordinates of the pole in recent years derived from the results of the 5 ILS stations are given and they are compared with the provisional ones by the IPMS stations. Polar motion analyzed into annual and Chandler terms is also demonstrated.

2. Polar Coordinates by the 5 ILS Stations

Coordinates of the instantaneous pole x, y and the non-polar variation of latitude, z are calculated by the equations

$$\Delta\varphi_i = x \cos\lambda_i + y \sin\lambda_i + z,$$
$$\Delta\varphi_i = \varphi_i - \Phi_i,$$

$$(1)$$

where φ_i is the observed latitude at the station i, the west longitude and mean latitude of which are λ_i and Φ_i respectively.

---

* *Annual Report of the IPMS* for the years 1962, 1963, 1964, 1965, 1966, 1967 have been published.
† Origin of the polar coordinates adopted at the XIII General Assembly of IAU (Prague 1967) and also at the XIV General Assembly of IUGG (Lucern, 1967)

L. Mansinha et al. (eds.), Earthquake Displacement Fields and the Rotation of the Earth, 45–53.
All Rights Reserved. Copyright © 1970 by D. Reidel Publishing Company, Dordrecht-Holland

Polar coordinates x, y referred to the CIO and z are practically calculated as the least square solutions of (1) by the formulae

$$x = - .4359\Delta\varphi_M + .1227\Delta\varphi_K + .4483\Delta\varphi_C + .1232\Delta\varphi_G - .2583\Delta\varphi_U,$$

$$y = - .2636\Delta\varphi_M - .3133\Delta\varphi_K - .0172\Delta\varphi_C + .3382\Delta\varphi_G + .2559\Delta\varphi_U, \quad (2)$$

$$z = + .2305\Delta\varphi_M + .2007\Delta\varphi_K + .1755\Delta\varphi_C + .1850\Delta\varphi_G + .2082\Delta\varphi_U.$$

Mean latitude of each of the 5 ILS stations defined in the CIO System and the adopted longitudes which are used in the calculation of coefficients of $\Delta\varphi_i$ in (2) are given in Table I.

TABLE I

Station	λ	Φ (CIO)
Mizusawa, Japan	$-9^h24^m31.^s4$	$+39°8'\ 3''.602$
Kitab, USSR	$-4\ 27\ 31.4$	1.850
Carloforte, Italy	$-0\ 33\ 14.9$	8.941
Gaithersburg, USA	$+5\ 8\ 47.8$	13.202
Ukiah, USA	$+8\ 12\ 50.3$	12.096

Values of x, y and z calculated for every month are given in Table V and Figure 1.

3. Polar Coordinates by the 26 IPMS Stations

Coordinates of the instantaneous pole are also calculated from the results of latitude observations at the 26 IPMS stations (see Table II), including the 5 ILS stations, which afforded the continued data from the beginning of 1962.0.

TABLE II

No.	Station	Instrument	No.	Station	Instrument
1	Mizusawa	F.Z.T.	14	Greenwich	P.Z.T.
2	Blagoveshchensk	ZTL-180	15	Ottawa	P.Z.T.
3	Irkutsk	ZTL-180	16	Washington	P.Z.T.
4	Engelhardt	ZTL-180	17	Richmond	P.Z.T.
5	Poltava	V.Z.T.	18	Potsdam	Astrolabe
6	Warsaw	V.Z.T.	19	Besançon	Astrolabe
7	Belgrade	V.Z.T.	20	Alger	Astrolabe
8	Borowiec	V.Z.T.	21	Paris	Astrolabe
9	Mt. Stromlo	P.Z.T.	22	Mizusawa	V.Z.T.
10	Mizusawa	P.Z.T.	23	Kitab	V.Z.T.
11	Tokyo	P.Z.T.	24	Carloforte	V.Z.T.
12	Hamburg	P.Z.T.	25	Gaithersburg	V.Z.T.
13	Neuchâtel	P.Z.T.	26	Ukiah	V.Z.T.

The results are given in Table V and Figure 1. Differences between the results by the 5 ILS and the 26 IPMS are also given in Table V.

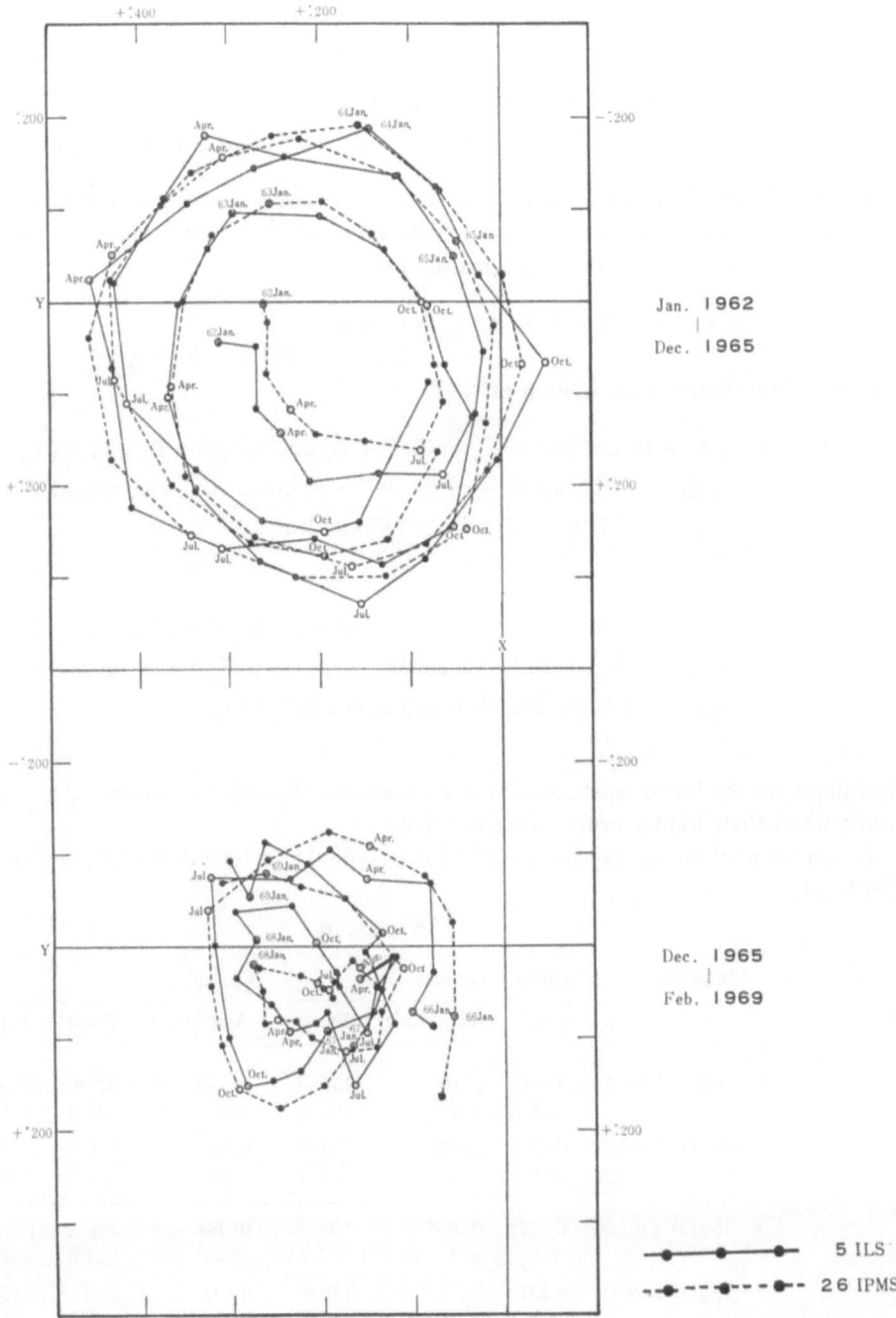

Fig. 1.   Polar orbit.

4. Annual and Chandler Terms

Coordinates of the instantaneous pole are analysed into the forms

$$x = A_x + B_x \cos 2\pi T + C_x \sin 2\pi T + D_x \cos 2\pi F_x T + E_x \sin 2\pi F_x T,$$
$$y = A_y + B_y \cos 2\pi T + C_y \sin 2\pi T + D_y \cos 2\pi F_y T + E_y \sin 2\pi F_y T, \quad (3)$$

where T is the epoch in years, A, B, C, D, E and F are the unknowns to be determined.

Let A', B', C', D', E', F' be the approximate values of the 6 unknowns and a, b, c, d, e, f be their respective small corrections, that is,

$$A = A' + a, \quad B = B' + b, \quad C = C' + c,$$
$$D = D' + d, \quad E = E' + e, \quad F = F' + f,$$

then the Equations (3) are written as

$$x = (A'_x + B'_x \cos 2\pi T + C'_x \sin 2\pi T + D'_x \cos 2\pi F'_x T + E'_x \sin 2\pi F'_x T)$$
$$+ \{(a_x + b_x \cos 2\pi T + c_x \sin 2\pi T + d_x \cos 2\pi F'_x T + e_x \sin 2\pi F'_x T)$$
$$+ (- 2\pi T D'_x \sin 2\pi F'_x T + 2\pi T E'_x \cos 2\pi F'_x T) f_x\}$$
$$+ \cdots$$

$$(4)$$

$$y = (A'_y + B'_y \cos 2\pi T + C'_y \sin 2\pi T + D'_y \cos 2\pi F'_y T + E'_y \sin 2\pi F'_y T)$$
$$+ \{(a_y + b_y \cos 2\pi T + c_y \sin 2\pi T + d_y \cos 2\pi F'_y T + e_y \sin 2\pi F'_y T)$$
$$+ (- 2\pi T D'_y \sin 2\pi F'_y T + 2\pi T E'_y \cos 2\pi F'_y T) f_y\}$$
$$+ \cdots$$

and these are the linear equations for the unknowns of small corrections a, b, c, d, e and f when their higher orders are neglected.

Results of analyses during the period 1962.0–1968.0 and 1963.0–1969.0 are given in Table III.

TABLE III

	Interval	Const. part	Annual Amplitude	Phase	Chandler Amplitude	Phase	Period
x	1962.0–1968.0	0″.034 ±2	0″.101 ±4	235°.3 ±2.0	0″.146 ±4	80°.9 ±2.9	1^{yr}.206 ±4
	1963.0–1969.0	0.025 ±3	0.099 ±4	238.9 ±2.4	0.142 ±6	138.1 ±2.6	1.208 ±4
y	1962.0–1968.0	0″.225 ±1	0″.076 ±2	135°.6 ±1.7	0″.146 ±2	349°.0 ±1.8	1^{yr}.205 ±2
	1963.0–1969.0	0.231 ±2	0.071 ±3	140.6 ±2.7	0.137 ±5	51.4 ±2.3	1.198 ±4

For the comparison of the results obtained with those during the past, polar coordinates from 1890 are analysed by a similar method. Results are shown in Figures 2 and 3. Their mean values over the whole period are given in Table IV.

TABLE IV

		x-comp.	y-comp.
Annual	Amplitude	0″.097	0″.080
Chandler	Amplitude	0″.151	0″.150
	Period	1ʸʳ.177	1ʸʳ.177

Each component of the constant part in the analysed results are plotted on the *x*-*y* diagram as shown in Figure 4.

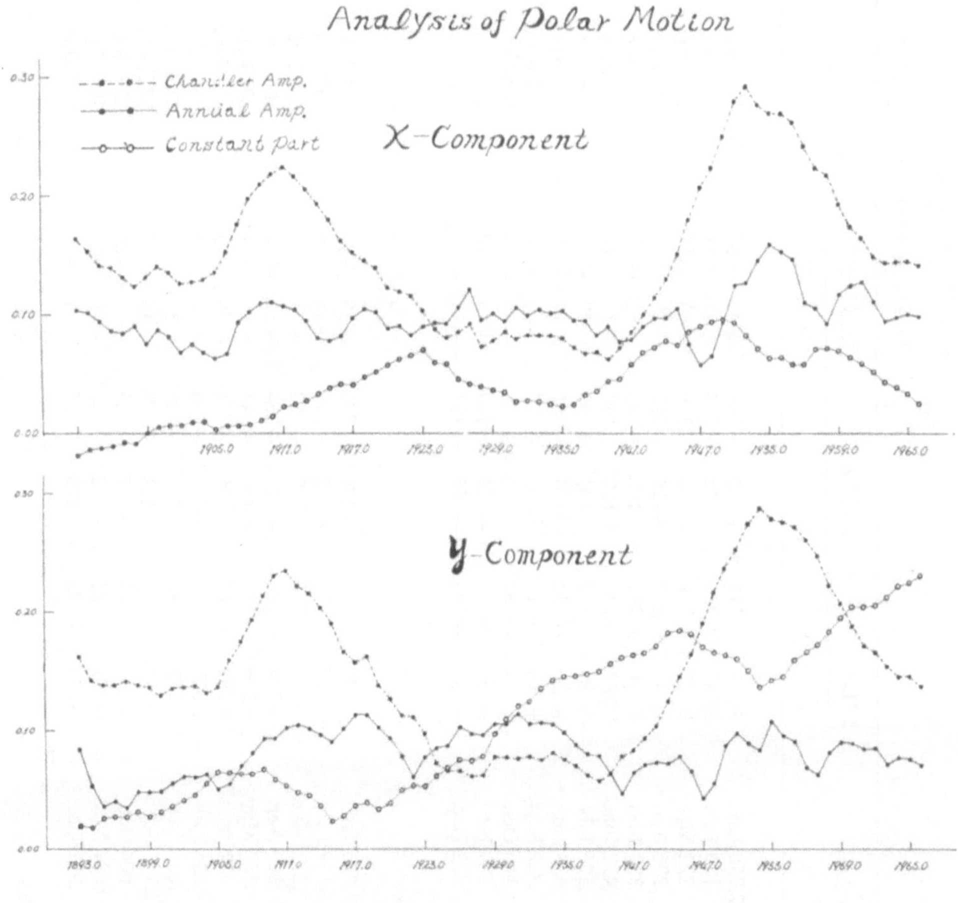

Fig. 2. Polar motion analysed into annual, Chandler and constant parts.

TABLE V

Polar coordinates (unit: 0".001)

Besselian year	5 ILS x	5 ILS y	26 IPMS x	26 IPMS y	(ILS-IPMS) Δx	(ILS-IPMS) Δy
1962.056 Jan.	+ 43	+ 311	+ 2	+ 261	+ 41	+ 50
.139 Feb.	+ 49	+ 269	+ 22	+ 256	+ 27	+ 13
.222 Mar.	+ 117	+ 269	+ 78	+ 258	+ 39	+ 11
.306 Apr.	+ 143	+ 242	+ 118	+ 231	+ 25	+ 11
.389 May	+ 196	+ 211	+ 145	+ 203	+ 51	+ 8
.472 June	+ 187	+ 136	+ 152	+ 150	+ 35	− 14
.556 July	+ 189	+ 66	+ 162	+ 90	+ 27	− 24
.639 Aug.	+ 126	+ 33	+ 109	+ 65	+ 17	− 32
.722 Sept.	+ 69	+ 64	+ 69	+ 74	0	− 10
.806 Oct.	+ 4	+ 82	0	+ 88	+ 4	− 6
.889 Nov.	− 57	+ 128	− 73	+ 142	+ 16	− 14
.972 Dec.	− 93	+ 198	− 108	+ 195	+ 15	+ 3
Mean					+ 25	0
1963.056 Jan.	− 96	+ 294	− 106	+ 253	+ 10	+ 41
.139 Feb.	− 58	+ 322	− 72	+ 317	+ 14	+ 5
.222 Mar.	+ 3	+ 356	0	+ 350	+ 3	+ 6
.306 Apr.	+ 104	+ 368	+ 92	+ 364	+ 12	+ 4
.389 May	+ 206	+ 338	+ 190	+ 347	+ 16	− 9
.472 June	+ 282	+ 266	+ 256	+ 271	+ 26	− 5
.556 July	+ 329	+ 157	+ 288	+ 166	+ 41	− 9
.639 Aug.	+ 279	+ 86	+ 264	+ 83	+ 15	+ 3
.722 Sept.	+ 173	+ 5	+ 183	+ 16	− 10	− 11
.806 Oct.	+ 66	− 49	+ 67	− 23	− 1	− 26
.889 Nov.	− 29	+ 25	− 30	− 1	+ 1	+ 26
.972 Dec.	− 124	+ 72	− 120	+ 67	− 4	+ 5
Mean					+ 10	+ 3
1964.056 Jan.	− 187	+ 144	− 191	+ 155	+ 4	− 11
.139 Feb.	− 144	+ 271	− 179	+ 250	+ 35	+ 21
.222 Mar.	− 106	+ 345	− 139	+ 339	+ 33	+ 6
.306 Apr.	− 24	+ 454	− 51	+ 429	+ 27	+ 25
.389 May	+ 71	+ 430	+ 39	+ 455	+ 32	− 25
.472 June	+ 224	+ 409	+ 172	+ 431	+ 52	− 22
.556 July	+ 269	+ 308	+ 254	+ 342	+ 15	− 34
.639 Aug.	+ 258	+ 206	+ 299	+ 227	− 41	− 21
.722 Sept.	+ 286	+ 132	+ 298	+ 129	− 12	+ 3
.806 Oct.	+ 245	+ 54	+ 248	+ 40	− 3	+ 14
.889 Nov.	+ 122	+ 29	+ 133	+ 17	− 11	+ 12
.972 Dec.	+ 54	+ 20	+ 26	+ 9	+ 28	+ 11
Mean					+ 13	− 2
1965.056 Jan.	− 50	+ 53	− 66	+ 47	+ 16	+ 6
.139 Feb.	− 136	+ 112	− 137	+ 115	+ 1	− 3
.222 Mar.	− 157	+ 237	− 176	+ 221	+ 19	+ 16
.306 Apr.	− 179	+ 324	− 156	+ 304	− 23	+ 20
.389 May	− 111	+ 371	− 104	+ 374	− 7	− 3
.472 June	+ 21	+ 427	− 23	+ 431	+ 2	− 4
.556 July	+ 110	+ 414	+ 85	+ 428	+ 25	− 14
.639 Aug.	+ 183	+ 337	+ 200	+ 363	− 17	− 26
.722 Sept.	+ 238	+ 263	+ 262	+ 276	− 24	− 13
.806 Oct.	+ 250	+ 195	+ 277	+ 195	− 27	0
.889 Nov.	+ 240	+ 158	+ 259	+ 126	− 19	+ 32
.972 Dec.	+ 87	+ 80	+ 163	+ 71	− 76	+ 9
Mean					− 11	+ 2

Date						
1966.056 Jan.	+ 72	+102	+ 77	+ 57	− 5	+45
.139 Feb.	+ 28	+ 79	− 26	+ 58	+54	+21
.222 Mar.	− 67	+ 82	− 77	+ 87	+10	− 5
.306 Apr.	− 73	+152	−109	+148	+36	+ 4
.389 May	−105	+192	−123	+193	+18	− 1
.472 June	− 73	+236	− 94	+259	+21	−23
.556 July	− 74	+323	− 39	+327	−35	− 4
.639 Aug.	+ 1	+319	+ 43	+323	−42	+ 4
.722 Sept.	+100	+305	+107	+313	− 7	+ 8
.806 Oct.	+151	+285	+155	+294	− 4	+ 9
.889 Nov.	+146	+255	+176	+248	−30	+ 7
.972 Dec.	+135	+225	+152	+201	−17	+24
Mean					0	+ 4
1967.056 Jan.	+ 93	+196	+108	+167	−15	+29
.139 Feb.	+ 73	+146	+ 71	+145	+ 2	+ 1
.222 Mar.	+ 11	+124	+ 47	+135	−36	−11
.306 Apr.	+ 35	+161	+ 25	+161	+10	0
.389 May	+ 12	+119	+ 15	+168	+ 3	−49
.472 June	+ 45	+142	+ 38	+188	+ 7	−46
.556 July	+ 94	+153	+ 48	+194	+46	−41
.639 Aug.	+ 44	+182	+ 57	+190	−13	− 8
.722 Sept.	+ 29	+186	+ 46	+200	−17	−14
.806 Oct.	− 5	+207	+ 40	+206	−45	+ 1
.889 Nov.	− 43	+234	+ 32	+224	−75	+10
.972 Dec.	− 38	+296	+ 23	+270	−61	+26
Mean					−17	− 9

Date						
1968.056 Jan.	− 7	+274	+ 20	+277	−27	− 3
.139 Feb.	+ 35	+295	+ 49	+266	−14	+29
.222 Mar.	+ 64	+257	+ 84	+263	−20	− 6
.306 Apr.	+ 93	+237	+ 80	+250	+13	−13
.389 May	+ 84	+209	+ 82	+230	+ 2	−21
.472 June	+ 73	+196	+ 99	+212	−26	−16
.556 July	+151	+166	+114	+177	+37	−11
.639 Aug.	+ 85	+122	+110	+142	−25	−20
.722 Sept.	+ 6	+154	+ 71	+136	−65	+18
.806 Oct.	− 14	+135	+ 24	+111	−38	+24
.889 Nov.	− 88	+217	+ 51	+176	−37	+41
.972 Dec.	−113	+262	+ 65	+223	−48	+39
Mean					−21	+ 5
1969.056 Jan.	− 54	+281	− 79	+262	+25	+19
.139 Feb.	− 92	+302	− 69	+311	−23	− 9

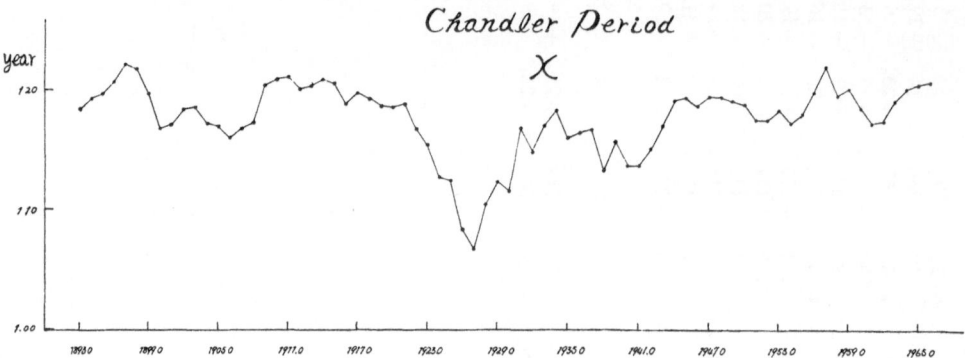

Fig. 3. Chandler period of the polar motion.

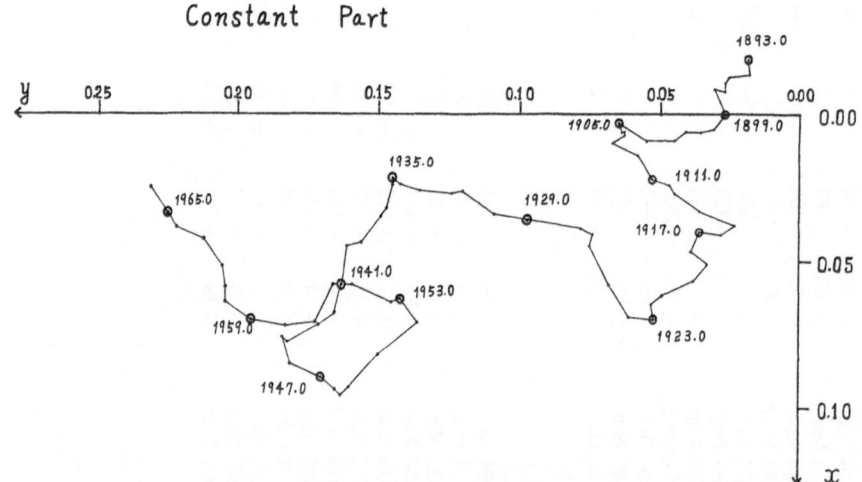

Fig. 4. Locus of the constant part in the analysed results of polar motion.

Discussion

Major: You are working to a 0.001 sec of arc, is that right?

Yumi: I give values of the polar coordinate to 0".001 which is probably the accuracy when comparing positions in a very short interval. When you compare them over a long period the accuracy probably falls to 0".01.

Major: The point that interests me is that in the measurements themselves the tilt due to the earth's tide at the observing station is apparently not important, is that right?

Yumi: It is very important and I will talk about that in the next two papers I give.

Smylie: With an annual term, and a Chandler period of about 1.2 years do you not need a longer time record than 6 years to accurately resolve these periods?

Yumi: Yes.

Markowitz: I would like to comment on a couple of questions that have been asked because I think it will help geophysicists if they know what astronomers are doing. The question of earth tides was raised. Since the results that Dr. Yumi has presented were taken from the ILS monthly means, the earth tide effects will tend to disappear. If you want to look for earth tide effects the data will have to be analysed differently.

Anonymous questioner: How frequently is an observation made?

Markowitz: Once a night if the weather is clear, which would be about 16 times a month. Some places might observe about 300 nights a year (such as Ukiah) whereas others might be as low as 150, but you have the results of five stations which are tied together.

Swetnick: How well known are the positions of the observation stations and are they tied to a common global coordinate system or to a local datum?

Yumi: The polar coordinates are calculated assuming that the positions of the stations are known on a global system. However, there might be some influence due to crustal movement since a tilting of the earth's surface would change the position of the plumb line. The IAU recommended simultaneous geophysical and astronomical stations to check this point.

Mansinha: Could you or Guinot comment on the reason for the large phase shift for 1926; how many stations were in operation in 1926 and how reliably established is the sudden change of phase?

Yumi: The polar motion was observed at only three stations.

Mansinha: So that this large change of phase might not be real.

Yumi: Yes.

Guinot: I made the analysis on the three stations individually, as well as on *x* and *y*, and the large shift of phase appeared on the three station results as well.

O'Hora: This was also shown quite independently of the ILS results by Markowitz when he combined the results of the Washington PZT with the Greenwich floating zenith tube.

Runcorn: Like Smylie, I am a little disturbed at using such small lengths of data for analysis and I wonder whether you have thought of just using ordinary spectral analysis?

Yumi: I did not attempt any spectrum analysis but I think it would be necessary to make more detailed analyses in the future.

Haubrich: A couple of the questions have centred on how to separate two closely spaced peaks in frequency when you have only a limited length of record. A few years ago it was shown by Munk and Hasselman that it is possible, in a method called super resolution, to separate frequencies which are less than the reciprocal record length. One is not limited if the method is used properly as in the present work.

Markowitz: I would like to clarify the answer to Swetnick's question. The coordinates of an ILS station do not have to be known. You adopt a purely arbitrary position which is regarded as a constant, so what astronomers are really dealing with is a purely arbitrary origin given by the adopted constants. Any attempt to tie these positions into other stations would be a different matter.

WORK OF THE BUREAU INTERNATIONAL DE L'HEURE
ON THE ROTATION OF THE EARTH

BERNARD GUINOT

Bureau International de l'Heure, Paris, France

Abstract. The methods used by the BIH are described: the polar motion and UT1 are deduced from 58 series of universal time observations and 44 series of latitude observations (80 instruments); precautions are taken to keep an homogeneous system in spite of the changes in the list of the participating observatories. Recent results are presented. The accuracy on x and y is $0''.01$; on UT1, $0^s.001$ (standard errors).

During the last few years, the Bureau International de l'Heure devised an entirely new method to obtain UT1 and simultaneously the coordinates of the pole. A complete description of this method appears in the *Annual Report* of the BIH (Guinot and Feissel, 1969). In the present paper, our aim is to describe the salient features of this method and to show what contribution the BIH results may bring in astronomical and geophysical research.

1. The Data

Our purpose is to compute the coordinates of the pole and the universal time UT1 from all available measurements. Measurement of lunar distances and long baseline interferometry bring the hope of an improved accuracy; but nowadays we have only at our disposal angular measures of latitude and UT0 and they lead to two types of equations:

$$x \cos L_0 + y \sin L_0 = \varphi - \varphi_0 \tag{1}$$

$$-x \operatorname{tg}\varphi_0 \sin L_0 + y \operatorname{tg}\varphi_0 \cos L_0 + UT1 - UTC = UT0 - UTC \tag{2}$$

where φ_0 and L_0 are the initial latitude and longitude of the astronomical instrument, φ the measured instantaneous latitude, UTC a common time scale (broadcast by time signals); UT0 is computed from the observed local sidereal time with the use of L_0.

The well-known solution obtained since 1900 by the ILS/IPMS rests upon a very small number of type (1) equations where an auxiliary unknown z is introduced to take into account the common errors of the data (presently 5 equations). This solution ignored a considerable amount of valuable data: in 1968, the BIH used 102 equations [44 of type (1) and 58 of type (2)], obtained with the instruments of Table I. However, one must note that the 18 visual transit instruments and the 2 circumzenithals received very small weights. In the near future, the list of Table I will be increased by 2 PZT's (Argentina, Canada) and by several astrolabes (Argentina, Belgium, Canada, Italy, Turkey, U.S.A.).

L. Mansinha et al. (eds.), Earthquake Displacement Fields and the Rotation of the Earth, 54–62.
All Rights Reserved. Copyright © 1970 by D. Reidel Publishing Company, Dordrecht-Holland

TABLE I

Instruments included in the BIH solution in 1968

Instrument	Number of instruments	
	for φ	for UT0
Photographic zenith tube	9	9
Astrolabe	15	19
Zenith telescope	20	
Visual transit instrument		18
Photoelectric transit instrument		10
Circumzenithal		2

2. The Elimination of Systematic Errors

The errors in the right hand terms of Equations (1) and (2) impose some precautions before solving the system. These errors are the following:

(i)   errors on initial latitudes and longitudes, φ_0 and L_0;

(ii) errors due to the erroneous positions and proper motions of stars;

(iii) periodic annual errors of various sources (meteorological, instrumental);

(iv) irregular systematic errors (personal, instrumental,...); and

(v)   accidental errors.

The global error E, computed by difference between the observed values (of φ or UT0) and the computed values from the BIH solution, is represented by a constant term, annual and semi-annual components:

$$E = a + b \sin 2\pi t + c \cos 2\pi t + d \sin 4\pi t + e \cos 4\pi t$$

(t expressed in years). Inasmuch as errors (iv) and (v) are comparatively small, the coefficients a, b, c, d, e will be nearly constant, or very slowly changing with time, due to the errors of proper motions, as long as the observation program remains unchanged. Table II gives a few examples of the variation of these coefficients. The predominance of periodic errors on irregular errors is apparent. Generally speaking, the fluctuation of the coefficients is of the order of $\pm 0\overset{''}{.}01$ to $\pm 0\overset{''}{.}02$ (or ± 1 to ± 2 ms) for the PZT's and the astrolabes, somewhat larger for other types of instruments.

The stability of these coefficients allows us to compute them from the past observations, then to use them as corrective terms for the present observations. In practice, to keep the coefficients up to date, they are computed twice a year, over one year intervals. After this correction, results of all instruments are freed from all the constant and periodic errors (i), (ii) and (iii), and it is permissible to alter the list of the instruments entering in the solution and their weights. These changes do not alter the reference system for the coordinates of the pole and UT1. When a new instrument, or a new program of observations on the same instrument, is put into operation, the BIH uses its data only after one year has elapsed and its coefficients are known. Similarly, these changes do not alter the measured annual terms of the polody and of UT1.

BERNARD GUINOT

TABLE II

Systematic errors of the stations (typical examples)

$$E = a + b \sin 2\pi t + c \cos 2\pi t + d \sin 4\pi t + e \cos 4\pi t$$

(t in years)

Instrument	Latitude (0″.001)					Universal time (0ˢ.0001)				
Interval of time	a	b	c	d	e	a	b	c	d	e
PZT (Washington)										
1966.00–1966.95	+ 5	− 105	+ 57	+ 9	− 8	+ 4	0	+ 4	− 11	− 35
1966.50–1967.45	0	− 116	+ 40	− 11	+ 2	0	− 13	+ 23	+ 14	− 13
1967.00–1967.95	+ 20	− 144	+ 19	+ 7	− 2	− 2	− 1	+ 20	+ 7	+ 11
1967.50–1968.45	+ 30	− 129	+ 27	+ 14	− 5	− 22	− 28	+ 33	+ 11	+ 9
Astrolabe (Paris)										
1966.00–1966.95	− 6	− 2	+ 35	+ 37	− 7	− 5	− 26	+ 22	+ 13	+ 25
1966.50–1967.45	0	+ 12	+ 18	+ 19	− 21	0	− 18	+ 22	+ 2	+ 18
1967.00–1967.95	+ 7	0	+ 40	− 6	− 25	− 3	− 12	+ 14	+ 6	+ 26
1967.50–1968.45	+ 23	+ 13	+ 46	− 4	− 7	− 10	− 14	+ 19	+ 17	+ 10
Zenith telescope (Pulkovo)										
1966.00–1966.95	− 1	− 32	+ 22	− 4	+ 8					
1966.50–1967.45	0	− 34	+ 14	− 16	+ 15					
1967.00–1967.95	− 7	− 33	+ 27	− 20	+ 7					
1967.50–1968.45	+ 5	− 9	+ 20	− 23	− 11					
Visual transit instrument (Buenos Aires Nav.)										
1966.00–1966.95						− 7	− 13	− 76	− 55	− 36
1966.50–1967.45						0	+ 6	− 59	− 42	− 67
1967.00–1967.95						− 35	+ 54	− 38	− 67	− 58
1967.50–1968.45						− 67	− 3	+ 32	+ 14	− 10
Photoelectric transit instr. (Pulkovo)										
1966.00–1966.95						+ 32	+ 2	+ 9	− 1	+ 29
1966.50–1967.45						0	− 31	− 22	− 16	+ 18
1967.00–1967.95						− 6	− 36	+ 1	− 34	− 13
1967.50–1968.45						0	− 42	+ 3	− 30	+ 23

3. Influence of the Errors on the Proper Motions of Stars. The 1968 BIH System

A. ORIGIN OF THE COORDINATES OF THE POLE

It is often said that the only way to get rid of the errors on the polody caused by the errors on the proper motions is to observe the same stars on the same parallel. Strictly speaking, this is correct. However, it is easy to demonstrate that a statistical method of elimination is possible with negligible errors.

A first attempt to estimate the influence of the proper motions is to suppose that the errors they cause on φ and UT0 are distributed at random among the 102 equations. If we suppose that the drift on φ or $(UT0 - UTC) \cos\varphi_0$ caused by the errors on proper motion is characterized by a standard error of $0''003$ (order of magnitude confirmed by catalogue comparisons and modern observations), the fictitious drift of the pole thus introduced is about $0''0005$ per year. But, in fact, the drifts of every instrument are not independent, because the errors on the proper motions are systematic and slowly changing with the declination. Thus the fictitious drift of the pole becomes smaller than $0''0002$ per year (Guinot and Feissel, 1968). This drift is small compared to the uncertainties due to real drifts of the stations: for instance, when discussing the same data from ILS/IPMS, Markowitz (1968) and Yumi (1968) obtained a difference in the drift of the pole which amounted to $0''0012$ per year (Markowitz: $0''0035$ per year. Yumi: $0''0022$ per year).

Owing to the large number of cooperating stations, the BIH pole is not very much affected by a perturbation in one station; for instance an error of $0''1$ on the latitude of one station would give rise to an error of $0''04$ on the pole as obtained by the IMPS, of $0''005$ on the BIH pole.

If there are local drifts of the verticals (Yumi, 1968), only the statistical method can be used to maintain a constant origin of the polody. The statistical method is then comparable to the method used for deriving the precession of the celestial pole from the proper motion of stars. A large number of cooperating stations is necessary.

An initial set of values of φ_0, L_0 was chosen so that the BIH origin was in coincidence with the Conventional International Origin (CIO), as determined by the IPMS, for the period 1964–1966. But, at subsequent dates, the BIH origin is obtained independently from that of the IPMS. The comparison of the mean pole obtained by both organisations will certainly lead to interesting conclusions.

B. REFERENCES FOR UT1

Similarly, the longitudes L_0 define a reference point on the equator of the BIH origin of the pole: it is defined statistically if there are drifts of the verticals.

The reference point on the celestial sphere is the equinox, but as defined by the catalogues of stars. There is no possibility of eliminating the movement of the catalogue equinox with respect to the real equinox. Therefore, the rotation of the Earth is referred to a slowly and uniformly rotating frame of reference; the order of magnitude of this rotation is $0''01$ per year. This fact has no practical consequence if the same catalogue is used for computing UT1 and for referring the observations of artificial or natural celestial bodies. This is done in principle by the use of the FK4.

C. THE 1968 BIH SYSTEM

The reference system formed by the initial values of φ_0 and L_0 for an initial set of observatories is called the *1968 BIH system*.

4. Methods of Computation. Presentation of the Results

The methods of computation are described in the annual reports of the BIH. The list of observatories, their coordinates, their systematic corrections, their data and residuals are given. We will only recall here that the BIH gives:

(i) a solution of Equations (1) and (2) for every 1/20th of a year. No smoothings are applied to the station data, nor to the published results (published in the annual report);

(ii) a solution for every 5-day interval, computed directly from the observed values of φ and UT0, for every star group or night. No smoothings are applied (published in the annual report);

(iii) smoothed values of x, y, UT1–UTC, obtained from the solution for the 5-day intervals (published in the monthly circulars D, with one month delay and in the annual report).

The values obtained in (iii) fulfill the current needs. Their standard errors are about $0\rlap{.}''01$ for x and y, $0\rlap{.}^{s}001$ for UT1. However, in some exceptional cases, the values of (iii) may differ significantly from the best smoothing of (ii) by an amount of $0\rlap{.}''02$ or $0\rlap{.}^{s}002$. This difference is due to the very short delay available to smooth the raw data and also to amendments to the transmitted data. For these reasons, we strongly recommend to use, for astronomical and geophysical research, the raw data under the forms (i) or (ii).

Besides the systematic corrections described above, we applied no other corrections to the data (no correction for diurnal nutation and for the periodic displacement of the vertical).

Figure 1 shows the polody from 1964.00 to 1968.95 (results under the form (i), without smoothings).

Figure 2 shows the raw values of x, y, UT2–UTC for every 5-day interval, from 1967.0 to 1969.4.

5. Revision of Past Data

We will apply the methods described above to the data transmitted to the BIH since 1955 (beginning of the atomic time). Results will be published as soon as possible; those of 1962 and 1963 will be available in 1970.

6. Some Studies of the BIH Results

In this field of the rotation of the Earth, the duties of the BIH end with the publication of UT1, x and y. Various studies on these results are in process at the Paris Observatory. I briefly present them here, not for the intrinsic value, but only to show the possibilities offered by the BIH data.

(1) Short term irregularities of x, y, UT1. The deviations of individual values for 5-day reductions from the smoothed curve are about 1.5 larger than expected from the standard deviations; it is perhaps due to some short term irregularities. The most remarkable irregularity happened on UT2. It is shown in Figure 3. It seems that UT1 (or UT2) suffered a jump of -10 ms between April 8 and 13, 1969.

(2) Length of the day. The variation in the length of the day, measured in seconds

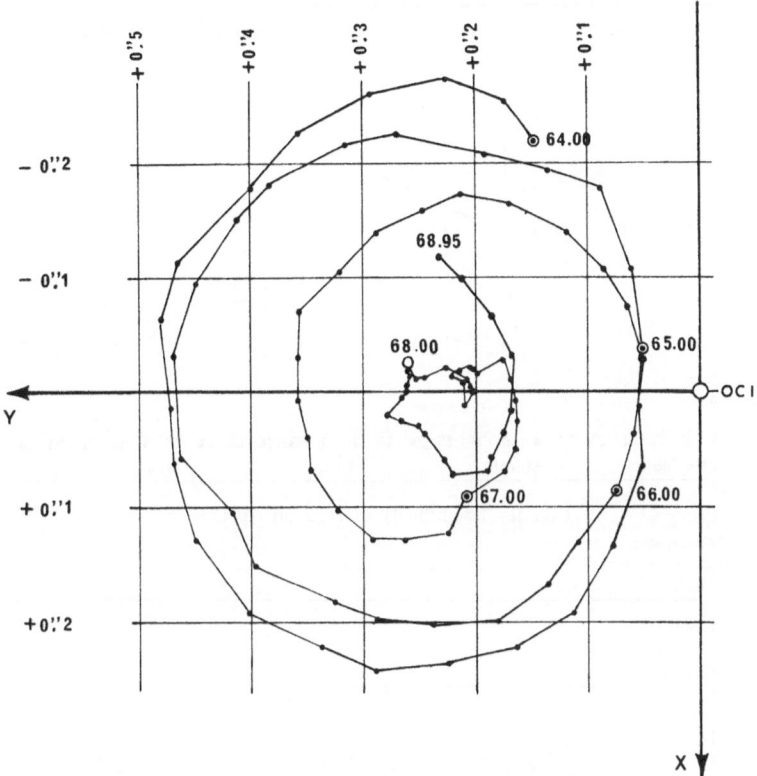

Fig. 1.   Path of the pole from 1964.00 to 1968.95
(unsmoothed values computed for every 1/20th year).

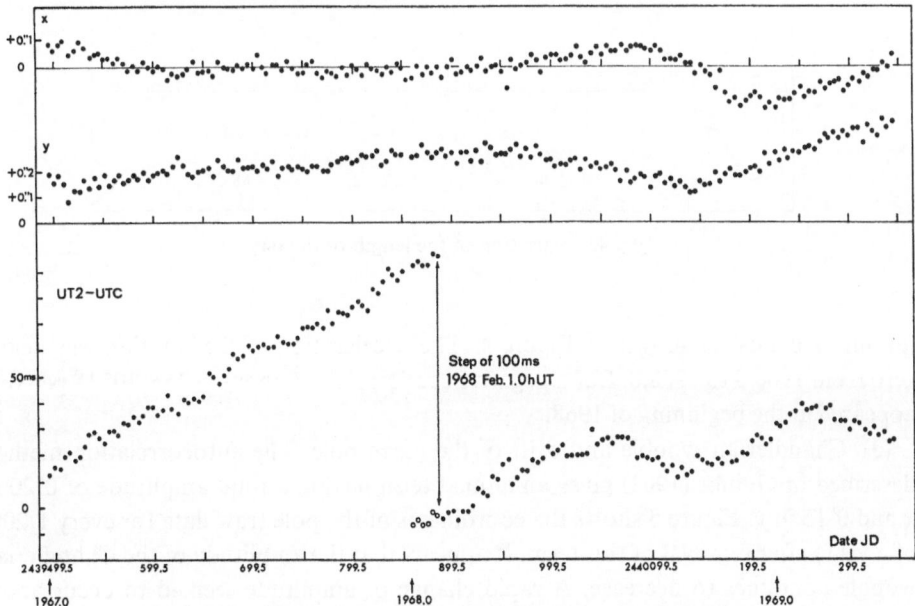

Fig. 2.   Values of x, y, UT2 – UTC for every 5-day interval.

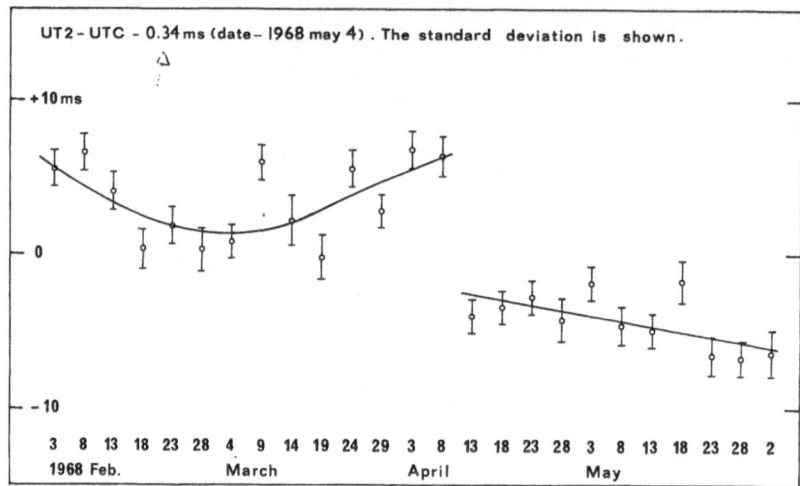

Fig. 3.   Irregularity of UT2 on April 1968.

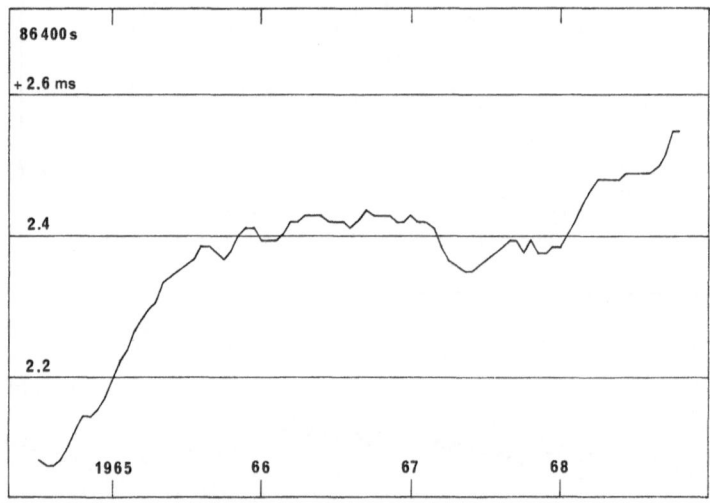

Fig. 4.   Variation of the length of the day.

(atomic seconds) is shown in Figure 4. The acceleration of the rotation was nearly zero from 1966.0 to 1968.0. But a negative acceleration of about -25×10^{-10} per year appeared at the beginning of 1968.

(3) Chandlerian wobble and drift of the mean pole. The autocorrelation method described in Guinot (1963) gives an annual term having a total amplitude of $0.''20$ in x and $0.''15$ in y. Figure 5 shows the coordinates of the pole (raw data for every 1/20th of a year) after removal of this term. It appears that the amplitude of the Chandlerian wobble continues to decrease. A rapid change of amplitude seemed to occur about

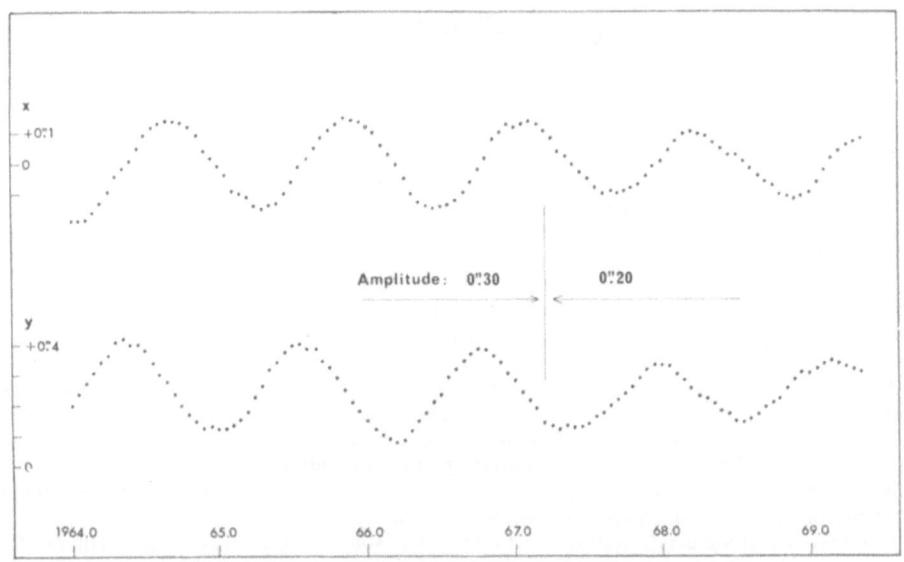

Fig. 5.   Coordinates of the pole corrected for the annual variation.

1967.2. The present amplitude is about 0″20 both in x and y. No significant drift is shown. If there was a drift of the pole between 1964.0 and 1969.3, the rate was less than 0″004/year.

(4) A Fourier analysis of the differences between the raw 5-day results and the smoothed results was performed for the years 1967 and 1968. No significant terms are clearly shown, except in UT1 where the following terms are found:

period	semi-amplitude
13.6 days	0.8 ms
14.2 –	0.8 –
14.8 –	0.9 –

Researches on this problem are in progress.

References

Guinot, B.: 1963, *Bull. Astron.* **24**, 461.

Guinot, B. and Feissel, M.: 1968, in *Continental Drift, Secular Motion of the Pole, and Rotation of the Earth* (ed. by Wm. Markowitz and B. Guinot), IAU Symposium no. 32, D. Reidel, Dordrecht, p. 63.

Guinot, B. and Feissel. M.: 1969, *Annual Report for 1968*, Paris.

Wm. Markowitz: 1968, in *Continental Drift, Secular Motion of the Pole, and Rotation of the Earth* (ed. by Wm. Markowitz and B. Guinot), IAU Symposium no. 32, D. Reidel, Dordrecht, p. 25.

Yumi, S.: 1968, in *Continental Drift, Secular Motion of the Pole, and of the Earth* (ed. by Wm. Markowitz and B. Guinot), IAU Symposium no. 32, D. Reidel, Dordrecht, p. 33.

Discussion

Haubrich: I am a bit surprised at the results if you have used two years of data with five day windows. I have done some four day spectra on short sets of pole data. For a seven year set of data all the power outside the Chandler peak is completely obliterated by the side bands of the Chandler peak. How did you avoid this problem in your two year set of data?

Guinot: I cannot say much except that these three lines appear very sharply in the results.

Haubrich: I would like to suggest that perhaps these three lines are due to the side bands of the Chandler peak. How high above the neighboring frequencies were they?

Guinot: They were about twice the amplitude.

Haubrich: Then I would suggest again that they could be questioned because the factor of two above neighboring frequencies is not significant for data which originally have a Gaussian distribution.

Guinot: Well, you must also take into account the fact that the lines are very narrow; they are not bands.

Markowitz: I would like to go back to the point about the motion of the mean pole. Yumi's figure agrees with mine because we used identical data. The difference is in interpretation of how much is due to polar motion and how much to station motion. Yumi puts 50% and I put 0% in station as moving; some people put 100%. The important point is that although the absolute motion of the pole depends upon the interpreter, the apparent motion is the same. The 0".0002 figure represents accidental error; it does not represent the systematic error.

Mansinha: The IPMS uses 26 stations and the ILS uses 5 stations; does the IPMS use all the available data from the 26 stations to reduce pole positions and find the x and y coordinates?

Tanner: My impression is that Yumi is using 26 for the period from 1962 because those were the ones which were continuous.

Yumi: That is right. We have a total of about 50 stations available but some of them discontinued their observations in the past six or seven years.

Guinot: I used the same data plus some other series because with my method, as soon as we have a series lasting more than a year they can be used. For instance, I used the astrolabe data from Spain, and similar other ones, but Dr. Yumi did not.

Bender: Why didn't the periodic motions show up in x and y as well as in time?

Guinot: I am not sure but one possibility is that if you have an effect due either to the diurnal nutation or to the movement of the vertical, this affects similarly all the latitude measurements so that it vanishes through the z terms introduced. This effect, if it is visible in latitude observations, would appear in the z term but not in the x and y. I think it completely disappeared because it is the same in all the stations on the same parallel but this is not the case for time.

Markowitz: I would like to clarify something raised by Mansinha. There are three different polar motions under discussion. The BIH which uses about 60 stations, the IPMS, which uses 36 stations and the ILS, based on only 5 stations. The polar motion curve which is shown in each annual report of the IPMS is based on only the 5 stations.

STABILITY OF STATION AND THE EARTH'S FIGURE

S. YUMI

International Latitude Observatory, Mizusawa-shi, Japan

Abstract. From the analyses of the residual latitudes of the 26 IPMS stations during six years from 1962 to 1967, a local trend in latitude variation by other than the polar motion was obtained for each station. They are found to be most likely to depend on the terms of 3λ, which might rather be attributed to an assumed gradual deformation of the Earth or a variation of geopotential surface than to the astronomical origins.

Coordinates of the pole calculated from the results of the 26 IPMS stations are compared with those obtained from the five ILS stations and the systematic difference between them is discussed from the viewpoint of the deformable earth or the variable geopotential surface.

1. Magnitude of Errors of the Coordinates of the Pole

Errors in the mean latitude, averaged over 100–400 observations during a month and, consequently, those of the coordinates of the pole are reasonably considered to be in the order of $\pm 0''.01$–$0''.02$, since an error of a single observation of latitude at the ILS stations with the zenith telescope is estimated as $\pm 0''.2$–$0''.3$. In practice, however, the estimated errors in the values of the polar coordinates have been gradually increasing during the past sixty years and may even amount to $\pm 0''.04$–$0''.08$ in recent years. As an example, the estimated probable errors in the x-component of the polar motion which is calculated from the ILS data with reference to the CIO are shown in Figure 1. Probable errors were also estimated for the values of coordinates referred to the revised origin which was defined by the mean latitudes of the 5 ILS stations during these 6 years. Results are also given in Figure 1.

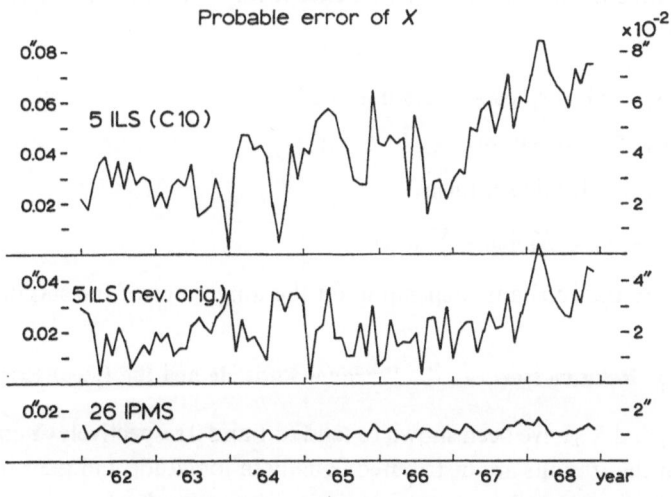

Fig. 1.   Probable error in the x-component of polar motion.

L. Mansinha et al. (eds.), Earthquake Displacement Fields and the Rotation of the Earth, 63–68.
All Rights Reserved. Copyright © 1970 by D. Reidel Publishing Company, Dordrecht-Holland

2. Residual Latitudes

As the errors of the coordinates of the pole are usually estimated by the use of the residual latitudes of the stations concerned, gradual increases in the former are assumed to be due to those in the latter. The residual latitude, which is supposed to be of the same order of $0''.01-0''.02$ as the errors in the monthly mean latitudes, has been gradually increasing during the past sixty years; it amounts even to $0''.13$ for a certain station of the ILS. As a yearly mean it varies from $0''.01$ to $0''.10$ during these years.

Yumi and Wako (1966a) supposed that the gradual increase in residual latitude might be attributed to the southward secular displacement of Mizusawa and to the northward one of Ukiah almost of the order of $0''.001$ per year. But no discussions on the fluctuation superposed on the secular drift were made in their paper. This fluctuation will be discussed in this paper.

Residual latitude $(o-c)$ at the station i is usually expressed as

$$(o - c)_i = \Delta\varphi_i - (x\cos \lambda_i + y\sin \lambda_i + z). \tag{1}$$

As x, y and z are

$$\begin{aligned} x &= \sum a_i \Delta\varphi_i, \\ y &= \sum b_i \Delta\varphi_i, \\ z &= \sum c_i \Delta\varphi_i, \end{aligned} \tag{2}$$

(1) is written as

$$(o - c)_i = (1 - a_i\cos \lambda_i - b_i\sin \lambda_i - c_i)\Delta\varphi_i - \sum{}'(a_j\cos\lambda_i$$
$$+ b_j\sin \lambda_i + c_j)\Delta\varphi_j, \tag{3}$$

where $\sum'$ means a summation on all i's except for $i=i$ and j stands for the stations other than $i=i$.

Put

$$\begin{aligned} K_{ii} &= (1 - a_i\cos \lambda_i - b_i\sin \lambda_i - c_i), \\ K_{ij} &= -(a_j\cos \lambda_i + b_j\sin \lambda_i + c_j), \end{aligned} \tag{4}$$

then the formula (3) takes a form

$$(o - c)_i = K_{ii}\Delta\varphi_i + \sum{}'K_{ij}\Delta\varphi_j. \tag{5}$$

K_{ii} and K_{ij} are the constants depending on the longitudes of the stations concerned.

3. Relation Between the Residual Latitude and the Local Error

Values of K_{ii} and K_{ij} have been shown to tend to 1 and 0 respectively (Yumi and Wako, 1966b) when the stations are distributed equally in longitude and their number grows large. Even in the case of unequal distribution, however, K_{ii} becomes very near to 1 and K_{ij} to 0, if the number of stations is large.

K_{ii} for each of the 5 ILS stations is 0.3–0.4 and K_{ij} is nearly $+0.3$ or -0.3. This means that the residual latitude at any station calculated from only a few stations can not represent a supposed local error of the station. Whereas the values of K_{ii} and K_{ij} for each of the 26 IPMS stations, which provided continuous data from 1962.0 and were used for the derivation of polar coordinates, are almost 0.9 and nearly $+0.1$ or -0.1 respectively. Therefore the residual latitude of the station calculated from the results of a large number of stations is considered to include almost all of the local error.

4. Analyses of Residual Latitudes

Monthly residual latitudes calculated from the results of the 26 stations inclusive of the 5 ILS stations during 6 years from 1962 to 1967 were analysed into a secular part and an annual one.

Annual part might be attributed to the declination errors of stars, instrumental error, meteorological effects and so on. But what cause can be supposed for the secular part in the residual latitude?

Annual amount of increment or decrement in the residual latitude is shown in Figure 2 which suggests a dependence of each annual amount on terms of 3λ. Similar tendency is also obtained from the 13 stations of comparatively equal distribution as shown in Figure 3.

Though it might not be very easy to give a physical explanation on terms of 3λ from the poor data of only 6 years, the writer supposes that it might be attributed to the gradual deformation of the Earth or the variation of geopotential surface rather than to an astronomical origin.

Fig. 2. Annual rate in residual latitude (26 stations).

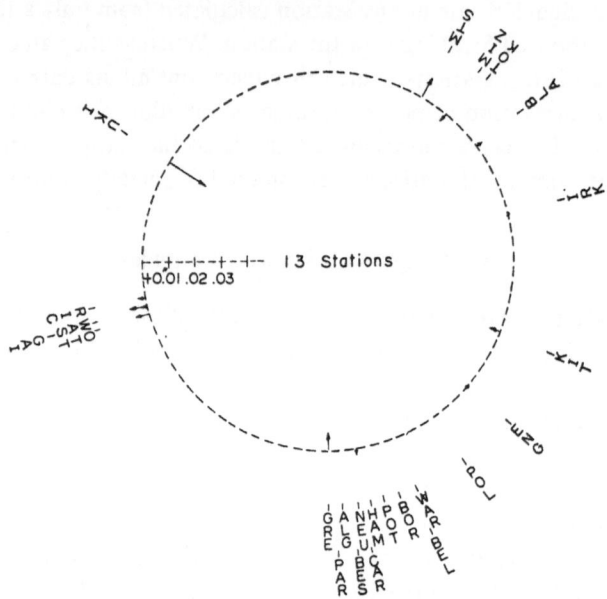

Fig. 3.   Annual rate in residual latitude (13 stations).

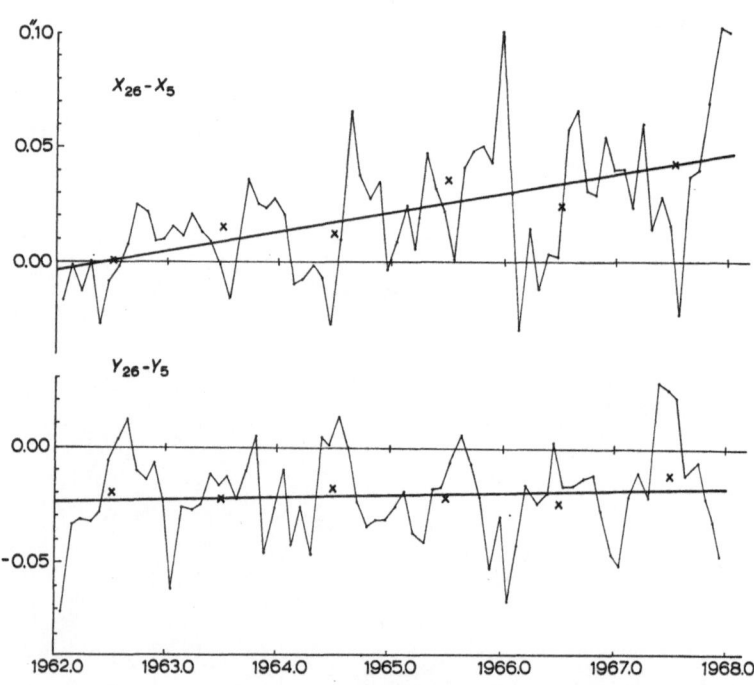

Fig. 4.   Difference between IPMS and ILS coordinates.

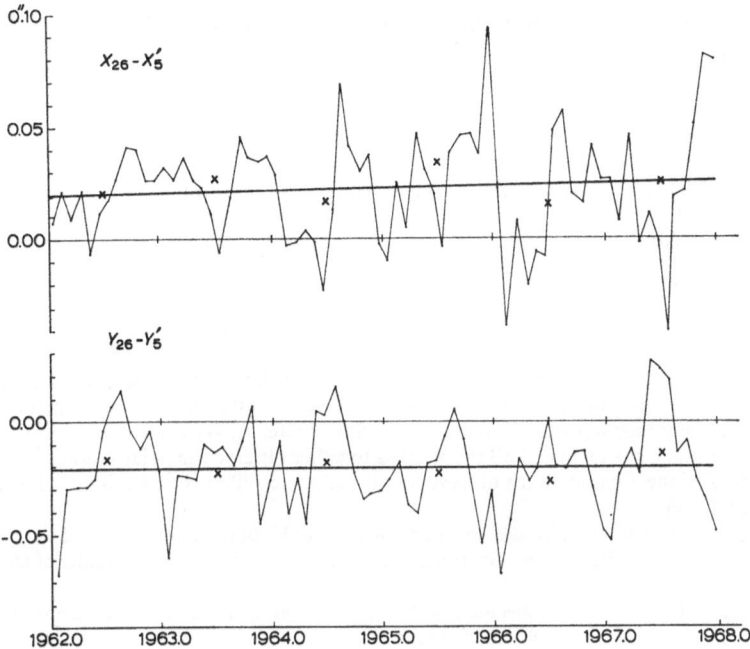

Fig. 5.   Differences between IPMS and ILS coordinates after correction for
annual changes in residual latitudes.

5. Coordinates of the Pole and z

Coordinates of the pole were calculated from the results of the 26 stations and the
13 stations respectively. No appreciable difference was found between the two. It is
noteworthy that nearly the same amount of z as that calculated from the results of the
5 ILS stations, corrected for their declination errors by the chain method, was obtained. This suggests that the Kimura term, amplitude of which is nearly $0''.03$ appears
not only for the ILS stations but also for the world-wide distributed stations with
different instruments which observe different stars independently. Probable error in
x-component of the polar motion by the 26 stations is shown in Figure 1.

Coordinates of the pole from the data of the 26 IPMS stations are compared with
those from the 5 ILS stations; Figure 4 shows the differences in the coordinates. Coincidence is faily good in y, but the difference in x increases year by year.

Differences of polar coordinates between the two results is found to be eliminated
almost completely when the observed values of latitude variation of the 5 ILS stations
are corrected for the annual changes in residual latitudes which are supposed to be
almost a real local error, and it is probable that the polar coordinates from only a few
stations are apt to be affected strongly by a local error of a certain station. Corrected
difference is shown in Figure 5.

From the results of several discussions in this paper, it is highly recommended that
the cooperative observation of latitude with the same instrument at large number of

stations on the same parallel of latitude should be made to find polar coordinates with high accuracy. At least fifteen stations are desirable.

References

Yumi, S. and Wako, Y.: 1966a, 'On the Secular Motion of the Mean Pole', *Publ. Int. Latit. Observ. Mizusawa* **5**, 61.

Yumi, S. and Wako, Y.: 1966b, 'On the Secular Motion of the Mean Pole', *Publ. Int. Latit. Observ. Mizusawa* **5**, 80.

Discussion

Rochester: Isn't there something rather anomalous about the large change in amplitude of the annual component of polar motion which is a factor of three over the amplitude of the annual motion?

Markowitz: May I answer that. It doesn't make any difference. We correct the observed $UT0$ for a $\Delta\lambda$. If the amplitude has increased all the stations together observe the polar motion. Guinot puts in the correction on the λ based on the observation. It makes no difference what it is; it could be a mile, a foot or a kilometer.

Guinot: I gave the total amplitude, not the semi-amplitude, of the annual term of the pole path.

Bender: What values did you use for the polar motion in correcting your results of the motion of the pole?

Markowitz: I did not use any specific results. The BIH publishes a table for each observatory $\Delta\lambda$. This was used, of course, at the Naval Observatory.

Bender: When you converted into angular measure it seemed that you could get systematic errors of $0''.2$ over a period of a month.

Markowitz: I would have to check the numbers but certainly systematic errors can and do exist. There are two PZT's and if you use the data taken over a period of a week to a month you are in the noise level; if you go for three months you are clear of it. If you use 50 or 60 instruments, as the BIH is doing, the error will be smaller.

SUDDEN CHANGES IN ROTATIONAL ACCELERATION
OF THE EARTH AND SECULAR MOTION OF THE POLE

WM. MARKOWITZ

Marquette University, Milwaukee, Wis., U.S.A.

Abstract. Development of the atomic clock in 1955 has made possible the detailed determination of irregular variations in speed of rotation. Sudden changes do not occur in speed but do in acceleration, about every 4 years. The motion of the mean pole consists of a progressive component of 10 cm/yr plus a 24-year oscillation. A deviation for 1966.0 is noted. No correlation is found between earthquakes and changes in rotational speed or acceleration.

1. Introduction

The study of the rotation of the earth, considered as a geophysical phenomenon, proceeds along two lines, observational and theoretical. Analysis of the observations provides a mathematical description of the rotation of the earth. This provides constraints which theories should fit.

Variations in speed of rotation of the earth, secular and irregular, were found by comparing rotational time (UT) with ephemeris time (ET) obtained from the moon. The irregular variation presented an enigma. Whereas the secular retardation could be accounted for by a lunar-tidal couple, there was no explanation for the irregular variation, whose detailed nature could not be determined with the moon as a reference. A matter of particular interest was that of the turning points in speed. It was generally assumed that the speed changed abruptly and remained constant between changes.

In 1952, however, D. Brouwer (1952) made the hypothesis that sudden changes in speed do not occur but sudden changes do occur in acceleration. He represented $\Delta T = ET - UT$ by arcs of parabolas. The probable error of a single annual mean of ΔT was 260 millisec, so that the moon could not be used to verify this hypothesis. Quartz clocks could not be used because their frequencies change with age. However, the cesium-beam atomic clock, developed in 1955.5, provides a reference with the requisite accuracy. The probable error of an annual mean of $UT - AT$, where AT is atomic time, is reduced to 1.2 millisec.

The precision of determining UT has increased in the last few decades through introduction of the PZT and the astrolabe. The PZT's of the U.S. Naval Observatory, at Washington, D.C., and Richmond, Fla., are the only such instruments which have been in continuous use since 1955.5. They have therefore been used to obtain analyses based on homogeneous data from 1955.5 to 1969.5. Comparison of Naval Observatory time determinations with those of other observatories, via time signals, shows that the various determinations of UT are coherent.

The Naval Observatory PZT's are separated by about 1500 km. They are similarly affected by changes in speed of rotation of the earth, but not by atmospheric anoma-

L. Mansinha et al. (eds.), Earthquake Displacement Fields and the Rotation of the Earth, 69–81.

lies, instrumental effects, or errors in the star catalogs. Hence, the use of two PZT's not only provides increased accuracy but enables observational errors to be separated from changes in speed. The atomic time system used is A.1, based on the frequency 9192 631 770 cycles/sec for cesium (U.S.N.O. Time Service Notice, 1959). This value was obtained in a joint experiment which utilized the cesium clock and the dual-rate moon camera (Markowitz *et al.*, 1958).

It is found that sudden changes in speed do not occur but that rapid changes in acceleration have occurred about every 4 years. Since acceleration is proportional to torque it follows that rapid changes in torque have occurred about every 4 years. The occurrence of such rapid changes is interesting, but unexplained.

Data of geophysical interest also come from observations of the polar motion. The International Latitude Service (ILS) stations provide the fundamental polar motion. The observed variations in latitude provide a check on the hypothesis of rotation of plates as a mechanism for continental drift.

2. Rotation of the Earth

In comparing theory with observation it should be clear just what quantities are obtained from the observations. When using the moon the observed quantity is $\Delta T = ET - UT$. With the atomic clock it is $H = UT1 - A.1$. A derived quantity, $K = UT2 - A.1$, is obtained through the relation $UT2 = UT1 + \Delta SV$, where ΔSV is a correction for seasonal variation. The adopted BIH formula is

$$\Delta SV = 25.1 \text{ ms } \sin(2\pi t - 0.499) - 9.2 \text{ ms } \sin(4\pi t - 0.862),$$

where t is the fraction of the year.

A. TIME

The heavy line in Figure 1 shows the mean monthly values of H, for Washington and

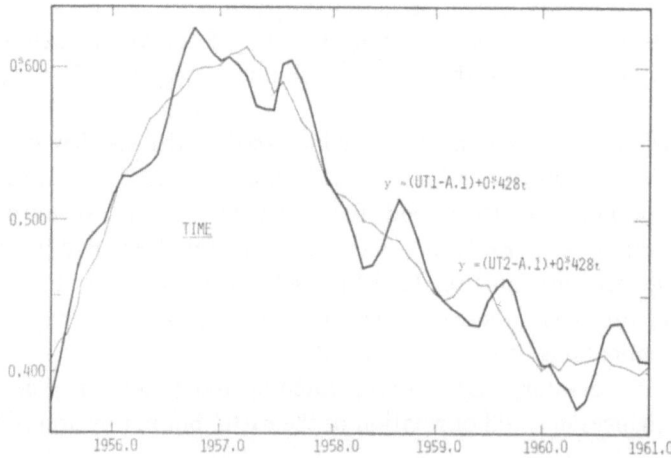

Fig. 1. Monthly means of Washington and Richmond, plus a linear term; t is in years from 1959.0. Heavy line, UT1–A.1; light line, UT2–A.1.

Richmond combined, from 1955.5 to 1961.0. The variations in speed of rotation of the earth are given by the variations in H. The irregular and 12-month variations are clearly seen. A 6-month variation is evidenced from the asymmetry of the annual variation. To study the irregular variation it is advantageous to use K, given by the light line. K can be represented by two parabolic arcs which are tangent just before 1958.0.

B. SPEED

Let L be the length of the day in seconds of atomic time, and let $dL = L - 86400$. Then $\sigma = -dL/L$ is the deviation in speed of rotation, expressed as a fraction. If $dL = +1$ millisec then $\sigma = -116 \times 10^{-10}$. Figure 2 shows mean monthly values of σ, for Washington and Richmond separately, from 1962.0 to 1966.0. These were obtained from K, so that σ represents the speed of rotation when corrected for seasonal variation by a formula with fixed coefficients, and not the actual speed. The graphs show a generally continuous deceleration, with a pronounced deviation in 1963. A strong correlation is evident between the decelerations, and the deviations therefrom, shown by Washington and Richmond.

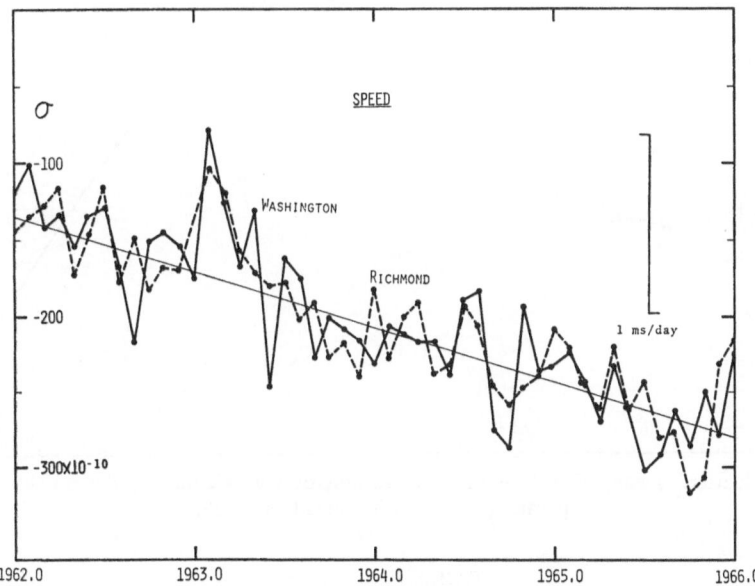

Fig. 2. Mean monthly deviations in speed of rotation for Washington and for Richmond; 1962.0 to 1966.0.

The observational probable error of σ for the mean of 2 PZT's is 14×10^{-10} for a monthly mean and 4×10^{-10} for a quarterly mean. We wish to study variations in σ of 10×10^{-10} so that quarterly means were used in the subsequent analysis.

The quarterly means showed an annual systematic effect. To eliminate this, corrections were added to the quarterly values of K as follows: $+2$, -5, -4, and $+7$ ms, respectively to obtain K^*. The systematic effect could be due to an error in the formula for ΔSV or to errors in the star positions of the PZT catalogs. In effect, ΔSV was replaced by a quantity ΔSV^* to obtain K^*.

Figure 3 shows K^* from 1955.625 to 1969.375. The nominal quarterly epochs are: 0.125, 0.375, 0.625, and 0.875. K^* is well represented by 4 parabolas with common tangents at the points of contact. The coefficients are given in Table I. Figure 4 shows the mean values of σ for each quarter. The speed of rotation is well represented by segments of straight lines. The turning points are discussed later.

Sudden changes in speed of rotation equivalent to a step change in L of 1 to 3 ms, or 116 to 348×10^{-10} in σ, have been announced at various times. No such sudden changes are shown by the atomic clock.

Figure 5a shows the speed of rotation from 1820.5 to 1969.5. Values up to 1954.5 are based on the moon and those from 1955.5 on are based on the atomic clock.

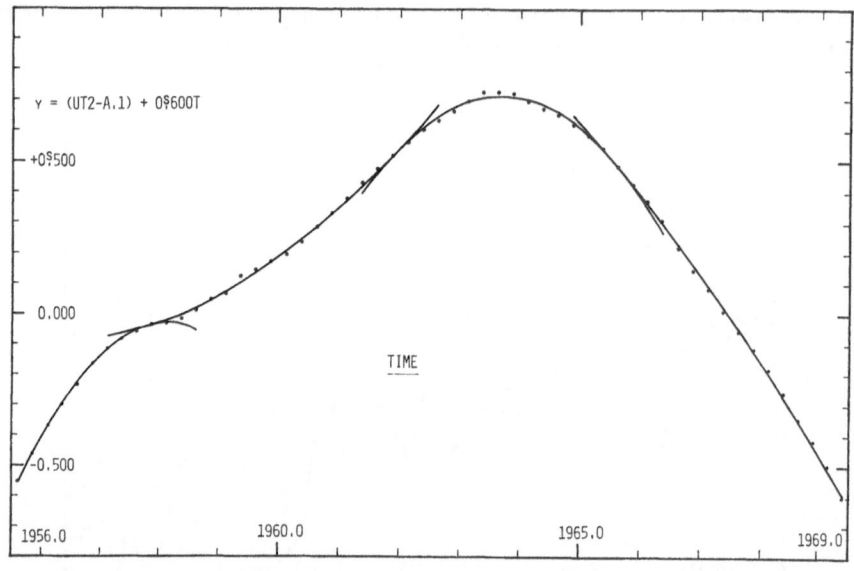

Fig. 3.   Quarterly means of $(UT2 - A.1)$ for Washington and Richmond, plus a linear term, and parabolic arcs; T is in years from 1958.0.

TABLE I

Coefficients of parabolas and accelerations;
unit $= 1$ sec; t is in years from 1960.0

	a	b	c	α
P_1	-1.486	-0.888	-0.0800	-51×10^{-10}/yr
P_2	-1.012	-0.457	$+0.0176$	$+11$
P_3	-1.305	-0.154	-0.0608	-38
P_4	$+0.207$	-0.693	-0.0128	-8

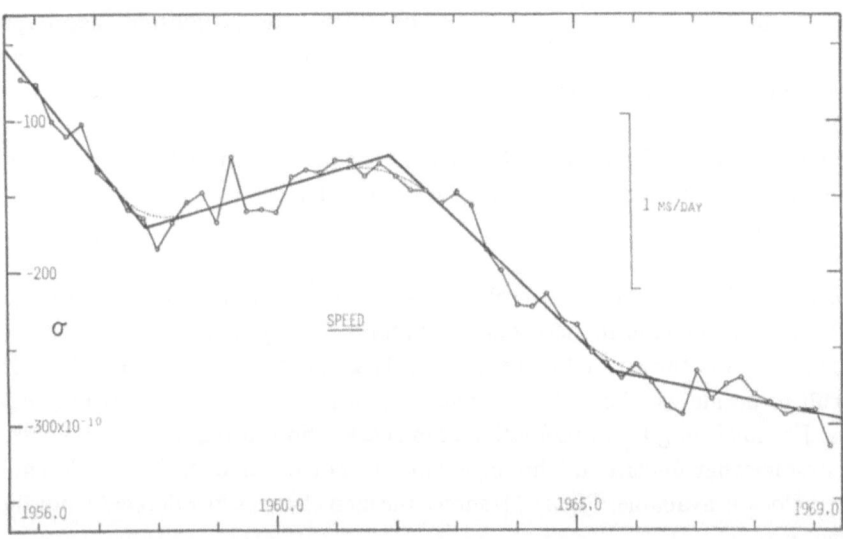

Fig. 4. Quarterly deviations in speed of rotation; mean of Washington and Richmond. Presumed variations near turning points are shown by dotted lines.

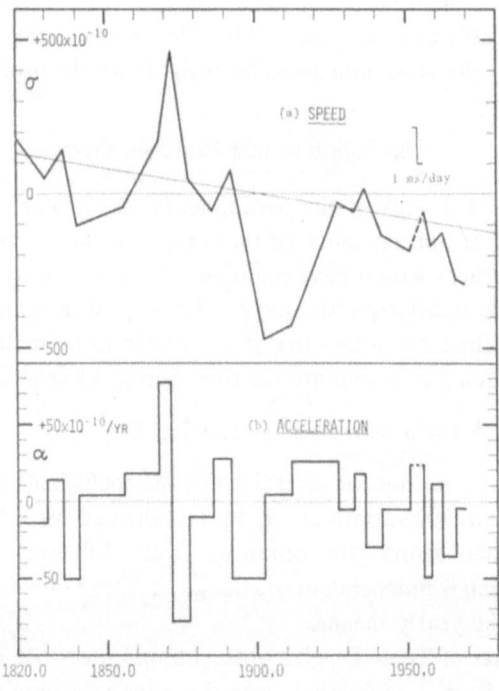

Fig. 5. (a) Deviations in speed of rotation. – (b) accelerations; 1820–1969.

The values up to 1950.5 are those given of Brouwer (1952) in Table VIIIa. The values from 1955.5 to 1969.5 are those obtained here. A constant acceleration is assumed between 1950.5 and 1955.5. The connection between ET and A.1 is based on moon camera observations from 1955.5 to 1958.5 (Markowitz et al., 1958; Markowitz, 1959).

The variation in σ between 1871 and 1902 is about 1 part in 10^7. We do not know if the large deviations near these times are correlated.

C. ACCELERATION

Let $\alpha = d\sigma/dt$ be the rotational acceleration. Values of the secular retardation derived several authors indicate an increase in L of about 1.4 ms/century, or $\alpha = -1.6 \times 10^{-10}/$ yr. This slope is shown in Figure 5a. The large excursions in σ show why it is so difficult to decide whether α has changed significantly during the past few thousand years. For analyzing past observations the acceleration of the moon is involved, and this is somewhat uncertain. This problem will not occur in the future because the atomic clock is available. Figure 5b shows the step changes in α derived from the data shown in Figure 5a.

A significant fact revealed with the atomic clock is that the changes in α occur about every 4 years, instead of about every 10 years as found by Brouwer. This difference, of course, is due to the reduction in observational probable error of an annual mean from 260 ms with the moon to 1.2 ms with the atomic clock. Indeed, it is remarkable that Brouwer was able to determine sudden changes in acceleration with the moon. That he was able to do so demonstrates his insight into the problem of the rotation of the earth.

3. Solutions and Probable Errors

Table I gives the coefficients of the parabolas $P_i = a + bt + ct^2$, and values of α. The parabolas have 3 common tangents, so that only 9 of the 12 coefficients are independent. Four straight lines would have required 8 constants, only 1 less.

Table II gives the quarterly residuals $R = K^* - P_i$, observed minus computed, and the yearly means. The table shows that it is possible to represent the observed difference between rotational time and atomic time during 14 years by the formula

$$UT1 - A.1 = a + bt + ct^2 - \Delta SV^* + R,$$

where a, b, and c are constant for several years, the coefficients in ΔSV^* are constants, and the residual quarterly variations, R, are less than 20 ms.

The observational errors are obtained from difference $S = K$ (Washington)– K (Richmond), which is independent of changes in speed of rotation. The last column of Table II gives the yearly means.

The first four lines of Table III give observational probable errors based on values of S for intervals from 1 night to 1 year. Line (a) was obtained directly from the differences, S; the value for 1 night was obtained previously (Markowitz, 1960a). The internal error, line (b), was computed from the entry above and to the left in (a) by

TABLE II

Residuals: R, based on the mean of Washington plus Richmond,
and S, the difference; unit $= 1$ millisec

			R			S
	0.125	0.375	0.625	0.875	yr	yr
1955			+ 5	0		
1956	+ 3	− 3	− 7	+ 5	0	− 6
1957	+ 2	+ 1	− 2	+ 1	0	− 2
1958	− 13	− 15	− 8	+ 1	− 9	− 2
1959	− 7	+ 17	+ 11	+ 2	+ 6	0
1960	− 10	− 7	− 1	+ 1	− 4	− 5
1961	+ 7	+ 11	+ 6	+ 4	+ 7	− 5
1962	− 4	− 6	− 13	− 11	− 8	+ 6
1963	+ 1	+ 17	+ 14	+ 11	+ 11	+ 3
1964	− 4	− 11	− 5	− 4	− 6	0
1965	+ 2	+ 2	+ 3	+ 1	+ 2	0
1966	+ 7	+ 7	− 5	− 18	− 2	− 2
1967	− 10	− 14	− 8	+ 4	− 8	+ 1
1968	+ 7	+ 9	+ 2	+ 5	+ 6	− 2
1969	+ 8	− 8				

TABLE III

Observational errors and time variations, V; unit $= 1$ millisec

	1 night	1 mo(16)	1 qtr(3)	1 yr(4)
(a) 1 PZT, external	4.8	3.8	3.0	1.7
(b) 1 PZT, internal		1.2	2.2	1.5
(c) Systematic error		3.6	2.0	0.8
(d) Obs. error, 2 PZT's		2.7	2.1	1.2
(e) P.E. of R			7.1	4.8
(f) Variation in time, V			6.8	4.7

dividing it by the square root of the number in parentheses. The systematic error is (c) $=$ $[(a)^2 - (b)]^{1/2}$. It is less than 1ms for the yearly mean of one PZT. Line (d) gives the total observational error for the mean of 2 PZT's.

Line (e) gives the probable errors of R. Correcting for (d) we obtain the quarterly and yearly deviations in time, V, expressed as a probable error. The values of V are several times as large as those in (d), so that R represents real variations in speed of rotation of the earth.

The probable error of an observed σ, derived from the deviations from the segments of straight lines of Figure 4, is 11×10^{-10}. The observational probable error in R is 2.1 ms for 2 PZT's for one quarter. In consequence, we derive probable errors of 3.8×10^{-10} for the observational and 10×10^{-10} for the real variations of σ.

4. Variations in Acceleration, Speed, and Time

Values of α were given in Table I. Values of σ at the turning points in acceleration and at the observational ends are as follows:

$$\sigma$$

(1955.5)	$- 53 \times 10^{-10}$
1957.79	$- 170$
1961.93	$- 123$
1965.61	$- 265$
(1969.5)	$- 297$

Variations in speed of rotation are classified as (a) secular, due chiefly to tidal friction, (b) irregular, due to coupling of mantle and core or to changes in moment of inertia, and (c) seasonal, due chiefly to winds. Table IV lists maximum effects for (b) and (c) and, for (a), the cumulative effects in speed and time during the past 2000 years.

Rotational torque is proportional to α. What is surprising is the large torque due to winds, about 8 times that of the irregular variation and 40 times that of the lunar tidal couple.

TABLE IV

Variations in acceleration, speed, and time

	Secular	Irregular	Seasonal
α	-1.6×10^{-10}/yr	$\pm 80 \times 10^{-10}$/yr	$\pm 650 \times 10^{-10}$/yr
σ	(-3×10^{-7})	$\pm 500 \times 10^{-10}$	$\pm 70 \times 10^{-10}$
Time	$(-10000$ sec$)$	± 30 sec	± 0.03 sec

5. Discussion of Rotation

Accelerations are produced by torques, whose sources are internal or external to the earth. Numerous possible sources are discussed by Munk and MacDonald (1960). As a working hypothesis we assume that (a) the variations represented by parabolas are due to internal effects, and (b) the deviations from the parabolas given by R in Table II are due to meteorological effects.

A. ACCELERATION TURNING POINTS

The observations have been represented by 4 parabolas with common tangents at the points of contact. This leads to discontinuities in the accelerations, and torques, which is physically impossible. The discontinuities can be avoided by using a single polynomial of high degree to represent the observations. However, it is doubtful if such representation can be used for studying the turning points. If there is a change in the physical conditions, e.g., a different element of the core acts on the mantle, then the same analytical formula will not represent the conditions before and after the change.

Hence, a single polynomial would provide illusory information on what occurs at the turning points. For this reason I have based the analysis on the parabolic representations of Figure 3.

The suddenness with which changes in α occur are determined best when the change in slopes of the straight line segments is large, as at 1957.8. By superimposing the three changes in acceleration and assigning various weights I find that the change in acceleration appears to take place within 6 months to 1 year. We cannot make a more precise determination at present. An increase in the accuracy of the determination of σ would not help because the deviations represent variations in speed of rotation, probably of meteorological origin, and not observational effects. Unless the deviations can be calculated from observations of winds, which seems unlikely, it will be necessary to accumulate about 25 to 50 years of observations of σ to obtain precise information on the turning points.

B. 2-YEAR CYCLE

Iijima and Okazaki (1966) announced the existence of a nearly biennial term in the earth's rotation, based on observations from 1956.9 to 1964.9 and later to 1967. They obtained a period of 26 months, which they ascribed to the effect of winds of this period.

Table II confirms the existence of a biennial effect, which, however is not strictly periodic. R was decidedly negative in the even years (earth slow) and positive in the odd years from 1958 to 1964. However, the reverse was true for 1967 and 1968. There was no decided trend in the pairs 1956, 1957 and 1965, 1966. A 26-month period is not found. Instead, there is a 2-year cycle which occurs for some pairs of years, or a succession, but not for all pairs of years.

The formula for the 2-year cycle, obtained from the years 1958 to 1964 inclusive, and 1967 and 1968, is

$$A = + 10.7 \text{ ms } \sin \pi\tau + 1.5 \text{ ms } \cos \pi\tau,$$

where A is the amount by which the earth is ahead and τ is the time in years since the beginning of the year in which A is positive. The coefficients obtained by Iijima and Okazaki were similar, $+8.6$ ms and $+2.7$ ms, respectively.

C. EARTHQUAKES AND THE ROTATION OF THE EARTH

The possibility of correlating earthquakes and changes in speed of rotation was discussed in 1927 by Lambert (1927), who used the moon as a reference. No definite results were obtained. The moon, of course, is not a sufficiently precise reference. However, the atomic clock now provides the requisite accuracy.

Two large earthquakes occurred in the interval shown in Figure 2, at 1963.78 in the Kurile Islands, and at 1964.24 in Alaska. No changes in acceleration or abrupt changes in speed of rotation are noted at these times. Thus, no correlation is observed between earthquakes and changes in rotational speed of the earth.

6. Motion of Mean Pole

The fundamental determination of the position of the pole of rotation is derived by the
International Polar Motion Service (IPMS) from observation of the 5 stations of the
International Latitude Service (ILS). The same stars are observed at all stations. The
method of control latitudes (Markowitz, 1961, 1968) shows that the ILS polar motion
is independent of errors in star positions, proper motions, and micrometer scale
values, or of changes in the observing lists. The position of the pole is obtained from
the variations in latitude. The position of the mean pole is obtained from 6-year means

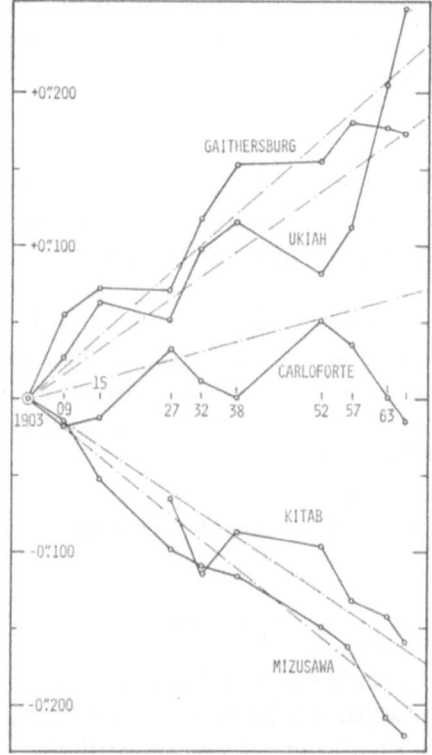

Fig. 6.   Variations in latitude of ILS stations; 6-year means. The dashed lines represent the variations which would have occurred if the mean pole had moved 0″.210 in 60 years along longitude 65°W.

since these are effectively free of the 12- and 14-year variations. Figure 6 shows the
variations in latitude and Figure 7 the position of the pole for 6-year normal points up
to 1966.0. The last point was kindly furnished by Dr. S. Yumi.

The motion of the ILS mean pole was found in 1960 to consist of a progressive
component of about 0″.0035/yr, or 10 cm/yr, along the meridian 65°W and a librational component (oscillation) of 24-year period along the meridian 122°W (Markowitz, 1960b).

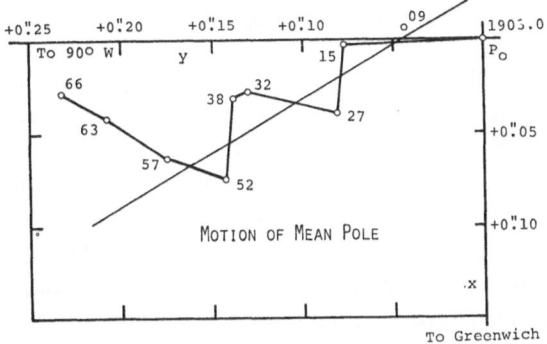

Fig. 7.   Motion of mean pole, ILS; 1903.0 to 1966.0. Each point is a 6-year mean.

The 24-year term is empirical; we do not know whether it will continue. Indeed, the position for 1966.0 deviates from the previous pattern. However, between 1932 and 1938 the motion also did not follow the usual pattern. We must wait to see whether the 24-year motion continues.

7. Plate Rotations and Continental Drift

The hypothesis of plate rotations is very attractive as a mechanism for continental drift. However, neither plate rotations nor continental drift have been confirmed by direct measurement as yet.

It has been suggested that the observed variations in latitude of the ILS stations could be due to plate rotations rather than to the motion of the pole. We consider this point. The variations in latitude shown by Figure 6 have oscillatory and progressive components. The phases agree closely with those computed on the assumption that the mean pole moves. No plausible theory of plate rotations can account for the coherence of the observed oscillations.

Figure 6 shows that the progressive components of the observed variations agree closely, both in sign and in magnitude, with that computed on the assumption that the mean pole moves. It would require a set of very fortuitous circumstances for several plates to rotate in such a way as to give the observed variations. The rotations derived by Whitten (1970) were obtained from the observed variations of latitude, taking two stations at a time. If plate rotations are to explain the observed variations, then the rotation rates and centers for all plates must be provided by geophysical data which are independent of observed variations in latitude. Thus far, I have been unable to obtain such a set of rotation rates and centers.

Within the errors of measurement, about 1 cm/yr, the observed variations of latitude of the ILS stations are accounted for by the assumption that the mean pole is moving. No set of homogeneous data is available which would account for the motions of all 5 stations on the assumption that the variations are due to plate rotations.

8. Discussion

For intervals of 3 months or longer, present astronomical techniques determine the speed of rotation of the earth with requisite accuracy. For intervals of 1 month or less, newer techniques, such as the use of the moon laser and radio interferometers, may provide accuracies needed for observing high frequency components of speed variations, due probably to winds. Whether significant correlations can be established between observed changes in the speed of rotation and winds remains to be seen.

Irregular variations in speed arise, presumably, from changes in core-mantle coupling. The associated torque remains fairly constant for several years and then changes suddenly, within 6 months or a year, to a new value.

Observed variations in latitude of the ILS stations, progressive and oscillatory, are accounted for by the simple hypothesis that only one quantity is changing, namely, the position of the mean pole. No similar simple explanation of the variations has been provided by the hypothesis that plates are rotating.

Acknowledgements

I thank Dr. R. G. Hall, with whom I began these studies at the U.S. Naval Observatory some years ago, for helpful discussions and for providing recent results. Also, I am indebted to Dr. G. M. R. Winkler, Director of the Time Service Division, and Mr. Don R. Monger, Director of the Richmond station, for discussions.

Mr. R. W. Tanner of the Dominion Observatory, Ottawa, kindly furnished determinations of time made with the Ottawa PZT for 1960 to 1968.

The Time, Frequency, and Polar Motion Laboratory of Marquette University was established with grants from the National Science Foundation, the University Committee on Research, and the Wehr Science Center Endowment. Operational support is provided by the Office of Naval Research, on behalf of the Naval Observatory, and by NASA.

References

Brouwer, D.: 1952, *Astron. J.* **57**, 125.
Iijima, S. and Okazaki, S.: 1966, *J. Geodetic Soc. Japan* **12**, 91.
Lambert, W. D.: 1927, *Wash. Acad. Sci.* **17**, 133.
Markowitz, W.: 1959, *Astron. J.* **64**, 106.
Markowitz, W.: 1960a, in *Methods and Techniques in Geophysics* (ed. by S. K. Runcorn), Interscience Publishers, New York, p. 325.
Markowitz, W.: 1960b, in *Telescopes* (ed. by G. P. Kuiper and Barbara Middlehurst), University of Chicago Press, Chicago, Ill., p. 88.
Markowitz, W.: 1961, *Bull. Géodésique* **59**, 29.
Markowitz, W.: 1968, in *Continental Drift, Secular Motion of the Pole, and Rotation of the Earth* (ed. by W. Markowitz and B. Guinot), Reidel Publishing Co., Dordrecht, p. 25.
Markowitz, W., Hall, R. G., Essen, L., and Perry, J. V. L.: 1958, *Phys. Rev. Letters* **1**, 105.
Munk, W. H. and MacDonald, G. J. F.: 1960, *Rotation of the Earth*, Cambridge University Press, Cambridge, Chapter 11.
U.S. Naval Observatory, *Time Service Notice No. 6*, 1 Jan. 1959, Washington, D.C.
Whitten, C. A.: 1970, this volume, page 255.

Discussion

Guinot: Lack of homogeneity in the past BIH results is not sufficient to reject any possibility of correlation with earthquakes.

Markowitz: Despite attempts to produce homogeneity, when one introduces a new station one may get a break. When correlation is found with one set of data and not the other, we have to be suspicious. For the ILS there has been no change of stations since 1931.

Mansinha: We were aware that the BIH started with a small number of stations in 1955 and the number was being slowly increased. Therefore, we ignored BIH data of 1955 and 1956. As for the ILS, I believe there was a period during which only 3 stations were actually observing.

Markowitz: No, there were 5. Carloforte had a reduced number of observations from 1939 to 1946, but not since then.

Smylie: The ILS data does not disagree with the earthquake hypothesis. It's just that we could not establish a significance to the correlation. Fifteen earthquakes out of 22 correlate with a break. The wider spacing of data points was partially responsible for the lack of significance.

SECULAR MOTION OF THE POLE

S. YUMI and Y. WAKO

International Latitude Observatory, Mizusawa-shi, Japan

Abstract. Disaccordance among the coordinates of the pole which are derived from several kinds of combination of the ILS stations suggests a possibility of secular changes in mean latitudes of the cooperative stations on $+39°8'$ irrespective of the effect by the secular drift of the mean pole. Local drifts of $-0''.00156$/year for Mizusawa and of $+0''.00105$/year for Ukiah were derived from the analyses of the notable increases in the residual latitudes of the five ILS stations during the period 1900–65. Subtraction of the apparent motion of the mean pole due to the local drifts of the stations from the total motion makes the secular motion of the mean pole as $0''.00220$/year in the direction $77°7$ W and this seems to be the real one.

1. Motion of the Mean Pole

Change of the position of the mean pole has hitherto been studied and discussed by many researchers, and the reality of secular change and libration were suggested by Sekiguchi (1954), Hattori (1959), and Markowitz (1960) on the assumption that the relative positions of the ILS stations did not change with each other.

Notwithstanding the fact that the position of the mean pole should be determined uniquely, different kinds of combination of the stations give the respective positions different from each other as demonstrated in Figure 1. This leads to a supposition that a real change in mean latitude of any station which is expected from a supposed displacement related with a crustal movement or a change in the plumb line at the station would produce an apparent change of the mean pole.

2. Secular Variation of Mean Latitude

Mean latitude of each of the ILS stations was calculated as a running mean through six years to eliminate the Chandler term, the assumed period of which is 1.2 years. Results are illustrated in Figure 2. Inclined straight lines in the figure show the variation of mean latitude of each station which would be expected from the secular variation of the mean pole obtained so far from the results of the five ILS stations. It can be seen in Figure 2 that the mean latitude of each station fluctuates with a long period around its mean direction. In addition, the mean latitudes of Mizusawa and Ukiah appear to have a tendency of deviating gradually from their respective mean directions.

3. Residual Latitude

Error of monthly mean latitude as an average of 100–400 observations during a month, the accuracy of which is estimated to be $\pm 0''.2$–$0''.3$ for a single observation at the ILS station, and therefore that of the coordinates of the pole is reasonably assumed to be an order of $\pm 0''.01$–$0''.02$. Consequently a value of the same order is expected in

L. Mansinha et al. (eds.), Earthquake Displacement Fields and the Rotation of the Earth, 82–87.
All Rights Reserved. Copyright © 1970 by D. Reidel Publishing Company, Dordrecht-Holland

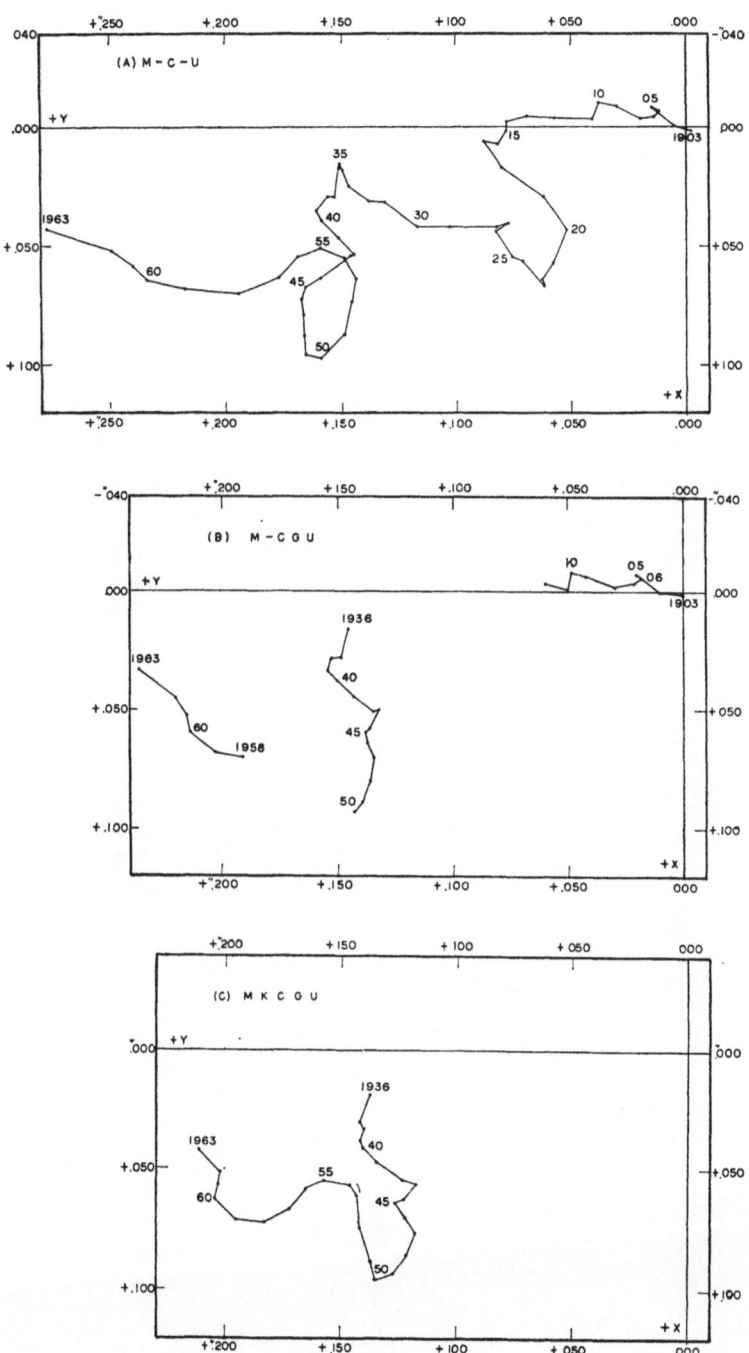

Fig. 1. Motion of the mean pole.

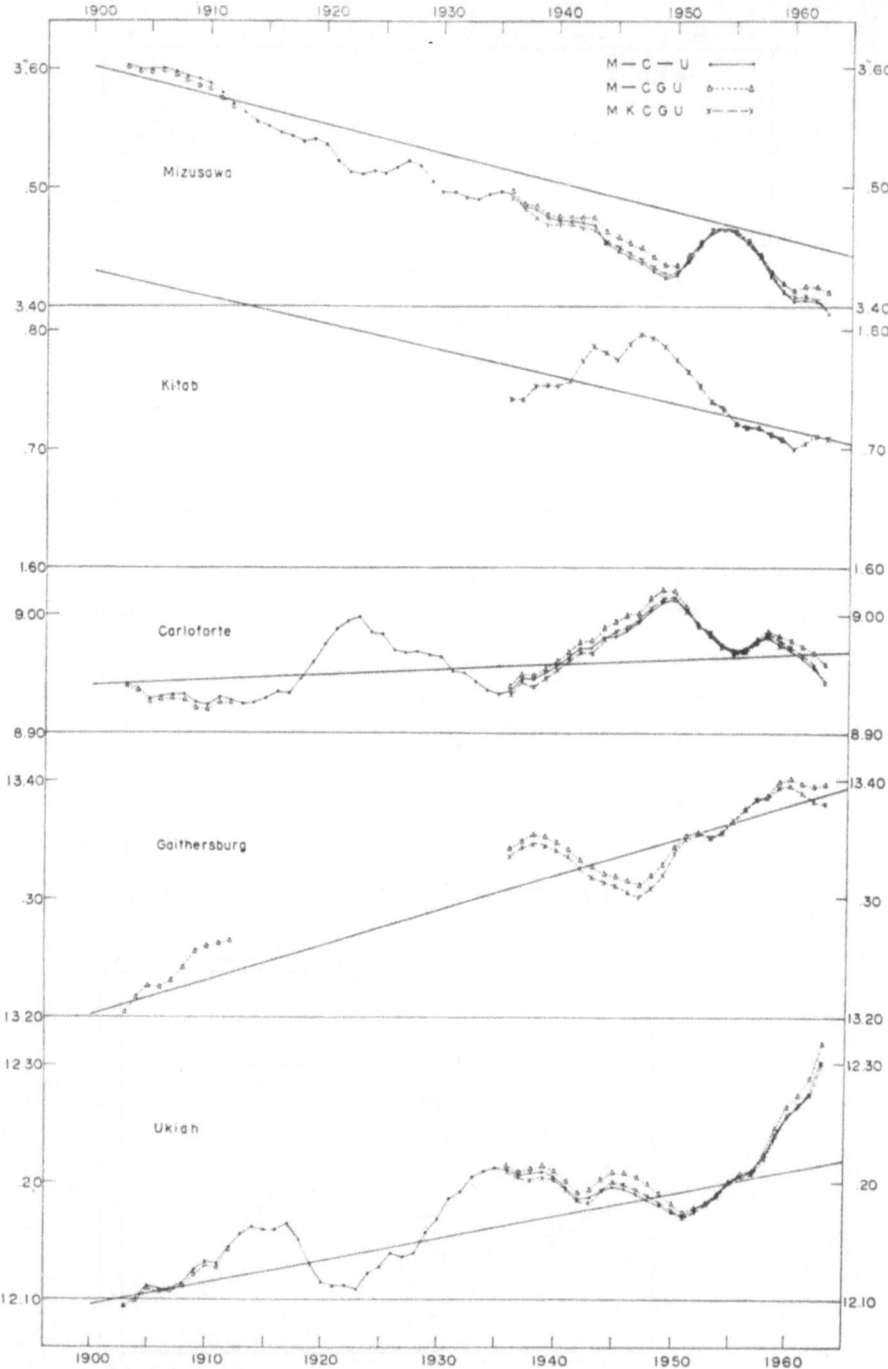

Fig. 2. Variation of mean latitude.

residual latitude, $(o - c)$, the difference between the observed latitude and the expected one calculated from the derived polar coordinates. In practice, however, the residual latitude gradually increased during the past sixty years. It amounts even to $0.''13$ for a certain station. As a yearly mean it varies from $0.''03$ to $0.''07$ in these years. Residual latitudes are illustrated in Figure 3 for the cases of M-CGU and MKCGU.

This ambiguity in residual latitude suggests the reality of local errors or local secular drift accompanied by fluctuations at any station and the resulting apparent motion of the mean pole. In this paper, however, only the secular part is dealt with, because the number of stations which afford the data are considered to be too few to discuss the fluctuation.

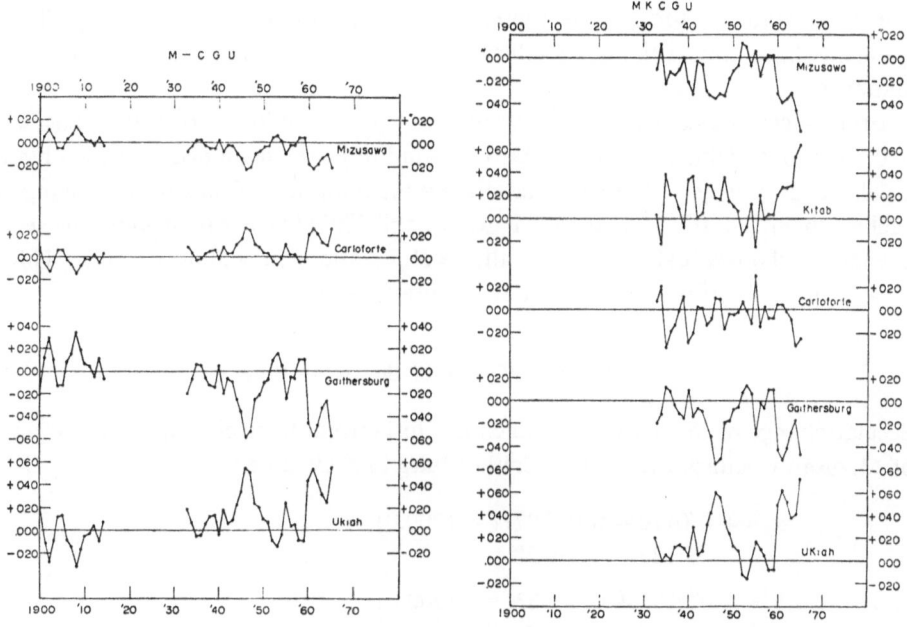

Fig. 3.   Change of $(o-c)_t$.

4. Local Error in Latitude

Residual latitude is customarily expressed as

$$(o - c)_i = \Delta\varphi_i - (x \cos \lambda_i + y \sin \lambda_i + z), \tag{1}$$

which can be rewritten as

$$(o - c)_i = K_{ii}\Delta\varphi_i + \sum{}' K_{ij}\Delta\varphi_j, \tag{2}$$

where i stands for the station concerned, and j the other stations, (Yumi and Wako, 1966).

K_{ii} and K_{ij} are the constants depending only on λ_i and λ_j. Assume that the observed latitude variation consists of true $\Delta\varphi$ and local error in it, that is

$$(\Delta\varphi_{\mathrm{obs}})_i = (\Delta\varphi_{\mathrm{true}} + \mathrm{d}\varphi_{\mathrm{local}})_i, \tag{3}$$

then (2) becomes

$$(o - c)_i = K_{ii}(\Delta\varphi_{true} + d\varphi_{local})_i + \sum{}' K_{ij}(\Delta\varphi_{true} + d\varphi_{local})_j$$
$$= K_{ii}(d\varphi_{local})_i + K_{ij}(d\varphi_{local})_j. \tag{4}$$

It is proved that K_{ii} and K_{ij} tend to 1 and 0 respectively, that is to say, the local error at any station i is represented almost by $(o-c)_i$ when the number of equally distributed stations becomes large. Actual distribution and the number of ILS stations, however, are so far from the ideal case that $d\varphi_{local}$ at the station i is not always calculated by $(o-c)_i/K_{ii}$ but is affected by the local errors in the other stations.

On the assumption that $(d\varphi_{local})_i$ is large enough compared with the local errors of the other stations, $(d\varphi_{local})_i$ is represented by $(o-c)_i/K_{ii}$ or $(o-c)_j/K_{ij}$. If these quotients have almost the same value, this is considered to be the local error of the station i.

From such an examination, the secular change in the local error at Mizusawa of $-0{.}''0312$ or at Ukiah of $+0{.}''00209$ was found to be the main origin of the obtained secular change in $(o-c)$ at each station. By the adoption of half secular change for each of them, $-0{.}''00156$/year for Mizusawa and $0{.}''00105$/year for Ukiah, the secular changes in the residual latitudes of all the ILS stations may almost vanish. This indicates the possibility of local drift at Mizusawa and Ukiah.

5. Secular Motion of the Mean Pole

Secular change of the mean pole which has hitherto been calculated from the results of Mizusawa, Kitab, Carloforte, Gaithersburg and Ukiah as

$$x = + 0{.}''0170 + 0{.}''00087(t - 1900.0),$$
$$\pm 139 \qquad \pm 28$$
$$y = + 0{.}''0105 + 0{.}''00283(t - 1900.0),$$
$$\pm 146 \qquad \pm 29 \tag{5}$$
$$S = 0{.}''00296/\text{year},$$
$$\theta = 72{.}°9 \text{ W},$$

will be reduced to about $\frac{1}{2}$ in its x-component and $\frac{3}{4}$ in the y-component by the adoption of the local drifts at Mizusawa and Ukiah described in the preceding section. The resulting magnitude of the reduced secular change is

$$S'_{x5} = S_{x5} - \Delta S_{x5} = + 0{.}''00087 - 0{.}''00040 = + 0{.}''00047,$$
$$S'_{y5} = S_{y5} - \Delta S_{y5} = + 0{.}''00283 - 0{.}''00068 = + 0{.}''00215, \tag{6}$$

that is

$$S' = 0{.}''00220/\text{year},$$
$$\theta' = 77{.}°7 \text{ W}, \tag{7}$$

and this change seems to be the real one.

This assumption also leads to another conclusion that the probable errors of the coordinates of the barycentres for the latest years which amount to about $\pm 0''.03$ in the current system of MKCGU are reduced to a small amount of $\pm 0''.01$.

In order to detect more precisely the reality of the local drifts of the stations and of the resulting secular motion of the mean pole, additional stations on the same parallel should be established to provide a nearly equal distribution in longitude. On the other hand, as the astronomical time and latitude are observed with reference to the plumb line at the station on the crust, change of plumb line and the deformation of the earth crust which might be related with the continental drift should be observed geographically and geodetically in parallel with the astronomical observations.

Details are described in Yumi and Wako (1966).

References

Hattori, T.: 1959, *Publ. Int. Latit. Observ. Mizusawa* 3, No. 1.
Markowitz, Wm.: 1960, *Methods and Techniques in Geophysics*, p. 350.
Sekiguchi, N.: 1954, *Publ. Astron. Soc. Japan* 5, 109.
Sekiguchi, N.: 1956, *Publ. Astron. Soc. Japan* 8, 13.
Yumi, S. and Wako, Y.: 1966, *Publ. Int. Latit. Observ. Mizusawa* 5, 74.

THE DYNAMICAL COUPLING BETWEEN INNER CORE AND
MANTLE OF THE EARTH AND THE 24-YEAR LIBRATION
OF THE POLE

F. H. BUSSE

Max-Planck-Institut für Physik, Munich, W. Germany

Abstract. The free nutations of the earth mantle and the rigid inner core are coupled by the pressure reactions in the fluid outer core. In the case that motions in the outer core are responsible for the excitation of the wobble of the earth, a period corresponding to the natural frequency of the inner core should be observable in the motions of the pole. It is proposed that this period corresponds to the 24-year libration of the pole found by Markowitz (1960).

1. Introduction

The moment of inertia of the inner core of the earth represents only a fraction of about $\frac{1}{500}$ of the total moment of inertia of the earth. For this reason it seems at first sight unlikely that the inner core should be relevant in the discussion of the wobble of the earth. Because of its rigidity, however, the inner core offers the possibility of an additional degree of freedom in the theory of the free nutations of the earth system. Thus it is an attractive idea to attempt the explanation of several puzzling features of the Chandlerian nutation by a dynamic coupling between inner core and mantle of the earth.

One of the unresolved problems is the rather high damping rate suggested by the broad peak corresponding to the Chandlerian period in the power spectrum of the librational motions of the pole. It is difficult to explain the broadening in terms of dissipative processes which are compatible with the knowledge about the properties of the earth system from other observations. In addition a high dissipation requires a correspondingly strong source of excitation, the identification of which has become the most challenging question in the problem of the Chandler wobble*. In this situation the assumption of a beating phenomenon caused by the coupling of the free nutations of the mantle and of the inner core would have obvious advantages. Unfortunately the large difference between the eigenfrequencies of the mantle and the inner core does not permit an explanation of the width of the Chandlerian peak along these lines. For this reason Colombo and Shapiro (1968) who originally considered this mechanism had to assume the hypothetical possibility of a coupling between two rather independent parts of the mantle as basis of a beating phenomenon.

It has to be concluded that the dynamic coupling between inner core and mantle is not likely to play an important role in the resolution of problems connected with the main peak in the spectrum of the Chandlerian nutation. Yet the possibility remains that the oscillation of the inner core could be identified in the spectrum of the ob-

---

* For a comprehensive discussion of the problem we refer to the monograph of Munk and Mac-Donald (1960).

L. Mansinha et al. (eds.), Earthquake Displacement Fields and the Rotation of the Earth, 88–98.

served motions of the pole. Markowitz (1960, 1968) has shown that a period of about 24 years appears to be distinguished by the analysis of the data. He has called this fact surprising "since we do not know of any geophysical phenomenon with this period". In the following we wish to lend support to the hypothesis that this period represents the period of the free nutation of the inner core.

It should be mentioned at this point that a different explanation of the 24-year libration of the pole has been proposed by Abraham (1968). From the short published note of Abraham, it cannot be understood, however, how the beating period of about 6 years between the annual and the Chandler wobble can cause a subharmonic response of the earth system with a period of 24 years even if nonlinear effects are invoked.

The theoretical analysis in the following two sections is intended to provide a basis for the qualitative discussion in the last section. For this reason a strongly simplified model is considered. Many physical properties, such as finite rigidity and stratification of the fluid core are neglected. They will become important, presumably, only when an increased knowledge about the earth core will permit a quantitative comparison between theory and observations.

2. A Simple Dynamical Model of the Earth

We shall base our discussion on a model of the earth consisting of an oblate spherical rigid shell filled with a homogeneous incompressible fluid. In the fluid an oblate homogeneous rigid spheroid is suspended with its center fixed in cocentric position. The effects of elasticity and gravity will be neglected. Let the basic state of this system be given by a constant angular velocity of all three parts about their common axis of symmetry. We use the rotational rate as scale of the time and describe the basic state of rotation by the unit vector $\mathbf{k}$. With respect to this rotating system the free nutation of the shell is given by

$$\frac{d}{dt}\delta_s - \frac{C_s - A_s}{A_s}\mathbf{k} \times \delta_s = 0 \tag{1}$$

in the case when the density of the fluid is vanishing. δ_s represents the small deviation of the angular velocity vector from its equilibrium value $\mathbf{k}$ and C_s, A_s are the moments of inertia about the figure axis and about an axis perpendicular to it. The unit vector describing the direction of the figure axis experiences the deviation γ_s from the equilibrium value $\mathbf{k}$. γ_s is determined by the assumption that the angular momentum of the oscillating shell is identical with its equilibrium value.

$$A_s\delta_s + (C_s - A_s)\gamma_s = 0. \tag{2}$$

It is well known (for reference see Lamb, 1932) that the frequency of nutation is increased when the cavity is filled by a fluid of finite density,

$$\frac{d}{dt}\delta_s - \frac{C_s + C_f - A_s - A_f}{A_s}\mathbf{k} \times \delta_s = 0. \tag{3}$$

The pressure balancing the centrifugal force in the fluid exerts a restoring torque as if the fluid were a rigid body with the moments of inertia C_f and A_f determined by the spheroidal shape of the cavity and the density ϱ_f of the fluid. The inertial response of the fluid, on the other hand, is negligible at least in the case when the ellipticity of the cavity is small.

The fluid has the opposite effect of decreasing the frequency of nutation of a rigid body submersed in it. When the moments of inertia of the rigid body and of the fluid replaced by it coincide, the frequency of nutation is zero, since the restoring torque vanishes. Accordingly the free nutation of the inner core is determined by

$$\frac{\mathrm{d}}{\mathrm{d}t}\delta_c - \frac{C_c - A_c}{A_c}\left(\frac{\varrho_c - \varrho_f}{\varrho_c}\right)\mathbf{k} \times \delta_c = 0, \tag{4}$$

where δ_c represents the small deviation of the rotation vector from $\mathbf{k}$. C_c and A_c are the moments of inertia of the inner core, ϱ_c is its density.

In the case of the earth the motions induced in the fluid outer core by the nutations (3), (4) are small because the frequencies

$$\omega_s = \frac{C_s + C_f - A_s - A_f}{A_s}$$
$$\omega_c = \frac{(\varrho_c - \varrho_f)(C_c - A_c)}{\varrho_c A_c} \tag{5}$$

are small compared with the natural frequency of the corresponding mode in the fluid which is of the order one. Accordingly, the motion of the fluid induced by the nutation of the shell will be of the order

$$r_s e_s |\delta_s \times \mathbf{k}|, \tag{6}$$

where the relation

$$\frac{\mathrm{d}}{\mathrm{d}t}\gamma_s = \delta_s \times \mathbf{k} \tag{7}$$

has been used. e_s and r_s denote the ellipticity and the equatorial radius of the cavity, respectively. In a spherical cavity no motions would be induced since viscous and hydromagnetic effects are neglected. Similarly the oscillation of the inner core produces motions of the order

$$r_c e_c |\delta_c \times \mathbf{k}|. \tag{8}$$

The torque exerted by the pressure associated with the fluid motion produces the dynamic coupling between mantle and inner core. The torques acting on the shell and the inner core are of the orders e_s and e_c, respectively, smaller than the pressure multiplied by the respective volumes. Thus it can be expected that the coupled system of the two oscillators (3) and (4) is described by equations of the form

$$\frac{d}{dt}\boldsymbol{\delta}_s - \omega_s \mathbf{k} \times \boldsymbol{\delta}_s + \mu_s \mathbf{k} \times \boldsymbol{\delta}_c = 0$$

$$\frac{d}{dt}\boldsymbol{\delta}_c - \omega_c \mathbf{k} \times \boldsymbol{\delta}_c + \mu_c \mathbf{k} \times \boldsymbol{\delta}_s = 0$$

(9)

with coupling constants

$$\mu_s \equiv e_c e_s \, 4g \, \frac{\varrho_f A_c}{\varrho_c A_s}, \quad \mu_c \equiv e_c e_s \, 4g \, \frac{\varrho_f}{\varrho_c}$$

of the order $e_c \, e_s$. It has been assumed that $\omega_s, \omega_c, e_s, e_c$ are small parameters in the case of the earth which can be neglected in comparison with unity. g represents a factor of the order one. The rather weak coupling has the consequence that the fundamental frequencies of the system (9) are essentially identical with ω_s, ω_c. Since dissipative processes have been neglected the kinetic energy of the system (9) is conserved,

$$\frac{d}{dt}(A_s \boldsymbol{\delta}_s \cdot \boldsymbol{\delta}_s + A_c \boldsymbol{\delta}_c \cdot \boldsymbol{\delta}_c) = 0.$$

(10)

Before we shall enter the discussion about the conclusions which can be drawn from the system (9) in the case of the earth we wish to give a mathematical formulation for the heuristic arguments used above. For this purpose we are going to solve the equations of motion in the outer core which will enable us also to calculate the number g.

3. The Coupling Mechanism

We consider the motion induced in the fluid core by the oscillations of the shell and the inner core. The equations of motion for a homogeneous inviscid and incompressible fluid in a rotating system are given by

$$\frac{\partial}{\partial t}\mathbf{v} + 2\mathbf{k} \times \mathbf{v} = -\nabla \pi,$$

(11)

$$\nabla \cdot \mathbf{v} = 0.$$

The time is measured again in units of the basic rotation rate. π denotes the pressure change induced by the motion and divided by ϱ_f. Hence the total pressure p is given by

$$p = \tfrac{1}{2}\varrho_f |\mathbf{k} \times \mathbf{r}|^2 + \varrho_f \pi,$$

(12)

where $\mathbf{r}$ denotes the position vector with respect to the center of the system. At the moving rigid boundaries of the shell and the inner core the velocity vector $\mathbf{v}$ has to satisfy the conditions

$$\mathbf{v} \cdot \nabla F_s + \frac{\partial}{\partial t} F_s = 0,$$

$$\mathbf{v} \cdot \nabla F_c + \frac{\partial}{\partial t} F_c = 0.$$

(13)

The function

$$F_s(\mathbf{r}, t) = \frac{1}{2r_s} (|\mathbf{r}|^2 + 2e_s|(\mathbf{k} + \boldsymbol{\gamma}_s(t)) \cdot \mathbf{r}|^2) \tag{14}$$

describes the core-mantle interface; the analogous expression with c replacing the index s describes the surface of the inner core. We are interested in the solution of (11), (13) in the case of slow movements of the boundaries. Accordingly, we shall neglect the time dependence in (11) and retain it only in the inhomogeneous part of the boundary conditions (13). This assumption is essentially identical with the neglect of e_c, e_s in comparison with one.

In the stationary case Equations (11) require

$$\mathbf{k} \cdot \nabla \pi = 0 \tag{15}$$

and the Taylor-Proudman theorem

$$\mathbf{k} \cdot \nabla \mathbf{v} = 0. \tag{16}$$

The general stationary solution of (11) can be written in the form

$$\mathbf{v} = \tfrac{1}{2}\mathbf{k} \times \nabla \pi + \mathbf{k}w.$$

The boundary conditions (13) yield

$$\begin{aligned}
\tfrac{1}{2}\mathbf{s} \times \mathbf{k} \cdot \nabla \pi + \mathbf{k} \cdot \mathbf{v}w &= -2e_s \mathbf{k} \cdot \mathbf{r}\,\dot{\boldsymbol{\gamma}}_s \cdot \mathbf{r}, \\
\tfrac{1}{2}\mathbf{s} \times \mathbf{k} \cdot \nabla \pi + \mathbf{k} \cdot \mathbf{v}w &= -2e_c \mathbf{k} \cdot \mathbf{r}\,\dot{\boldsymbol{\gamma}}_c \cdot \mathbf{r},
\end{aligned} \tag{17}$$

at the respective surfaces. For convenience, we have indicated the time derivative of the vectors $\boldsymbol{\gamma}_s$, $\boldsymbol{\gamma}_c$ by a dot. The vector $\mathbf{s}$ is defined by

$$\mathbf{s} \equiv (\mathbf{k} \times \mathbf{r}) \times \mathbf{k}.$$

In polar coordinates (s, φ, z) with respect to the axis of rotation the dependence of π and w on z vanishes according to (15), (16). When the Equations (17) are divided by $\mathbf{k} \cdot \mathbf{r}$ and subtracted at fixed values of s, φ the variable w is eliminated and a single equation for π in the region $s < r_c$ is obtained,

$$\tfrac{1}{4}\mathbf{k} \times \mathbf{s} \cdot \nabla \pi = (e_c \dot{\boldsymbol{\gamma}}_c \cdot \mathbf{r} - e_s \dot{\boldsymbol{\gamma}}_s \cdot \mathbf{r}) \frac{\sqrt{r_s^2 - s^2}\sqrt{r_c^2 - s^2}}{\sqrt{r_s^2 - s^2} - \sqrt{r_c^2 - s^2}} \operatorname{sign}(z).$$

The solution of this equation is

$$\pi = \mathbf{k} \times \mathbf{r} \cdot (e_s \dot{\boldsymbol{\gamma}}_s - e_c \dot{\boldsymbol{\gamma}}_c) \frac{4\sqrt{r_s^2 - s^2}\sqrt{r_c^2 - s^2}}{\sqrt{r_s^2 - s^2} - \sqrt{r_c^2 - s^2}} \operatorname{sign}(z) \tag{18}$$

since the vectors $\boldsymbol{\gamma}_s$, $\boldsymbol{\gamma}_c$ are always perpendicular to $\mathbf{k}$ according to their definition. By adding the two boundary conditions at the interface $F_s = \tfrac{1}{2}$ for $s > r_c$ it follows that $\mathbf{k} \times \mathbf{s} \cdot \nabla \pi$ and therefore π vanish for $s > r_c$.

Hence the nutations of the shell and the inner core do not induce a pressure variation outside the cylindrical surface $s = r_c$. The component of the velocity field perpendicular to $\mathbf{k}$ which is determined by π has been sketched in Figure 1. Owing to the stretching

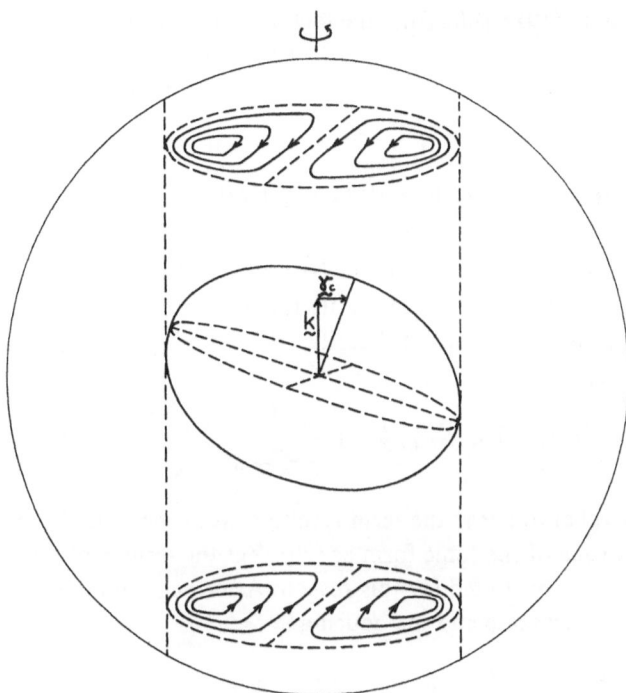

Fig. 1. The flow pattern of the velocity component parallel to the equatorial plane which is induced by the free nutations of the inner core and of the mantle. The motion of the fluid vanishes outside the cylinder $s = r_c$ when the mantle is at rest, with respect to the rotating system.

and compressing of vortex lines, flows of negative and positive vorticity with respect to the sense of rotation are induced, respectively. On the surface $s = r_c$ the pressure π vanishes like $\sqrt{(r_c^2 - s^2)}$ with the consequence that the azimuthal velocity component has a singularity at $s = r_c$. In reality thin shear layers will develop at $s = r_c$ owing to viscous effects in addition to the Ekman layers at the rigid boundaries. The corresponding modifications of the velocity field can be analysed by boundary layer techniques; we refer to Greenspan's (1968) book for details on this subject. For the calculation of the torque the modifications can be neglected since their influence vanishes like $E^{1/4}$ where E is the Ekman number,

$$E \equiv \frac{v}{\Omega(r_s - r_c)^2}.$$

v denotes the kinematic viscosity.

The torque $\mathbf{T}_c$ acting on the inner core is given by

$$\mathbf{T}_c = -\oint \mathbf{r} \times \mathbf{n} p \, d\Sigma_c = -\frac{\varrho_f}{2} \oint \mathbf{r} \times \nabla F_c |\mathbf{k} \times \mathbf{r}|^2 \, d\Sigma_c$$

$$- \varrho_f \oint \mathbf{r} \times \nabla F_c \pi \, d\Sigma_c. \tag{19}$$

The negative sign in (19) results from the fact that the normal unit vector **n** points into the fluid. **n** can be replaced by ∇F_c since the length of the vector ∇F_c differs only by a negligible amount of the order e_c from unity. The first integral on the right hand side of (19) yields the torque

$$\frac{\varrho_f e_c}{r_c} \oint \mathbf{r} \times (\mathbf{k} + \gamma_c) \, |\mathbf{k} \times \mathbf{r}|^2 \, (\mathbf{k} + \gamma_c) \cdot \mathbf{r} \, d\Sigma_c = - \mathbf{k} \times \gamma_c \frac{\varrho_f}{\varrho_c} (C_c - A_c) \quad (20)$$

which has been taken into account in the derivation of the Equation (4). In the second integral on the right hand side of (19) only the term proportional to $\dot{\gamma}_s$ is of interest. We denote this term by $\mathbf{T}_{cs}$ since it describes the coupling torque exerted by the shell on the inner core,

$$\mathbf{T}_{cs} = 2\varrho_f e_c e_s \oint \mathbf{k} \times \mathbf{r} \dot{\gamma}_s \cdot \mathbf{k} \times \mathbf{r} \, \frac{4(r_c^2 - s^2)\sqrt{r_s^2 - s^2}}{\sqrt{r_s^2 - s^2} - \sqrt{r_c^2 - s^2}} \, d\Sigma_c. \quad (21)$$

Because $\dot{\gamma}_c$ is parallel to $\mathbf{k} \times \gamma_c$ the term resulting from the nutation of the inner core gives rise to a torque of the same form as (20). Yet the term is of the order e_c smaller than (20) and can be neglected for this reason. After performing the integration with respect to the φ-coordinate and introducing

$$x \equiv \frac{z}{r_c}, \quad \alpha^2 = \frac{r_s^2}{r_c^2} - 1$$

we can rewrite the expression (21) in the form

$$\mathbf{T}_{cs} = 4\varrho_f e_c e_s \frac{8\pi}{15} r_c^5 \dot{\gamma}_s \int_0^1 \frac{15(1 - x^2)\sqrt{x^2 + \alpha^2}}{2(\sqrt{x^2 + \alpha^2} - x)} x^2 \, dx$$

$$= 4 e_c e_s \frac{\varrho_f}{\varrho_c} A_c \dot{\gamma}_s g \quad (22)$$

g is defined by the integral on the right hand side in the first line of (22). The evaluation of the integral yields

$$g = 1 + \tfrac{3}{7}\alpha^{-2} + \alpha^3 + \alpha^{-2}(\alpha^2 + 1)^{5/2} + \tfrac{4}{7}\alpha^{-2}(\alpha^7 - (\alpha^2 + 1)^{7/2}). \quad (23)$$

Using the relation (7) it is readily seen that $\mathbf{T}_{cs} A_c^{-1}$ is identical with the coupling term in the second equation of (9) which we have anticipated in the preceding section.

Obviously the torque $\mathbf{T}_{sc}$ exerted by the inner core on the shell is given by (22) after γ_s is replaced by $-\gamma_c$. In order to test the above derivation it may be of interest, however, to obtain $\mathbf{T}_{sc}$ by direct calculation,

$$\mathbf{T}_{sc} = - 4\varrho_f e_c e_s \frac{8\pi}{15} r_s^5 \dot{\gamma}_c \int_\beta^1 \frac{15\sqrt{\xi^2 - \beta^2}(1 - \xi^2)}{2(\xi - \sqrt{\xi^2 - \beta^2})} \xi^2 \, d\xi. \quad (24)$$

This time the variable of integration is defined by

$$\xi \equiv \frac{z}{r_s}$$

and the positive constant β^2 is given by

$$\beta^2 \equiv \frac{r_s^2 - r_c^2}{r_s^2} = \frac{\alpha^2}{1 + \alpha^2} \tag{25}$$

The introduction of the new variable of integration

$$x = \sqrt{\frac{\xi^2 - \beta^2}{1 - \beta^2}}$$

transforms the integral in (24) into

$$(1 - \beta^2)^{5/2} \int_0^1 (1 - x^2) x^2 \frac{15 \sqrt{x^2 + \frac{\beta^2}{1 - \beta^2}}}{2 \left(\sqrt{x^2 + \frac{\beta^2}{1 - \beta^2}} - x \right)} dx$$

which is identical with $(r_c/r_s)^2 \, g$ according to (22) and (25). Hence $\dot{\gamma}s \cdot \mathbf{T}_{sc}$ is equal to $-\dot{\gamma}_c \cdot \mathbf{T}_{cs}$.

Another test for the treatment of the problem given here is the fact that it yields the correct results in the limit when r_c/r_s tends to zero. In this case the pressure induced by the free nutation of the shell vanishes. This result is compatible with the exact solution of the time dependent problem (11), (13) for $r_c = 0$,

$$\mathbf{v} = - 2e_s (\mathbf{k} \, \dot{\gamma}_s \cdot \mathbf{r} + \omega_s \dot{\gamma}_s \, \mathbf{k} \cdot \mathbf{r}) ((1 + e_s)(\omega_s + 1) + e_s)^{-1},$$
$$\pi = 2e_s ((\omega_s + 1)^2 - 1) \mathbf{r} \cdot \mathbf{k} \times \dot{\gamma}_s \, \mathbf{k} \cdot \mathbf{r} ((1 + e_s)(\omega_s + 1) + e_s)^{-1}. \tag{26}$$

Since the pressure π in (26) is of the order of frequency of nutation smaller than in the solution (18) at finite values of r_c neglecting it is consistent within the approximation used above.

4. Discussion

In order to draw conclusions from the model about the earth system we have to introduce dissipation and to consider forced oscillations. We do the first by adding the terms $\kappa_s \delta_s, \kappa_c \delta_c$ on the left hand side of Equations (9) without specifying the dissipative process. The action of external torques can be described by replacing the zeros on the right hand side of Equations (9) by

$$\mathbf{f}_s A_s^{-1} e^{i\omega t}, \quad \mathbf{f}_c A_c^{-1} e^{i\omega t} \tag{27}$$

respectively. Since the vectors $\mathbf{f}_s$ and $\mathbf{f}_c$ as well as δ_s and δ_c are supposed to lie in a plane perpendicular to **k** it is convenient to regard this plane as a complex plane in which $\mathbf{f}e^{i\omega t}$ represents a circularly polarized wave. The stationary state of forced oscillations can be described by

$$\delta_s = \mathbf{d}_s e^{i\omega t}, \quad \delta_c = \mathbf{d}_c e^{i\omega t}$$

with $\mathbf{d}_s$ and $\mathbf{d}_c$ satisfying the equations

$$
\begin{aligned}
i(\omega - \omega_s - i\kappa_s) A_s \mathbf{d}_s + i\mu_s A_s \mathbf{d}_c &= \mathbf{f}_s \\
i(\omega - \omega_c - i\kappa_c) A_c \mathbf{d}_c + i\mu_c A_c \mathbf{d}_s &= \mathbf{f}_c
\end{aligned}
\tag{28}
$$

The solution of this system of equation is

$$
\begin{aligned}
\mathbf{d}_s &= i\, \frac{\mathbf{f}_s(\omega - \omega_c - i\kappa_c) A_s^{-1} - \mu_s \mathbf{f}_c A_c^{-1}}{\mu_s \mu_c - (\omega - \omega_c - i\kappa_c)(\omega - \omega_s - i\kappa_s)}, \\
\mathbf{d}_c &= i\, \frac{\mathbf{f}_c(\omega - \omega_s - i\kappa_s) A_c^{-1} - \mu_c \mathbf{f}_s A_s^{-1}}{\mu_s \mu_c - (\omega - \omega_c - i\kappa_c)(\omega - \omega_s - i\kappa_s)}.
\end{aligned}
\tag{29}
$$

Since the ellipticities of the core-mantle interface and of the inner core are about $\frac{1}{400}$ (MacDonald, 1967) $\mu_s \mu_c$ is only about 10^{-12} and can be neglected for this reason in the denominators of (29). Of particular interest is the case when the inner core is excited at its natural frequency. Assuming $\mathbf{f}_s = 0$, $\omega = \omega_c$ we obtain as response of the mantle

$$\mathbf{d}_s \approx \frac{-\mu_s \mathbf{f}_c}{(\omega_c - \omega_s)\kappa_c A_c}. \tag{30}$$

Accordingly, the amplitude of $\mathbf{d}_s$ depends strongly on κ_c. We have mentioned in the introduction that the damping rate κ_s for the mantle which is of the order 10^{-4} is rather large. Probably the dissipation in the ocean is responsible for most of the damping. For the inner core we expect a much lower damping rate. The most likely cause of damping in this case is of hydromagnetic origin.

Rochester and Smylie (1965) have concluded that hydromagnetic dissipation at the core-mantle boundary can account for not more than $10^{-4} - 10^{-5}$ of the observed damping of the Chandler wobble. In the case of the inner core hydromagnetic dissipation will be more important since the moment of inertia of the inner core is relatively small in comparison with its surface. If we assume tentatively that the inner core does not differ from the lower mantle in other respects, we obtain a damping rate $\kappa_c \approx 10^{-7}$. In this case the response (30) of the mantle to a pressure variation inside the cylinder $s = r_c$ with the frequency ω_c would be about 10 times as large as the response to the pressure variation at the natural frequency ω_s.

It is very likely that the damping rate κ_c is much higher than assumed above, in particular because the electrical conductivity is likely to be larger by a few orders of magnitude in the inner core than in the lower mantle. On the other hand the analysis by Rochester and Smylie was intended to yield an upper bound on the dissipation and was based for this reason on the assumption of a rigid outer core. In addition the torques exerted by motions in the fluid core may be more powerful at periods of the order of decades than at shorter periods. It is difficult to obtain more reliable results at this point since the properties of the inner core and the forces which are responsible for the excitation of the wobble are not well known. We conclude that the possibility exists that the motion of the pole includes a distinguished long period libration which is excited at the natural frequency of the inner core by pressure variations in the fluid core inside the cylindrical surface $s = r_c$.

The observations show indeed a polar libration with a period of about 24 years and an amplitude of approximately $\frac{1}{7}$ of the Chandlerian nutation, as was pointed out by Markowitz (1960, 1968). According to (5) a period of 24 years would require a density contrast between inner and outer core of about 6%, a number which is well in accordance with density distributions calculated on the basis of seismic data. Theory and observations are less compatible with respect to the polarisation. Markowitz interprets the results of his analysis in the form of a linearly polarized libration while a libration of positive circular polarisation should be expected according to the theory. Because of the difficult reduction of the data and because the libration will be less well defined in the direction of the superimposed secular motion of the pole, it is not yet clear whether the apparent distinction of a direction represents a real effect. In the case of a nearly linearly polarized libration the theory outlined in the preceding sections remains applicable when a nonaxisymmetric inner core is assumed.

If the hypothesis proposed in this paper is correct, important information is obtained about the inner core of the earth which always has been a rather obscure entity in geophysics. A phenomenon which is related to the nutation of the inner core is the gravitational oscillation which also depends on the density contrast between inner and outer core. Slichter (1961) has proposed that the gravitational oscillation of the inner core can be excited by large earthquakes strongly enough to become observable by gravimeters. The interpretation of the observations is compatible with this hypothesis if it is assumed that the fluid outer core reacts partly elastically to the high frequency oscillation of the inner core.

References

Abraham, H. J.: 1968, in *Continental Drift, Secular Motion of the Pole, and Rotation of the Earth* (ed. by Wm. Markowitz and B. Guinot), D. Reidel Publishing Company, Dordrecht, The Netherlands, p. 98.

Colombo, G. and Shapiro, I. I.: 1968, *Nature* **217**, 156.

Greenspan, H. P.: 1968, *The Theory of Rotating Fluids*, Cambridge University Press.

Lamb, Sir Horace: 1932, *Hydrodynamics*, 6th ed., Dover Publications, p. 724.

MacDonald, G. J. F.: 1967, in *Advances in Earth Sciences* (ed. by P. M. Hurley), M. I. T. Press.

Markowitz, Wm.: 1960, in *Methods and Techniques in Geophysics* (ed. by S. K. Runcorn), Interscience Publ., New York, p. 325.

Markowitz, Wm.: 1968, in *Continental Drift, Secular Motion of the Pole, and Rotation of the Earth* (ed. by Wm. Markowitz and B. Guinot), D. Reidel Publishing Company, Dordrecht, p. 25.

Munk, W. H. and MacDonald, G. J. F.: 1960, *The Rotation of the Earth*, Cambridge University Press.

Rochester, M. G. and Smylie, D. E.: 1965, *Geophys. J. Roy. Astron. Soc.* **10**, 289.

Slichter, L. B.: 1961, *Proc. Nat. Acad. Sci.* **47**, 186, 416.

Discussion

Runcorn: Doesn't the ellipticity of the inner core depend very much on what density you ascribe to it?

Busse: No, the ellipticity is pretty well independent of the densities assumed. This is a property of Clairaut's equation, which I used.

Garland: Would you comment on your neglect of the hydromagnetic effect on the outer core?

Busse: It is known from other problems, for instance the precessional problem, that hydromagnetic coupling is rather small compared with the pressure coupling.

DECONVOLUTION OF THE POLE PATH

D. E. SMYLIE and G. K. C. CLARKE

Dept. of Geophysics, University of British Columbia, Vancouver, Canada

and

L. MANSINHA

Dept. of Geophysics, University of Western Ontario, London, Canada

Abstract. Since the observed pole path is the convolution of the excitation with the impulsive wobble response, the problem of recovering the excitation from pole path observations is one of deconvolution. In the presence of measurement noise this can be done in an optimum way by designing a Wiener deconvolution filter. Optimum Wiener filters are designed for both the acausal and causal cases on the basis of a particular statistical model for the excitation – that appropriate to the earthquake excitation hypothesis. Deconvolution is carried out in the frequency domain with both filters. The results are found to be comparable to those obtained previously by a method which involved both deconvolution and detection. Premonition of great earthquakes is once again observed but it is somewhat reduced in the case of the causal filter. Shifts in the secular pole of the order of 0″.1 arc are found, in agreement with recent dislocation theory results for real Earth models.

1. Introduction

The position of the north pole of rotation is conveniently given by the complex function of time

$$m(t) = m_1 + im_2,$$

where m_1, m_2 are the direction cosines the axis of rotation makes with equatorial axes through 0° and 90°E longitude, respectively. This quantity has been the subject of observation now for over eighty years (Smylie and Mansinha, 1968).

Geophysical interest lies not so much in the observed quantity $m(t)$ as in the 'modified excitation function' $\psi(t)$ which describes the driving force producing the polar motion (Munk and MacDonald, 1960, Ch. 6).

The Earth wobbles in response to this excitation as a damped harmonic oscillator, resonating at the natural or Chandler angular frequency ω_0 (5.24 rad/yr). The equation of motion is (Munk and MacDonald, 1960, Ch. 6)

$$\dot{m} - i\sigma_0 m = - i\sigma_0 \psi = f(t), \tag{1}$$

where $\sigma_0 = \omega_0 + i/\tau$ and $\tau(\sim 20 \text{ yr})$ is the damping time.

Integration of (1) shows that for a particular excitation $f(t)$ we should observe

$$m(t) = \int_{-\infty}^{t} e^{i\sigma_0(t-\lambda)} f(\lambda) \, d\lambda$$

$$= \int_{0}^{\infty} e^{i\sigma_0\lambda} f(t-\lambda) \, d\lambda$$

L. Mansinha et al. (eds.), Earthquake Displacement Fields and the Rotation of the Earth, 99–112.

$$= \int_{-\infty}^{\infty} g(\lambda) f(t - \lambda) \, d\lambda \qquad (2)$$

where

$$g(t) = \begin{cases} 0, & t < 0 \\ e^{i\sigma_0 t}, & t > 0 \end{cases}$$

is the impulsive wobble response of the Earth (Figure 1).

Thus, $m(t)$ is the result of *convolution* of the excitation $f(t)$ with the impulsive wobble response $g(t)$. The problem of recovering $f(t)$ from the observed quantity $m(t)$ is then one of *deconvolution*.

This paper is concerned with the construction of optimum Wiener deconvolution filters for an excitation with particular statistical properties – namely those appropriate to the earthquake excitation hypothesis (Mansinha and Smylie, 1967; Smylie and Mansinha, 1968).

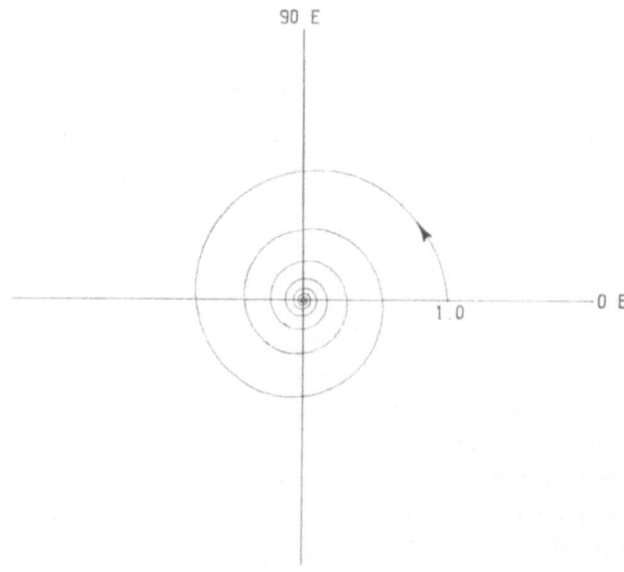

Fig. 1. Impulsive wobble response of the Earth. The pole path resulting from a unit impulse excitation is shown. For purposes of illustration the damping time has been reduced to two years.

2. The Wobble and Deconvolution System

Figure 2 shows a systems representation of the deconvolution problem. The excitation $f(t)$ acts on the Earth with impulsive wobble response $g(t)$, producing the polar motion $m(t)$. In the process of measuring $m(t)$ measurement noise $n(t)$ is added. A deconvolution filter with impulsive response $h(t)$ is to be constructed, such that with input

$$i(t) = m(t) + n(t),$$

the output $f_0(t)$ will be optimized with respect to a desired output $d(t)$.

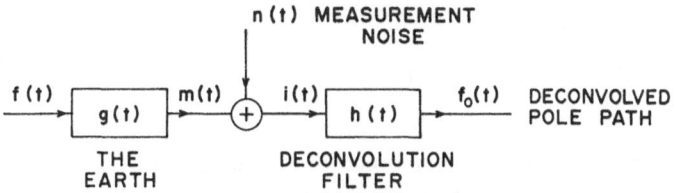

Fig. 2.   The wobble and deconvolution system.

The error in the optimization will be

$$\varepsilon(t) = f_0(t) - d(t).$$

When the optimization is such as to minimize the mean square of $\varepsilon(t)$, the filter is an optimum Wiener filter (Lee, 1960, Ch. 14).

The condition on $h(t)$ for this optimization is that

$$\phi_{di}(\tau) = \int_{-\infty}^{\infty} h(\lambda) \, \phi_{ii}(\tau - \lambda) \, d\lambda, \tag{3}$$

where

$$\phi_{di}(\tau) = E\left[d(t) \, i^*(t - \tau)\right]$$

and

$$\phi_{ii}(\tau) = E\left[i(t) \, i^*(t - \tau)\right]$$

are the crosscorrelation of $d(t)$ with $i(t)$ and the autocorrelation of $i(t)$, respectively. $E[...]$ denotes taking the expected value of the quantity in brackets and a superscript asterisk denotes complex conjugation.

Equation (3) is known as the Wiener-Hopf equation. It is written in the form applicable to stationary time series. In the case where the time series has already been completely sampled, the filter may be acausal and then Equation (3) holds for all values of τ. When only data prior to a certain time are to be processed, the filter must be causal and Equation (3) then holds only for $\tau \geqslant 0$.

The solution of the Wiener-Hopf equation is available for both the acausal and causal cases (Lee, 1960, Ch. 14).

In the acausal case, Equation (3) holds for all values of τ, allowing the solution to be obtained directly by Fourier transformation. In the frequency domain

$$H(\omega) = S_{di}(\omega)/S_{ii}(\omega), \tag{4}$$

where

$$H(\omega) = \int_{-\infty}^{\infty} h(t) \, e^{-i\omega t} \, dt,$$

$$S_{di}(\omega) = \int_{-\infty}^{\infty} \phi_{di}(t) \, e^{-i\omega t} \, dt$$

and

$$S_{ii}(\omega) = \int_{-\infty}^{\infty} \phi_{ii}(t)\, e^{-i\omega t}\, dt.$$

These are, respectively, the transfer function of the optimum filter, the cross power spectrum between $d(t)$ and $i(t)$ and the power spectrum of $i(t)$.

In the causal case, the solution may be obtained by factoring the input power spectrum into

$$S_{ii}(\omega) = S_{ii}^{+}(\omega)\, S_{ii}^{-}(\omega),$$

where $S_{ii}^{+}(\omega)$ contains all the poles and zeroes of $S_{ii}(\omega)$ that are in the upper half of the complex ω-plane and $S_{ii}^{-}(\omega)$ contains all those in the lower half. Then, the transfer function of the optimum filter is given by

$$H(\omega) = \frac{1}{S_{ii}^{+}(\omega)} \int_{0}^{\infty} q(t)\, e^{-i\omega t}\, dt, \qquad (5)$$

where

$$q(t) = \frac{1}{2\pi} \int_{-\infty}^{\infty} \frac{S_{di}(\omega)}{S_{ii}^{-}(\omega)}\, e^{i\omega t}\, d\omega.$$

For the deconvolution filter, the optimization of the output will be with respect to $f(t)$, so that

$$\begin{aligned}
\phi_{di}(\tau) &= E[f(t)\, i^{*}(t - \tau)] \\
&= E[f(t)\, \{m^{*}(t - \tau) + n^{*}(t - \tau)\}] \\
&= E\left[f(t) \left\{ \int_{-\infty}^{\infty} g^{*}(\lambda)\, f^{*}(t - \tau - \lambda)\, d\lambda + n^{*}(t - \tau) \right\} \right] \\
&= \int_{-\infty}^{\infty} g^{*}(\lambda)\, E[f(t)\, f^{*}(t - \tau - \lambda)]\, d\lambda + E[f(t)\, n^{*}(t - \tau)]
\end{aligned}$$

or

$$\phi_{di}(\tau) = \int_{-\infty}^{\infty} g^{*}(\lambda)\, \phi_{ff}(\tau + \lambda)\, d\lambda + \phi_{fn}(\tau), \qquad (6)$$

where

$$\begin{aligned}
\phi_{ff}(\tau) &= E[f(t)\, f^{*}(t - \tau)], \\
\phi_{fn}(\tau) &= E[f(t)\, n^{*}(t - \tau)]
\end{aligned}$$

are, respectively, the autocorrelation of the input excitation $f(t)$ and the crosscorrelation of $f(t)$ with the measurement noise $n(t)$.

Taking the excitation not to be correlated with the measurement noise, (6) reduces to

$$\phi_{di}(\tau) = \int_{-\infty}^{\infty} g^*(\lambda)\, \phi_{ff}(\tau + \lambda)\, d\lambda.$$

Because the autocorrelation is Hermitian, we may write

$$\phi_{di}(\tau) = \int_{-\infty}^{\infty} g^*(\lambda)\, \phi_{ff}^*(-\tau - \lambda)\, d\lambda$$

or

$$\phi_{di}^*(-\tau) = \int_{-\infty}^{\infty} g(\lambda)\, \phi_{ff}(\tau - \lambda)\, d\lambda$$

on taking complex conjugates and changing the sign of τ. Fourier transformation gives

$$S_{di}^*(\omega) = G(\omega)\, S_{ff}(\omega)$$

or

$$S_{di}(\omega) = G^*(\omega)\, S_{ff}(\omega), \tag{7}$$

where the real function $S_{ff}(\omega)$ is the power spectrum of the excitation $f(t)$.

Manipulations similar to those leading to (7) yield the relations

$$S_{ii}(\omega) = G(\omega)\, S_{di}(\omega) + S_{nn}(\omega) \tag{8}$$

and

$$S_{mm}(\omega) = |G(\omega)|^2\, S_{ff}(\omega) \tag{9}$$

where $S_{nn}(\omega)$ and $S_{mm}(\omega)$ are the noise power spectrum and the wobble power spectrum, respectively. We can combine (7), (8) and (9) to give

$$S_{di}(\omega) = S_{mm}(\omega)/G(\omega) \tag{10}$$

and

$$S_{ii}(\omega) = S_{mm}(\omega) + S_{nn}(\omega). \tag{11}$$

Given $S_{mm}(\omega)$ and $S_{nn}(\omega)$, we are now in a position to proceed to the transfer functions of the acausal and causal deconvolution filters via Equations (4) and (5), respectively.

3. Earthquake Excitation

A simplified model of the excitation due to earthquakes has been presented by Mansinha and Smylie (1967). According to this model the excitation $f(t)$ can be represented as a two-dimensional random walk in the complex plane.

Let a point in the complex plane be denoted by

$$z = z_0 + z_1 = z_0 + re^{i\theta}$$

where r, θ are its polar coordinates with z_0 as origin. Then, if z_0 denotes the initial value $f(0)$, the probability that $f(t)$ will lie between r and $r+dr$, θ and $\theta+d\theta$, after a time t, is

$$p(r, t) = \frac{1}{\pi \langle r^2 \rangle t} e^{-r^2/\langle r^2 \rangle t} r \, dr \, d\theta,$$

where $[\langle r^2 \rangle t]^{1/2}$ is the root mean square departure of $f(t)$ from $f(0)$.

The process generating $f(t)$ is clearly non-stationary. In this circumstance the autocorrelation and related quantities become functions of time (Bendat and Piersol, 1966, Ch. 9). The time dependent autocorrelation is defined to be

$$\phi_{ff}(\tau, t) = E[f(t + \tau/2) f^*(t - \tau/2)].$$

If we write

$$f(t - \tau/2) = z_0 + re^{i\theta},$$
$$f(t + \tau/2) = z_0 + re^{i\theta} + \varrho e^{i\alpha},$$

for $0 < \tau < 2t$, then

$$\phi_{ff}(\tau, t) = \int_0^{2\pi} \int_0^\infty \int_0^{2\pi} \int_0^\infty (z_0 + re^{i\theta} + \varrho e^{i\alpha}) \, p(\varrho, \tau)$$
$$\times (z_0^* + re^{-i\theta}) \, p(r, t - \tau/2) \, \varrho \, d\varrho \, d\alpha r \, dr \, d\theta$$
$$= |z_0|^2 + \langle r^2 \rangle (t - \tau/2).$$

For $-2t < \tau < 0$, we obtain

$$\phi_{ff}(\tau, t) = |z_0|^2 + \langle r^2 \rangle (t + \tau/2).$$

At $\tau = 0$, ϕ_{ff} is obviously just $|z_0|^2 + \langle r^2 \rangle t$.

Outside the range $-2t < \tau < 2t$ we may take $\phi_{ff} = |z_0|^2$. This implies that the process generating $f(t)$ was turned on at $t=0$ with initial value $f(0)$ and that $f(t) = f(0)$ for $t < 0$.

Hence,

$$\phi_{ff}(\tau, t) = \begin{cases} |z_0|^2, & |\tau| > 2t \\ |z_0|^2 + \langle r^2 \rangle (t - |\tau|/2), & |\tau| < 2t. \end{cases} \tag{12}$$

Figure 3 illustrates this autocorrelation function.

Having obtained the expression (12) for the instantaneous autocorrelation of the excitation, we can make use of Equation (2) to find the instantaneous autocorrelation of the excited wobble.

$$\phi_{mm}(\tau, t) = E[m(t + \tau/2) m^*(t - \tau/2)]$$
$$= E\left[\int_0^\infty \int_0^\infty e^{i\sigma_0 \lambda_1 - i\sigma_0^* \lambda_2} f(t + \tau/2 - \lambda_1) f^*(t - \tau/2 - \lambda_2) \, d\lambda_1 \, d\lambda_2 \right]$$
$$= \int_0^\infty \int_0^\infty e^{i\sigma_0 \lambda_1 - i\sigma_0^* \lambda_2} \phi_{ff}\left(\lambda_2 - \lambda_1 + \tau, t - \frac{\lambda_1 + \lambda_2}{2} \right) d\lambda_1 \, d\lambda_2.$$

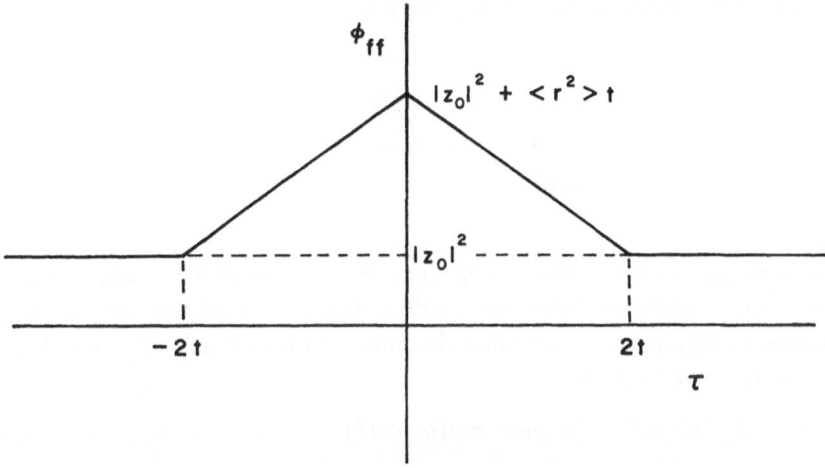

Fig. 3. The autocorrelation of the non-stationary earthquake excitation function. At later times the function is a photographic enlargement of that shown.

For $\tau \geqslant 0$, the latter integral can be evaluated to give

$$\phi_{mm}(\tau, t) = |z_0|^2/|\sigma_0|^2, \quad \tau > 2t$$

and

$$\phi_{mm}(\tau, t) = \frac{|z_0|^2}{|\sigma_0|^2} + \frac{\langle r^2 \rangle}{|\sigma_0|^2} \left[(t - \tau/2) - \frac{i}{\sigma - \sigma_0^*} \left(e^{i(\sigma_0 - \sigma_0^*)(t - \tau/2)} - 1 \right) e^{i\sigma_0\tau} \right.$$

$$\left. + \frac{i}{\sigma_0} \left(e^{i\sigma_0(t - \tau/2)} - 1 \right) e^{i\sigma_0\tau} - \frac{i}{\sigma_0^*} \left(e^{-i\sigma_0^*(t - \tau/2)} - 1 \right) \right],$$

$$0 < \tau < 2t. \quad (13)$$

The expression for $\tau < 0$ may be found by making use of the Hermitean property of ϕ_{mm}; that is

$$\phi_{mm}(-\tau, t) = \phi_{mm}^*(\tau, t).$$

The instantaneous power spectrum of the excited wobble is

$$S_{mm}(\omega, t) = \int_{-\infty}^{\infty} \phi_{mm}(\tau, t) e^{-i\omega\tau} \, d\tau.$$

Of more practical interest in the present problem is the average value of $S_{mm}(\omega, t)$ over a record. The smoothed power spectrum over a record of length T is

$$\bar{S}_{mm}(\omega) = \frac{1}{T} \int_0^T S_{mm}(\omega, t) \, dt.$$

Beginning with expression (13) we can show that

$$
S_{mm}(\omega) = 2\pi \frac{|z_0|^2}{|\sigma_0|^2} \delta(\omega) + \langle r^2 \rangle \left[\frac{1}{\omega^2(\omega - \sigma_0)(\omega - \sigma_0^*)} - \frac{1}{\omega^2 |\sigma_0|^2} \right.
$$
$$
\left. \times \left\{ \frac{\sin 2\omega T}{2\omega T} + \frac{(\omega_0 - 2\omega)\sin 2\omega T - (1/\tau)(\cos 2\omega T - 1)}{T[(\omega_0 - 2\omega)^2 + 1/\tau^2]} \right\} \right].
$$

$$(14)$$

The smoothed wobble power spectrum, as given by (14), is shown plotted for various record lengths in Figures 4a, b and c. For records of the order of ten years and greater in length it is evident that the smoothed power spectrum is well approximated by the stationary form

$$
S_{mm}(\omega) = \langle r^2 \rangle / \omega^2 (\omega - \sigma_0)(\omega - \sigma_0^*). \tag{15}
$$

4. Optimum Filters

It remains to specify the measurement noise power spectrum $S_{nn}(\omega)$ before we can proceed to the transfer functions of the optimum deconvolution filters as given by (4) and (5). In the absence of detailed information on the nature of the noise statistics, we take it to be white and write

$$
S_{nn}(\omega) = \langle r^2 \rangle N^2.
$$

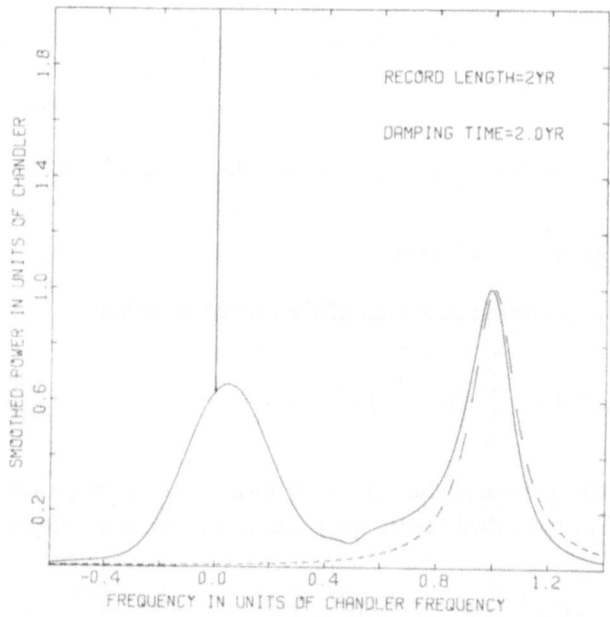

Fig. 4.a

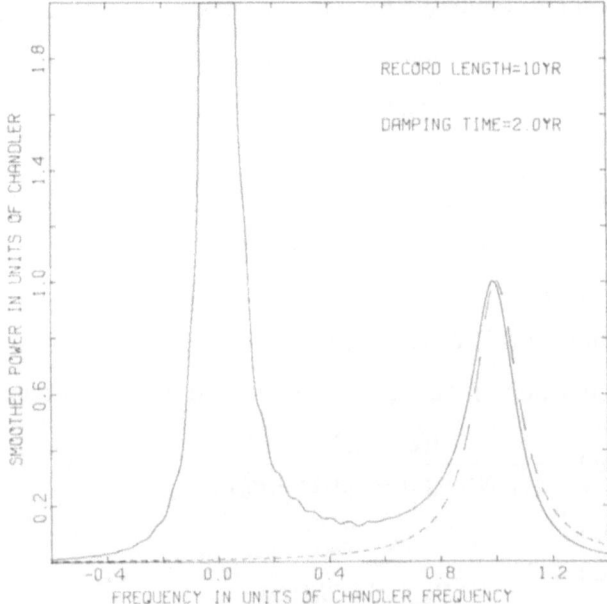

Fig. 4b.

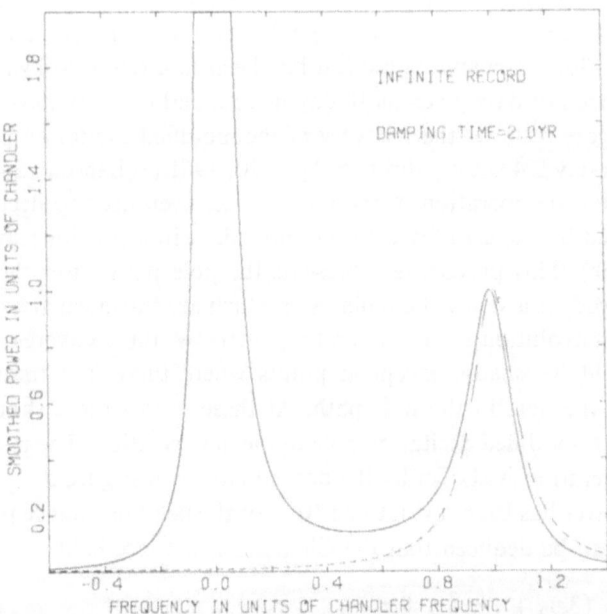

Fig. 4c.

Figs. 4a, b and c.   The smoothed wobble power as a function of record length. Again the damping time has been reduced to two years for illustration. The broken curve is the power spectrum obtained for a white excitation. Note that the Chandler resonance is slightly lowered in frequency and broadened. For an infinite record the spectrum is stationary everywhere but at zero frequency.

From (11) and (15) we have

$$S_{ii}(\omega) = \langle r^2 \rangle \frac{[1 + N^2\omega^2(\omega - \sigma_0)(\omega - \sigma_0^*)]}{\omega^2(\omega - \sigma_0)(\omega - \sigma_0^*)}. \tag{16}$$

Since

$$G(\omega) = \int\limits_{-\infty}^{\infty} g(t)\, e^{-i\omega t}\, \mathrm{d}t = 1/i(\omega - \sigma_0),$$

we obtain from (10)

$$S_{di}(\omega) = i\langle r^2 \rangle/\omega^2(\omega - \sigma_0^*). \tag{17}$$

The transfer function of the acausal filter derived from (4) is then

$$H(\omega) = \frac{i(\omega - \sigma_0)}{1 + N^2\omega^2(\omega - \sigma_0)(\omega - \sigma_0^*)}$$

or

$$H(\omega) = G^{-1}(\omega) \frac{S_{mm}(\omega)}{S_{mm}(\omega) + S_{nn}(\omega)}. \tag{18}$$

The latter form of the acausal deconvolution filter has been used in the frequency domain to filter the pole path produced by the Bureau International de l'Heure (BIH) for the period December 22, 1956 through January 4, 1968 (Bureau International de l'Heure, 1955–1967). The annual motion has been first removed by a least squares fit to the eleven years of data given as 10 day means, and the data have been reduced to zero mean. Figure 5 shows the recovery of the modified excitation pole from March 2, 1957 (Julian day 2,435,900) through April 10, 1961 (Julian day 2,437,400).

As a check on the operation of the filter, is has been used to deconvolve the pole path represented by the arcs fitted to the same data in a previous study (Smylie and Mansinha, 1968). This procedure represents the pole path, after the annual motion has been removed, as a series of circular arcs which are traversed at a constant angular rate. If the deconvolution filter is operating correctly, the recovered modified excitation pole should be steady, except at points where there is a transition from one circular arc to the next in the pole path. At these points there should be a sudden movement of the modified excitation pole to the new position. The result of this check on the filter operation is shown by the dashed curves in Figure 5.

The noise power has been taken to be 10^{-8} of the peak resonance power of $S_{mm}(\omega)$. From (15), it can be deduced that the Chandler resonance is at

$$\omega = (3\omega_0 + \sqrt{\omega_0^2 - 8/\tau^2})/4. \tag{19}$$

To obtain the transfer function of the causal filter, we need to factor (16). This results in

$$S_{ii}(\omega) = \langle r^2 \rangle\, N^2 \frac{(\omega - \omega_1)(\omega - \omega_1^*)(\omega - \omega_2)(\omega - \omega_2^*)}{\omega^2(\omega - \sigma_0)(\omega - \sigma_0^*)}, \tag{20}$$

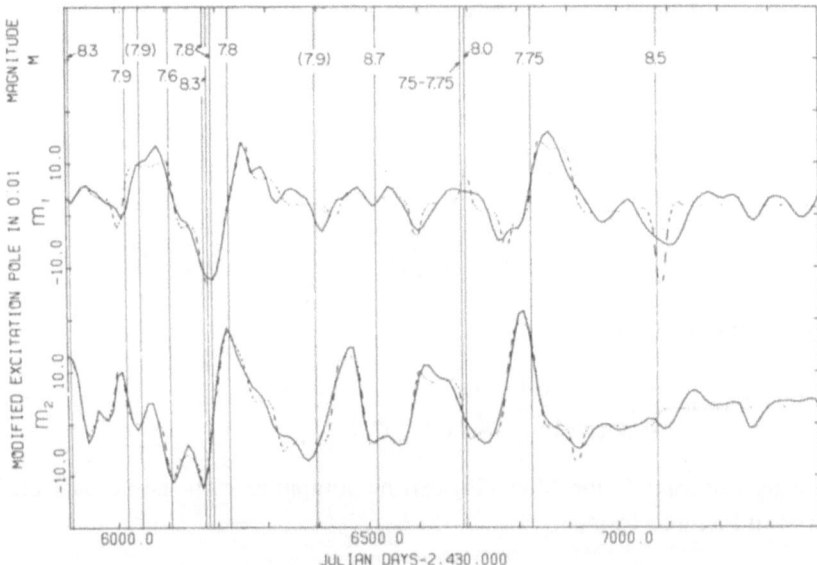

Fig. 5. The recovery of the modified excitation pole (solid curves) from March 2, 1957 to April 10, 1961 using the acausal filter in the frequency domain. The broken curves show the result of deconvolving the pole path represented by the arcs fitted to the same data in a previous study (Smylie and Mansinha, 1968).

where ω_1, ω_1^*, ω_2, ω_2^* are the two complex conjugate pairs of roots of the equation

$$\omega^4 - 2\omega_0\omega^3 + |\sigma_0|^2 \, \omega^2 + 1/N^2 = 0.$$

For a noise power level of 10^{-8} of peak resonance signal power (obtained by substituting (19) into (15)), we find numerically that

$$N^2 = 1.459 \times 10^{-7} \qquad \text{yr}$$
$$\omega_1 = -33.61 + i\,36.13 \quad \text{yr}^{-1}$$

and

$$\omega_2 = 38.84 + i\,36.13 \qquad \text{yr}^{-1}.$$

The input power spectrum (20) can be split into

$$S_{ii}^+(\omega) = \langle r^2 \rangle^{1/2} \, N \, \frac{(\omega - \omega_1)(\omega - \omega_2)}{(\omega - ia)(\omega - \sigma_0)}, \tag{21}$$

$$S_{ii}^-(\omega) = \langle r^2 \rangle^{1/2} \, N \, \frac{(\omega - \omega_1^*)(\omega - \omega_2^*)}{(\omega + ia)(\omega - \sigma_0^*)}, \tag{22}$$

where a is a real, positive quantity that will eventually be allowed to vanish. Similarly, we write

$$S_{di}(\omega) = \frac{i \langle r^2 \rangle}{(\omega - ia)(\omega + ia)(\omega - \sigma_0^*)} \tag{23}$$

in place of (17). The combination of (5), (21), (22) and (23) allows us to compute first.

$$q(t) = \frac{\langle r^2 \rangle^{1/2}}{N} \frac{e^{-at}}{(a + i\omega_1^*)(a + i\omega_2^*)}, \quad \text{for } t \geqslant 0,$$

and then the optimum filter transfer function

$$H(\omega) = \frac{-i(\omega - \sigma_0)}{N^2(a + i\omega_1^*)(a + i\omega_2^*)(\omega - \omega_1)(\omega - \omega_2)}.$$

Letting $a \to 0$, this becomes

$$H(\omega) = \frac{i(\omega - \sigma_0)}{N^2 \omega_1^* \omega_2^* (\omega - \omega_1)(\omega - \omega_2)}. \tag{24}$$

The causal deconvolution filter (24) can be compared with the acausal filter (18) expressed in factored form,

$$\frac{i(\omega - \sigma_0)}{N^2(\omega - \omega_1)(\omega - \omega_1^*)(\omega - \omega_2)(\omega - \omega_2^*)}.$$

In the time domain (24) is

$$h(t) = \begin{cases} 0, & t < 0 \\ \dfrac{(\sigma_0 - \omega_1) e^{i\omega_1 t} - (\sigma_0 - \omega_2) e^{i\omega_2 t}}{N^2 \omega_1^* \omega_2^* (\omega_1 - \omega_2)}, & t > 0. \end{cases}$$

Figure 6 shows the effect of deconvolving in the frequency domain with (24). The result may be compared with Figure 5, obtained by operating in the frequency domain with (18). The data and procedures used in the two deconvolutions are otherwise identical.

5. Discussion

The operation of the deconvolution filters has been verified by their performance on the uniform circular arcs previously fitted to the pole path (see broken curves in Figures 5 and 6). Ideally, we should see step changes in the level of the broken curves. Only approximations to step changes in level are found for two reasons. Firstly, the response of the filter cannot be made instantaneous owing to limitations set by the assumed noise level and the sampling interval of the data. Secondly, the fitted circular arcs were not constrained to intersect in such a way as to imply a pole motion which proceeds continuously from uniform circular movement along one arc to uinform circular movement along an intersecting succeeding arc. This introduces discontinuities which show up as departures from simple smoothed step changes.

The degree of similarity between the deconvolved real pole path (shown by the solid curves in Figures 5 and 6) and that resulting from the fitted arc path make it

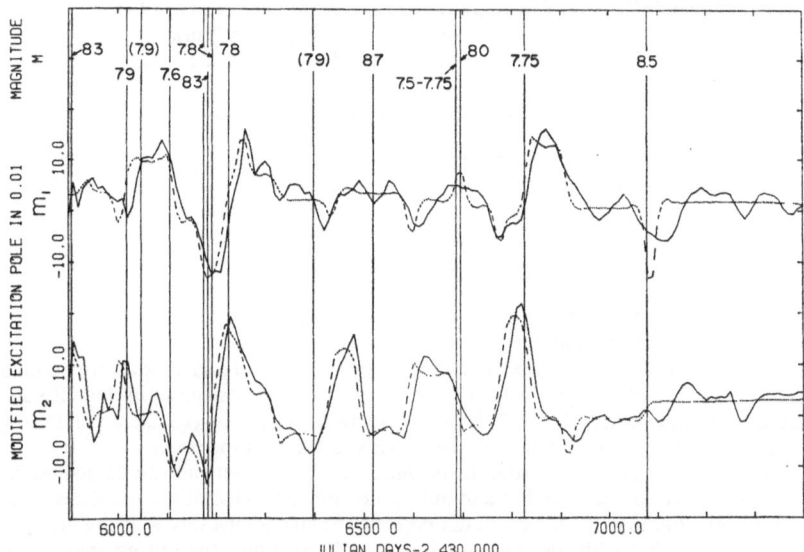

Fig. 6. The recovery of the modified excitation pole (solid curves) from March 2, 1957 to April 10, 1961 using the causal filter in the frequency domain. The broken curves in Figure 5 are repeated for reference.

extremely unlikely that the previous correlation (Smylie and Mansinha, 1968) found between major earthquakes and computed breaks in the pole path was spurious.

Premonition of great earthquakes is reduced in the case of the causal filter output but it is still evident, particularly in the case of the largest event in the series ($M = 8.7$).

The amplitudes of shifts in the modified excitation pole are of the order of $0''.1$ arc. Calculations based on dislocation theory in real Earth models (including the effects of sphericity, self-gravitation and radial variation of density, gravity and elastic constants, as well as a liquid core), presently in progress, indicate that pole shifts due to the mass rearrangement associated with major earthquakes are of this order.

We have discussed only deconvolution filters in this paper. Other Wiener filters could be easily designed following similar procedures. In particular, filters to extrapolate or smooth the pole path could be built. An alternative to the detection method used by Smylie and Mansinha (1968) would be to use the prediction error signal or the difference between the predicted pole path and the actual path as an indicator of changes in excitation.

Acknowledgement

The research reported in this paper was partly supported by the National Research Council of Canada.

References

Bendat, J. S. and Piersol, A. G.: 1966, *Measurement and Analysis of Random Data*, Wiley, New York.
Bureau International de l'Heure: *Bulletin Horaire, BIH,* Paris, bimonthly to 1967.

Bureau International de l'Heure: Circulaire D1 to D15, November 1966 to 1968.

Lee, Y. W.: 1960, *Statistical Theory of Communication*, Wiley, New York.

Mansinha, L. and Smylie, D. E.: 1967, 'Effect of Earthquakes on the Chandler Wobble and the Secular Polar Shift', *J. Geophys. Res.* **72**, 4731–4743.

Munk, W. H. and MacDonald, G. J. F.: 1960, *The Rotation of the Earth*, Cambridge University Press, London.

Smylie, D. E. and Mansinha, L.: 1968, 'Earthquakes and the Observed Motion of the Rotation Pole', *J. Geophys. Res.* **73**, 7661–7673.

Discussion

Haubrich: Did you assume a Poisson distribution for earthquakes?

Smylie: No. As we indicated in our 1967 paper, it is an inverse power law.

Haubrich: But the power law concerns the rate of earthquakes of a given size. If you take a given size earthquake and specify the rate you still have to specify how these occur in time. If you say they occur randomly in time, then this is a Poisson distribution. When you estimated the noise spectrum what was the frequency at which you took the power spectrum off the peak?

Smylie: We looked at everything from the Nyquist frequency down and you do not see very much anywhere except close to the Chandler and annual components. One of the problems is that if the noise level is raised too high the filter will not perform the step jumps very well and just gives a roll. This possibly means that with the noise level in the observations, the process may not be really feasible.

Haubrich: What noise level did you finally assume?

Smylie: 10^{-8} of the peak resonance. The peak was around 200 feet squared.

Haubrich: By peak do you mean the total power in the Chandler wobble.

Smylie: No. If you take the peak amplitude at the Chandler resonance as 1 then we assume the noise power to be 10^{-8}. Because of the nature of optimisation the power spectrum in the pole path is divided by the power spectrum in the pole path plus the power spectrum in the noise. So the level of the power spectrum of the pole path is not important.

Haubrich: That implies a Chandler amplitude of 10^{-5} sec which is smaller than the least count of the data. The data are quoted only to $0''.00025$ so your noise level is very small.

Smylie: This is a very hard thing to deal with and it is probably the weakest assumption which is made, but there is very little else one can do at the moment.

Stacey: Do you want an excitation statistically by large numbers of small earthquakes as well as the big ones?

Smylie: No, we took a mean slope, on the average polar shift vs. magnitude plot, of 2.5; in other words by going down one unit of magnitude you go down $10^{2.5}$ in contribution to the wobble. In effect, therefore, anything below about $7\frac{1}{2}$ would be negligible.

Stacey: I would suggest that if you have any $M = 8\frac{1}{2}$ earthquakes than anything below $M = 8.2$ is negligible.

Smylie: I haven't actually worked out the centre of gravity in the contribution but this could be done.

EXCITATION OF THE CHANDLER WOBBLE

SOLAR ACTIVITY AND THE ROTATION OF THE EARTH

R. J. JADY

Dept. of Mathematics, University of Exeter, England

Abstract. An estimate of the torque exerted by the solar wind on the Earth's magnetic dipole may be obtained in the simple case in which the magnetosphere is approximated by a thin plane perfectly conducting sheet. The effect of the lateral containment of the dipole is studied using an analytic solution of a model magnetosphere, in a form suitable for the automatic computation of the torque components in a frame of reference rotating with the Earth.

The results obtained are used in conjunction with the computed solar excitation spectrum, which has characteristic peaks at 11 years and 6 months, to discuss the influence of solar activity on the Earth's rotation. The application of this model to diurnal wobble is also considered.

1. Introduction

Evidence that solar activity may influence the Earth's rotation is given by Karklin (1967), following an analysis of the International Latitude Service data covering a period from 1900 through to 1959. The movement of the pole of rotation was analyzed using the method of moving harmonic analysis, and the amplitudes of the Chandlerian and annual oscillations obtained for this period. These amplitudes were further analyzed using periodogram analysis, and it was found that both the Chandler oscillation and the annual oscillation exhibit long period variations. The amplitudes of these variations are small and are shown in Table I.

TABLE I

Long period variations of the Chandler and annual oscillations
(after Karklin)

	Period of oscillations (years)	Amplitude	Phase relative to 1903	Probability of random amplitude
Chandler	10.3	0″.0200	194°	1:15
	12.3	0″.0244	219°	1:20
	17.7	0″.0421	215°	1:700
Annual	7.0	0″.0171	106°	1:1000
	10.7	0″.0172	193°	1:150
	17.8	0″.0186	173°	1:30

Apart from atmospheric processes induced by solar electromagnetic radiation, the suggestion has been made that the geomagnetic field will interact with the solar corpuscular flux to produce a torque capable of perturbing the pole of rotation. The average solar wind energy impinging on the magnetosphere is of order 10^{20} ergs/sec, a surprisingly large value (O'Brien, 1964). The torques produced by this interaction are examined here.

L. Mansinha et al. (eds.), Earthquake Displacement Fields and the Rotation of the Earth, 115–121.

2. Plane Sheet Approximation

The neutral ionized gas emitted by the sun is a good conductor of electricity, as it approaches the Earth; electric currents, the Chapman-Ferraro system (1931), are induced in the gas and these currents have the effect of shielding the interior of the gas from the geomagnetic field, and confining this field inside the magnetospheric surface. No exact solution has been found for this surface and one must rely on discussing simpler surfaces which approximate to the magnetosphere (Williams and Mead, 1965). An estimate of the torque exerted by the solar wind on the Earth's magnetic field is given in the simple case of a plane, perfectly conducting sheet placed a distance d from the Earth's dipole. The problem of the interaction of a thin plane conductor and a dipole field was first solved by Maxwell on the assumption that the currents induced in the neutral ionized stream perfectly shield the interior of the solar wind from the dipole field.

It is found that the magnetic field inside the cavity is the same as that field due to the original dipole and a correctly oriented image dipole placed at the same distance on the opposite side of the boundary. The magnitude of the extra field at the Earth's centre due to an image dipole parallel to the sheet is $M/(2d)^3$, where M is the moment of the geomagnetic dipole.

If now there is an increase in pressure due to a magnetic storm, the geomagnetic field is compressed, and there may be, during the first phase of the storm, an increase in the horizontal magnetic force of up to 100 γ. As shown by Chapman and Bartels (1940) approximately half of this increase is due to currents induced in the Earth by the sudden increase of the external field.

Using this model it is found that for an increase dH in the horizontal magnetic force of 20, 40, 50 γ, there corresponds a torque of approximately 1.7, 3.3, 4.1 × 10^{22} dyn cm.

3. The Effect of Lateral Containment

A more realistic model for the magnetosphere, which takes into account the lateral containment of the field, is required. Parker (1962) discussed the dynamics of the geomagnetic storm using a circular cylindrical cavity plugged at one end to correspond to the nose of the magnetosphere. For purposes of computation, however, it seems best to extend the method of images to the case of a rectangular cylindrical cavity blocked off at one end. The rear may be closed also but in view of its distance from the dipole the torque components are unlikely to be altered much.

The axis of the cavity will always lie parallel to the Sun–Earth line, and we take right-handed axes $OXYZ$ with OY along the axis of the cylinder in the anti-solar direction, OZ-perpendicular to the ecliptic plane, and origin of coordinates in the closed end of the cavity. The walls of the cavity correspond to $X = \pm a$, $Z = \pm c$. If the moment of the geomagnetic dipole with coordinates (o, b, o) has components (m_1, m_2, m_3), which are functions of time, the image system consists of one set of dipoles with coordinates $(2m\alpha, b, 2nc)$, and having components

$$\mathbf{M}_1(m, n) = \{(-1)^m m_1, m_2, (-1)^n m_3\},$$

where m, n take all integral values $m=n\neq0$, and a second set with coordinates $(2m\alpha, -b, 2nc)$ and with components

$$\mathbf{M}_2(m, n) = \{(-1)^m m_1, -m_2, (-1)^n m_3\}.$$

The total potential at a field point $P\,(XYZ)$ due to the image system is clearly

$$V = -\sum_{m=-\infty}^{\infty}\sum_{n=-\infty}^{\infty}$$

$$\times \left[\frac{\{(-1)^m m_1, m_2, (-1)^n m_3\}\cdot\{(2m\alpha - X), (b - Y), (2nc - Z)\}}{\{(2m\alpha - X)^2 + (b - Y)^2 + (2nc - Z)^2\}^{3/2}}\right.$$

$$\left. + \frac{\{(-1)^m m_1 - m_2, (-1)^n m_3\}\cdot\{(2m\alpha - X), (b + Y), (2nc - Z)\}}{\{(2m\alpha - X)^2 + (b + Y)^2 + (2nc - Z)^2\}^{3/2}}\right]$$

and the field is given by

$$\mathbf{H} = -\operatorname{grad} V.$$

The lines of force shown in Figure 1 have been computed in a plane containing the dipole in the case in which the dipole axis is perpendicular to the axis of the cavity.

In the motion of the Earth about the Sun, the axis of rotation remains fixed in space, over the time period under consideration; consequently the dipole axis will change its orientation inside the cavity. Further, since the dipole axis is inclined to the rotation axis by approximately $11\frac{1}{2}°$ the dipole field, which is constrained to rotate with the Earth, will have a diurnal variation.

If $\mathbf{M}$ is the dipole moment in $OXYZ$ axes and $\mathbf{M}^1$ is the dipole moment in a frame of reference rotating with the Earth, then

$$\mathbf{M}^1 = A\mathbf{M},$$

where A is the transformation matrix

$A =$

$$\begin{pmatrix} \cos\Omega\tau\cos\Phi - \cos\theta\sin\Phi\sin\Omega\tau & \cos\Omega\tau\sin\Phi + \cos\theta\cos\Phi\sin\Omega\tau & \sin\Omega\tau\sin\theta \\ -\sin\Omega\tau\cos\Phi - \cos\theta\sin\Phi\cos\Omega\tau & -\sin\Omega\tau\sin\Phi + \cos\theta\cos\Phi\cos\Omega\tau & \cos\Omega\tau\sin\theta \\ \sin\theta\sin\Phi & -\sin\theta\cos\Phi & \cos\theta. \end{pmatrix}$$

$\theta = 23\frac{1}{2}°$ is the obliquity of the ecliptic, Φ is the longitude of the Earth measured from the sun, and Ω denotes the mean diurnal rotation. The inverse transformation from geographic axes to $OXYZ$ axes is then

$$\mathbf{M} = A^{-1}\mathbf{M}^1 = \tilde{A}\mathbf{M}^1$$

in virtue of the orthogonality of A.

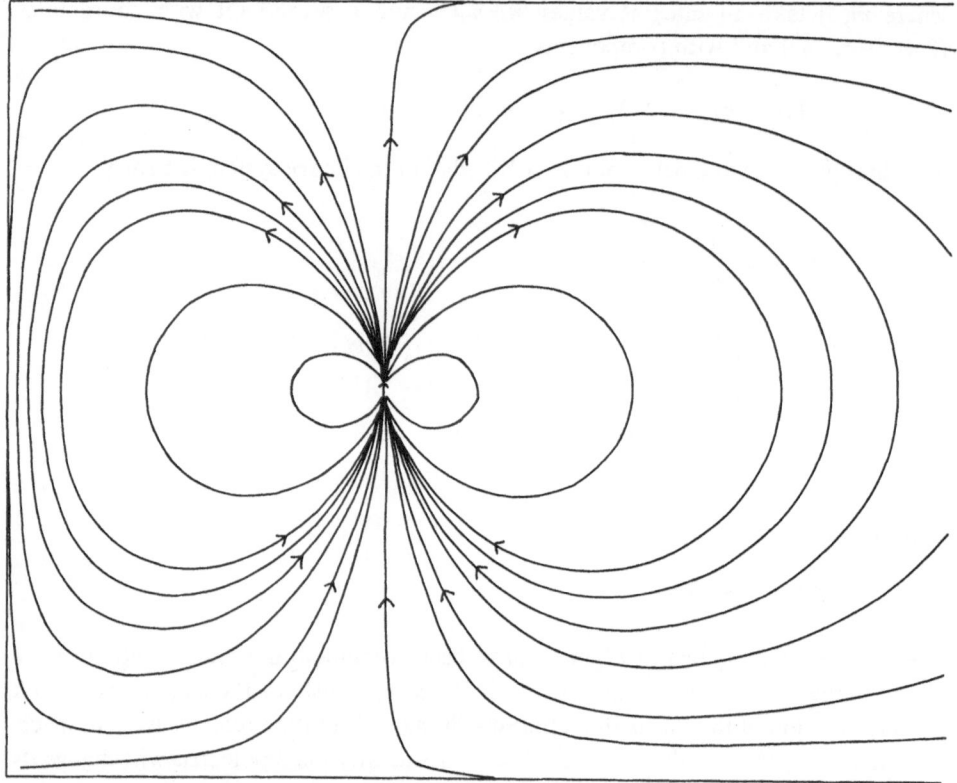

Fig. 1.   Lines of force of a magnetic dipole contained in a rectangular cylindrical cavity closed at
one end, computed in the case for which the dipole axis is perpendicular to the axis of the cylinder.
A section parallel to the sides of the cavity and containing the dipole axis is shown
$(m, n = 0, 1, 2)$.

The torque on the geomagnetic dipole is given by the usual expression

$$\mathbf{L} = \mathbf{M} \times \mathbf{H}^1,$$

where $\mathbf{H}^1$ is the field due to the set of image dipoles, excluding the geomagnetic dipole, at the Earth's centre.

The diurnal variation of one equatorial component of torque in the geographic frame has been computed for a typical case $a = b = c = 7\,R_e$ and is shown in Figure 2.

The maximum variation is at $\Phi = 0$ and π whilst at $\Phi = \pi/2$; $3\pi/2$ (the equinoxes) the variation is small and of semi-diurnal frequency. The maximum amplitude of torque in this case is 1.1×10^{23} dyn cm. During heightened solar activity the dimensions of the cavity will decrease, increasing the torque roughly proportional to the inverse cube of the cavity size. The axial component of torque remains small throughout the motion.

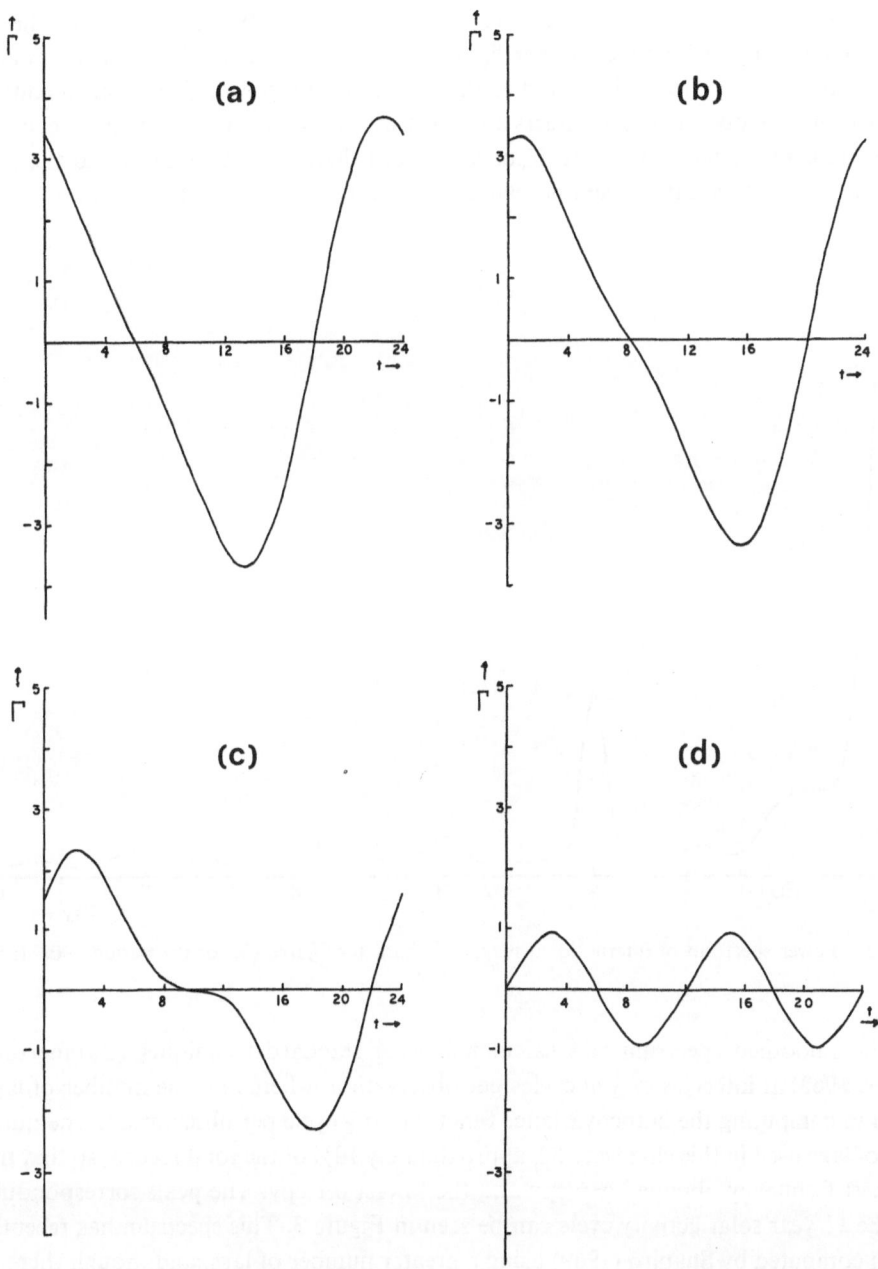

Fig. 2. The diurnal variation of one equatorial component of torque Γ (in units of 10^{22} dyn cm^{-1}), in a typical case at (a) $\Phi = 0$, (b) $\Phi = \pi/6$, (c) $\Phi = \pi/3$, and (d) $\Phi = \pi/2$.

4. The Excitation Spectrum

The power spectrum of magnetic disturbance using monthly mean values of the international daily magnetic character figure C_i for the period 1900–57 was computed and is shown in Figure 3. The C_i index, though rather subjective, gives firstly a concise indication of global magnetic effects and consequently of solar activity, and secondly a longer continuous record than K_p or a_p; indeed there is some doubt as to the adequacy of K_p to account for the solar wind momentum flux (Piddington, 1968).

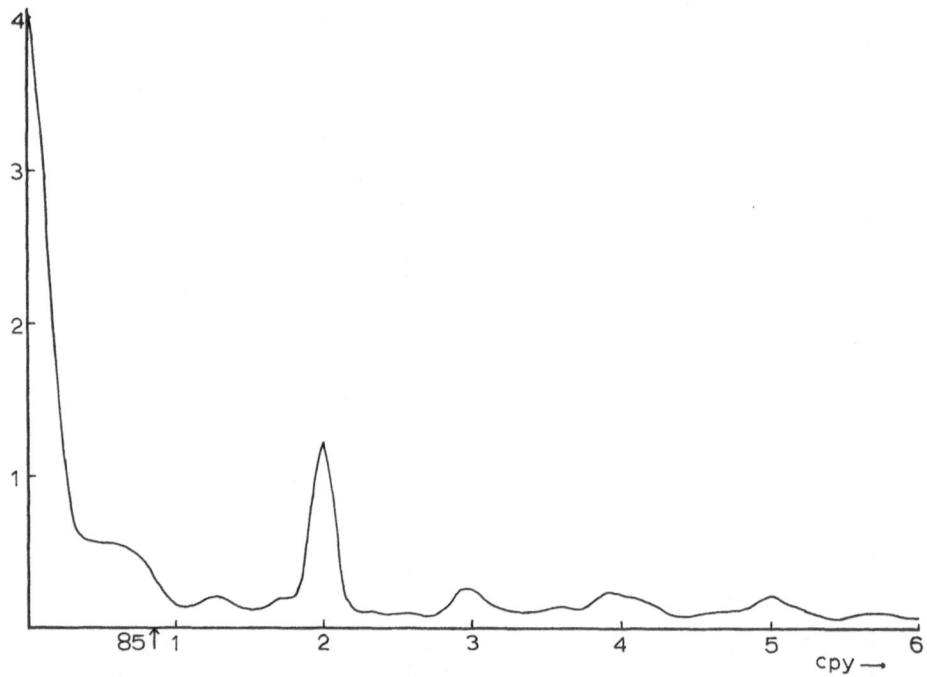

Fig. 3.   Power spectrum of international magnetic character figures C_i, for the period 1900–1957.

The smoothed spectrum was calculated using standard techniques (Swinnerton-Dyer, 1963) at intervals of $\frac{1}{2} m$ cycles per observation, where m is the number of lags used in computing the autocovariance function, to $\frac{1}{2}$ cycle per observation. The number of lags used in this case were 60, approximately 10% of the total record, so that the highest frequency obtained is 6 cpy and the lowest 0.1 cpy. The peak corresponding to the 11 year solar activity cycle can be seen in Figure 3. This spectrum has recently been computed by Shapiro (1969) using a greater number of lags, and though there is good general agreement Shapiro isolates two peaks at periods of 1.41, and 1.10 years.

The power of the Chandler frequency is small in both cases, and although any excitation would induce a resonance response there seems little doubt that there is insufficient power at the Chandler frequency to excite the Chandler wobble.

5. Discussion of Results

The magnitude of the torques computed are less than those required to account for the change in the length of day on the ms/decade scale (Rochester, 1960), though they are comparable to the lunar tidal torque ($\sim 2.7 \times 10^{23}$ dyn cm).

It is possible that the diurnal wobble mode (Rochester, 1968) may be excited by the solar wind torque, though this possibility has not been investigated fully. The amplitude of the diurnal wobble is small and more accurate observational results such as are envisaged will have to be awaited before being certain that this mode is indeed excited.

Acknowledgement

I would like to thank Professor A. T. Price for his helpful discussions on the problem.

References

Chapman, S. and Ferraro, V. C. A.: 1931, *Terrest. Magn. Atmos. Elect.* **36**, 77, 171.
Chapman, S. and Bartels, J.: 1940, *Geomagnetism,* Clarendon Press, Oxford.
Karklin, V. P.: 1967, *Geomagnetizm i Aeronomyia* **7**, 121.
O'Brien, B. J.: 1964, *J. Geophys. Res.* **69**, 13.
Parker, E. N.: 1962, *Space Sci. Rev.* **1**, 62.
Piddington, J. H.: 1968, *Geophys. J. Roy. Astron. Soc.* **15**, 39.
Rochester, M. G.: 1960, *Phil. Trans. Roy. Soc.* **A252**, 531.
Rochester, M. G.: 1968, *J. Geomag. Geoelect.* **20**, 387.
Shapiro, R.: 1969, *J. Geophys. Res.* **74**, 2356.
Swinnerton-Dyer, H. P. F.: 1963, *Computer J.* **5**, 16.
Williams, D. J. and Mead, G. D.: 1965, *J. Geophys. Res.* **70**, 3017.

Discussion

Rochester: I am a little bit disturbed about your use of the magnetostatic approach rather than a hydromagnetic approach. Are you familiar with the results of Beiser who obtained an interplanetary torque on the earth of an order of magnitude of 10^{23} dyne cm, and then this was cut down by a factor of 1000 by the more rigorous treatment of the problem by Maeda.

Jady: The crux of the matter is probably the approximation involved in assuming that the interplanetary fluid is a perfect conductor. I think this assumption is justified in taking a model of this sort but it is one which could very well have a consequence on the results.

SEISMIC EXCITATION OF THE CHANDLER WOBBLE

L. MANSINHA

University of Western Ontario, London, Canada

and

D. E. SMYLIE

University of British Columbia, Vancouver, Canada

Abstract. Results from the elasticity theory of dislocations as well as observational evidence are presented in support of the hypothesis of seismic excitation of the Chandler wobble. There appears to be a high degree of correlation between changes in the pole path and earthquakes with $M > 7.5$ in the 11 year period from 1957.0 to 1968.0. The probability of correlation by chance is shown to be very small for a range of values of two parameters.

1. Introduction

Periodic change in the geographic orientation of the earth's axis of rotation is termed 'wobble'. The Chandler wobble is the spectral component with a period of 438 days. A comprehensive history of polar motion is given by Lambert *et al.* (1931). The method of observation and reduction by the International Polar Motion Service (successor in 1962 to the International Latitude Service) is given by Yumi (1968). A second organization has been independently determining pole positions since 1955. The current methods of the Bureau International de l'Heure are given by Guinot and Feissel (1968, 1969).

The excitation of the wobble has remained problematic since its isolation from variation of latitude data in 1891 by S. C. Chandler. A relation between polar motion and earthquakes was suspected shortly thereafter by Milne (1893), who noticed that over a two year period more earthquakes appeared to occur around the time of sharp changes of curvature of pole path. But since the annual component of polar motion data was not removed, the relation is dubious. Milne continued to mention the relationship in subsequent works but rejected the idea of seismic excitation because he thought the displacements accompanying the earthquakes were of limited areal extent. (Milne, 1903, 1906, 1913). Larmor and Hills (1906) showed that a sudden displacement of 10^7 cubic miles of material on the surface by 10 ft due to an earthquake would change the wobble amplitude by $0\overset{''}{.}003$. Later Larmor (1909) concluded that the displacements involved in land earthquakes are insufficient but that submarine earthquakes may have a significant effect on the wobble. Munk and MacDonald (1960) quote Cecchini (1928) in connection with various attempts to link variations in the wobble to earthquakes and volcanic eruptions. Lambert (1926, 1931) considers and dismisses the idea that earthquakes may be triggered by the stresses due to variation of latitude. Stoyko (1952) reports a similarity in the plotted curves of Chandler amplitude and energy released by deep (focal depth $\geqslant 70$ km) earthquakes during the period 1908–1944. Munk and MacDonald (1960) dismiss seismic excita-

L. Mansinha et al. (eds.), Earthquake Displacement Fields and the Rotation of the Earth, 122–135.
All Rights Reserved. Copyright © 1970 by D. Reidel Publishing Company, Dordrecht-Holland

tion of the wobble from consideration of a block displacement model for earth-quakes. If a surface block with $100 \text{ km} \times 100 \text{ km} \times 30 \text{ km}$ dimensions is displaced by one meter, the contribution to the wobble is only 10^{-5} sec of arc. This is about 10^{-4} too small to account for the observed Chandler amplitude of the order of $0''.1$.

2. Theory

The theory of polar motion is given in the monograph by Munk and MacDonald (1960) and in this volume by Rochester (1970). For completeness the main expressions are reproduced here. The coordinate system is oriented with x_1 along the Greenwich meridian and x_2 along a meridian $90°E$. The x_3 axis passes through the Conventional International Origin (CIO). The position of CIO is defined through the mean latitudes of ILS observatories at Carloforte, Gaithersberg, Mizusawa and Ukiah during 1900 to 1905; and of Kitab during the two periods (1935 to 1940) and (1949 to 1954). The polar motion $m = m_1 + im_2$ is given by

$$m(t) = \frac{-e^{i\sigma_0 t}}{A} \int\limits_\infty^t e^{-i\sigma_0 t} \left[i\Omega(c_{13} + ic_{23}) + \dot{c}_{13} + i\dot{c}_{23} \right] dt, \tag{1}$$

where m_1 and m_2 are angular displacements of the rotation pole along x_1 and x_2 axes; Ω is the mean diurnal rotation rate of the earth; c_{13} and c_{23} are changes in the off-diagonal components of the inertia tensor and $\dot{c}_{13}$, $\dot{c}_{23}$ are the time derivatives; A is the equatorial moment of inertia; σ_0 is the complex Chandler angular frequency.

If the mass redistribution due to an earthquake occurs in a time short compared with the period of the wobble, one can assume a step function time dependence for c_{13} and c_{23}. Integration of Equation (1) gives the effect of a single earthquake on $m(t)$;

$$m(t) = \frac{-\Omega}{\sigma_0} \frac{(c_{13} + ic_{23})}{A} e^{i\sigma_0 t} + \frac{\Omega}{\sigma_0} \frac{(c_{13} + c_{23})}{A}. \tag{2}$$

The time dependent part is the contribution to the wobble whereas the time independent part is the shift of the secular pole. At $t = 0$, the two are equal in magnitude but opposite in sign.

To calculate c_{13} and c_{23}, the displacements u_1, u_2, u_3 need to be known throughout the crust and the mantle. Developments in the elasticity theory of dislocations by Steketee (1958), Chinnery (1961), Maruyama (1964) and Press (1965) made it possible to calculate the displacement fields in an elastic half-space due to slip on a fault. The half-space displacements were used to approximate corresponding terms in the real earth by wrapping them around a sphere (Mansinha and Smylie, 1967).

The amplitude $|m|$ for any earthquake is a function of geographic latitude as well as fault azimuth. The longitude of earthquake focus determines only the direction. In order to fix on the effect of earthquake magnitude (and corresponding fault parameters) the $|m|$ for each earthquake was averaged over latitude and fault azimuth.

While the locations of focii of all large earthquakes is known, the fault parameters are known for only a few. An empirical magnitude-fault length relationship
due to Press and Brace (1965) was used. The fault width was assumed to be 0.25 of
fault length. The slip on the fault was taken as 5 m. The computation for faults inter-
secting the surface are shown in Figure 1. On the same diagram $|m|$ for several earth-
quakes with known fault parameters are shown. Computations for shallow buried
faults were done, but the curves obtained were too close to the one shown in Figure 1
to be drawn separately. Hence the effect of depth of a fault on $|m|$ is not significant
at shallow depths.

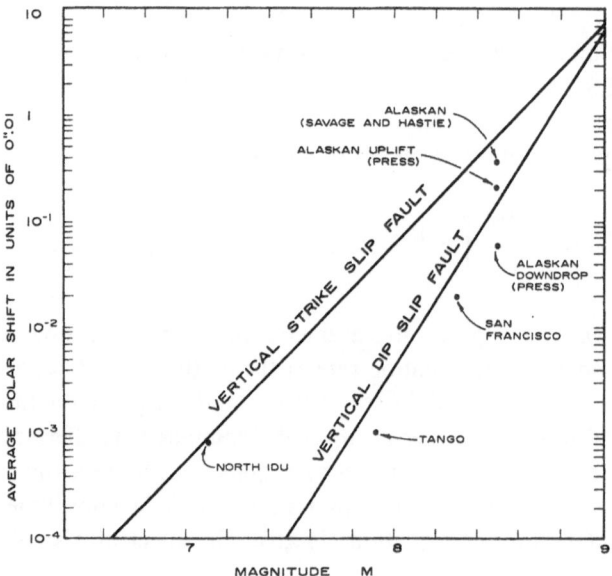

Fig. 1.   The dependence of average polar shift on magnitude.

3. Cumulative Effect

An estimated 10^6 earthquakes occur every year (Gutenberg and Richter, 1954). The
cumulative effect of this large number of events on the Chandler wobble is

$$\sum S_j e^{i[\sigma_0(t-t_j)+\pi]} \tag{3}$$

and on the secular polar shift is

$$\sum_j S_j$$

with

$$S_j = \frac{\Omega}{A\sigma_0}(\Delta c_{13} + i\Delta c_{23}). \tag{4}$$

The subscript j refers to the jth earthquake; i.e. t_j is the time of occurrence of the
jth earthquake. The time of occurrence of an earthquake as well as the longitude of

the focus serve as randomizing factors for Equations (3) and (4). Considering the longitude of focii as randomly distributed leads to an under-estimate of the secular polar shift. In the case of both the wobble and polar shift, a two dimensional random walk analysis is applied to estimate the seismic excitation.

From observational seismology, the empirical frequency-magnitude relationship of earthquakes is given by

$$\log_{10} N = A - bM \tag{5}$$

where N is the number of earthquakes of magnitude M or greater; A and b are constants. In a crude way one may give a relationship between S, the step-length (contribution) to the wobble or polar shift and M, the earthquake magnitude.

$$\log_{10} S = D + cM$$

Here D and c are constants. To eliminate D one can write

$$\log_{10} \frac{S}{S_1} = c(M - M_1) \tag{6}$$

when S_1 is the average step-length due to an earthquake of magnitude M_1.

Unlike A and b, the values of D and c cannot be established at the present time with any confidence. If $M_1 = 8.5$, the values of c and S_1 in Figure 1 are 2.03, $0''.0069$ for strike slip fault and 3.04, $0''.0015$ for dip slip fault respectively.

Combining expressions (5) and (6), the number of earthquakes per year with step length between S and $S + dS$ is given by

$$\frac{b}{c} \frac{N_1}{S_1} \left(\frac{S_1}{S} \right)^{b/c+1} dS . \tag{7}$$

Starting from the above equation, one finds the estimated rms annual seismic excitation to be

$$S_m \left[\frac{b}{2c - b} \left(\frac{S_0}{S_m} \right)^{b/c} N_0 \right]^{1/2} \tag{8}$$

where S_m is the maximum possible step length and N_0, S_0 refer to minimum allowable step-lengths. Taking into account the fact that the wobble decays between earthquakes, Equation (8) becomes

$$S_m \left[\frac{b}{2c - b} \left(\frac{S_0}{S_m} \right)^{b/c} N_0 \right]^{1/2} \left(\frac{\tau}{2} \right)^{1/2} \tag{9}$$

where τ is the damping time.

The estimated rms annual excitation computed from Equation (9) as a function of the parameters c and S_1 is shown in Figure 2. The following numerical values are used: M_0 and M_m are 7.0 and 8.9 respectively; M_1 is 8.5; $\tau = 20$ yr; $b = 1$; $N_0 = 23.5$/yr. For the value of $c = 2.03$, $S_1 = 0''.0069$ determined from Figure 1, we have an estimated

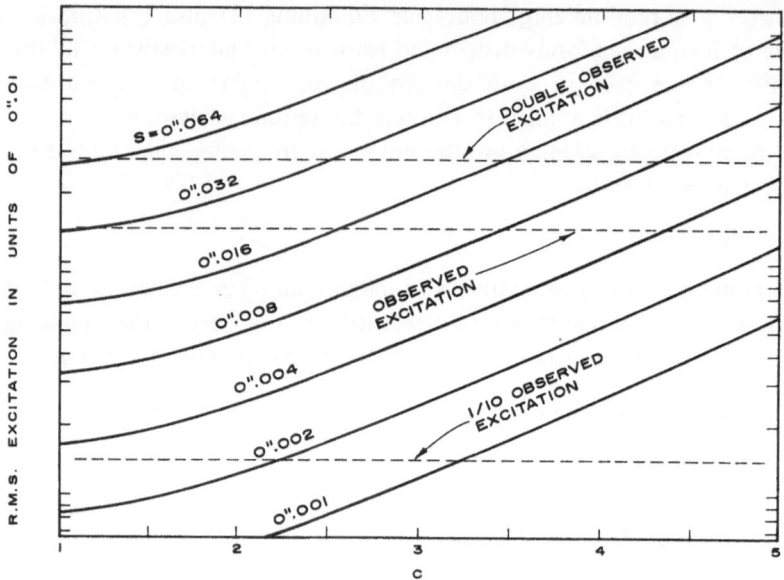

Fig. 2. The dependence of the rms Chandler wobble excitation on the parameters c and s_1, defined by Equation (9).

rms seismic excitation of 3 ft, roughly one-fifth of the observed rms Chandler wobble amplitude. The annual secular polar shift is given by $(1/\tau)^{1/2}$ of the wobble excitation and comes to about $0''.005$.

The above estimate of the excitation has been questioned by Haubrich (1970), who states that Equation (5) overestimates the number of earthquakes at the higher magnitude range. The number of earthquakes predicted from Equation (5) and the actual number occurring over 1904 to 1964 are shown in Table I. It is seen that

TABLE I

Table compares predicted number of earthquakes with the number of observed events during 1904 to 1967

Magnitude M	Predicted events 1904 to 1967	Observed events 1904 to 1967
8.9	4	2
8.8	5	0
8.7	6	9
8.6	8	10
8.5	10	2
8.4	12	12
8.3	16	37
8.2	19	1
8.1	25	25
Total	105	98

Equation (5) predicts 105 earthquakes for $M > 8.0$ while actually 98 earthquakes occur during the same period. It would be surprising indeed if one obtained a better agreement from an empirical relationship. Haubrich (1970) has used the magnitudes listed in Gutenberg and Richter (1954, photo reprinted 1967). Those magnitudes have since been revised as given in Richter (1958).

4. Observational Test

From Equation (2) it is seen that a sudden mass shift in the solid earth produces no immediate change in the position of the rotation pole. The only change is in the secular pole as well as in the radius of the pole path. An attempt was made to detect such changes in the pole path and to find any possible correlation with major earthquakes. The hypothetical pole path is shown in Figure (3).

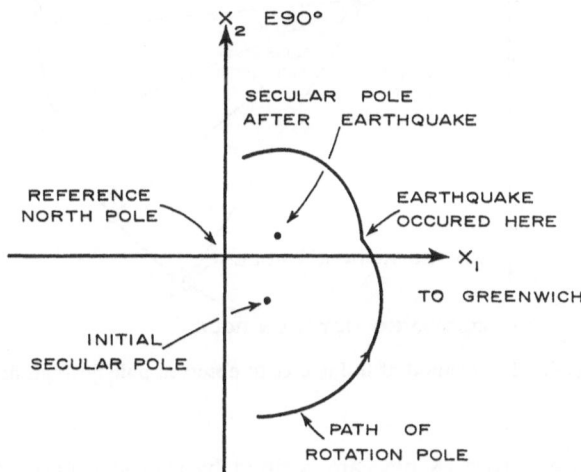

Fig. 3.   The effect of a sudden mass shift on the pole path and secular pole.

Let the time interval between published pole position data (with annual component removed) be δ days. Then the angle α through which the rotation pole advances with respect to the secular pole in δ days is

$$\alpha = \frac{2\pi}{438} \times \delta.$$

Knowing α, an exact arc can be fitted through two consecutive pole positions. The pole path is then extrapolated to predict the next point. If the pole path has changed to a different arc, the predicted and observed points will not agree. However, due to errors there may not be an agreement even without a change in pole path. To take into account the errors, an acceptance radius a is defined. If the difference between observed and predicted pole position is less than a, then no change in pole path is

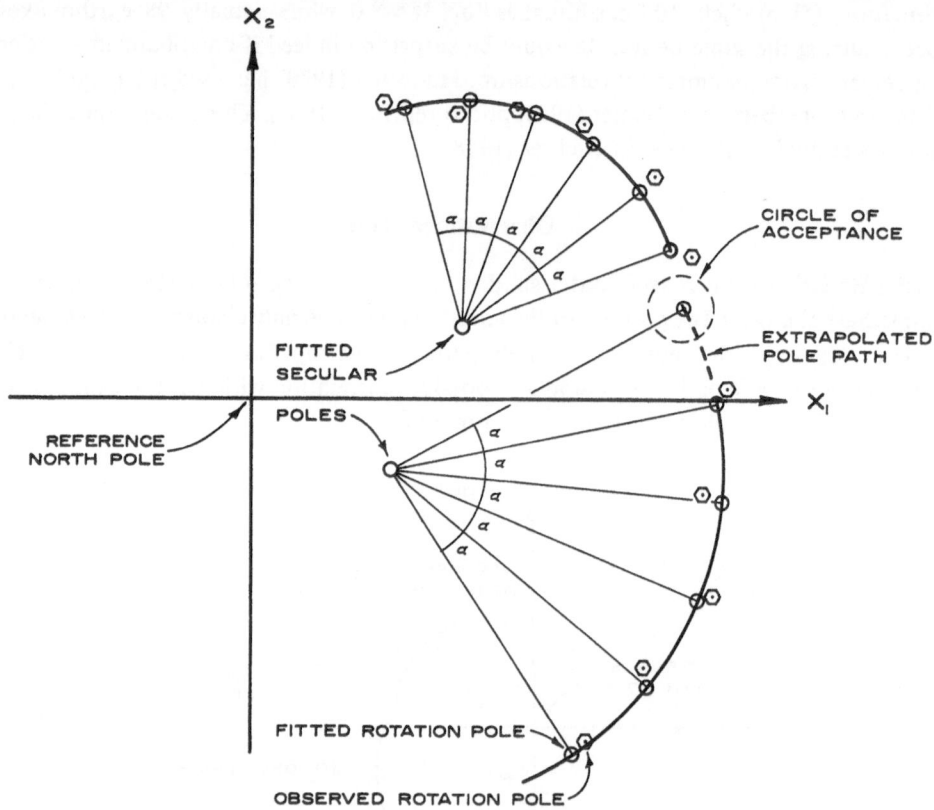

Fig. 4. The method of fitting arcs to observed pole position data.

assumed to have occurred. A new arc is fitted by a least squares procedure to the
three points. On the other hand if this difference exceeds the assigned value of a then
a change in pole path is assumed to have occurred. A new arc is then initiated. The
process is repeated for all subsequent points. Figure 4 illustrates the method of least
squares arc fitting.

The location of the breaks as well as the radius and center of fitted arcs are, to a
limited extent dependent on a. Any finite a may give a break one or more δ's away
from the true break, or even miss a break entirely. Therefore, a least square fitted arc
may be contaminated with data which should fall on neighbouring arcs. With a large
enough a the fitted arc will bear only a remote relation to the true pole path. The
minimum possible value of a is the uncertainty in data, 0″.01. At this value too many
spurious breaks will be caused by noise. In addition the spurious breaks will generate
spurious arc parameters.

Table II is reproduced from Smylie and Mansinha (1968). With $a=0″.02$, 32 breaks
were found in the BIH data from 1957.0 to 1968.0. For the ILS-IPMS data the corresponding figure is 41. The list of breaks was then compared with all earthquakes of

TABLE II

Complete listing of earthquakes with magnitude $M > 7.5$ which occurred over the period 1957.0 to 1968.0. The three nearest breaks in the arcs fitted to both the ILS-IPMS and BIH pole paths are shown for each earthquake.

BIH Three closest pole path breaks			Earthquakes 1957.0–1968.0			ILS-IPMS Three closest pole path breaks		
Days before quake	Days before quake	Days after quake	Date	Region	Mag.	Days before quake	Days before quake	Days after quake
				$M > 8.0$				
...	12	28	Mar 9 57	Aleutians	8.3	...	−32	124
72	2	38	Dec 4 57	Outer Mongolia	8.3	91	−55	164
79	19	81	Nov 6 58	Kurile Islands	8.7	63	−10	120
162	−8	358	May 22 60	Off Chile	8.5	78	42	50
261	1	149	Oct 13 63	Kurile Islands	8.25	57	2	71
168	18	242	Mar 28 64	Southern Alaska	8.5	60	5	68
Random probability = 0.09%						Random probability = 71%		
				$8.00 \geqslant M > 7.75$				
82	12	88	Jun 27 57	Lake Baikal, USSR	7.9	78	−14	69
113	43	57	Jul 28 57	Guerro, Mexico	7.9	109	17	38
67	−3	43	Nov 29 57	Southern Bolivia	7.8	86	−60	169
85	15	25	Dec 17 57	Santa Cruz Island	7.8	104	−42	151
48	8	32	Jan 19 58	Off Ecuador	7.8	137	−9	118
73	−40	100	Jul 10 58	SE Alaska	7.9	54	−1	56
98	−2	82	May 4 59	Off Kamchatka	8.0	60	−13	105
Random probability = 3.4%						Random probability = 60%		
				$7.75 \geqslant M > 7.50$				
101	1	69	Sep 24 57	Off Mindanao	7.6	75	20	126
90	−10	90	Apr 26 59	Off Formosa	7.5 − 7.75	52	−21	113
51	−9	89	Sep 14 59	Kermadec Islands	7.75	120	28	81
116	16	104	Sep 8 61	Sandwich Island	7.5 − 7.75	114	−50	306
202	−58	208	Aug 15 63	Peru-Bolivia	7.75	163	−2	57
227	77	183	May 26 64	S. Sandwich Is.	7.5 − 7.75	64	−9	100
320	60	150	Jan 24 65	Ceram Sea	7.6	88	−40	149
331	71	139	Feb 4 65	Rat-Aleutian Is.	7.75	98	−30	139
143	−177	197	Dec 28 66	Off N. Chile	7.75	133	79	140
Random probability = 29%						Random probability = 68%		

Magnitude $M > 7.5$ in the 11 year period. The list is taken from *Principal Earthquakes of the World* (1957 to 1965) and *Seismological Notes* (1966 to 1968).

One would not in general expect a perfect match between date of earthquakes and date of pole path breaks. As each published pole position is a mean over δ days, the time location of a break is uncertain by $\pm \delta$ days. In addition, the time location of a break is influenced by the value of a. This can be seen in Table III. As a increases, the

TABLE III

The effect of different acceptance radius (a) on breaks in arcs fitted to BIH pole path is shown. All earthquakes with $M > 7.5$ that occurred during 1957.0 to 1968.0 are listed chronologically. Two values of $\omega = \pm 10$ days and ± 20 days are used to calculate the random probability (RP)

Earthquakes

Date	Location	Mag	Acceptance radius (a) in ft						
			1.0	1.25	1.5	1.75	2.0	2.25	2.5
			Nearest break in days. Positive sign means before earthquake						
1 Mar 9 57	Aleutians	8.3	12	12	12	12	12	12	12
2 Jun 27 57	Lake Baikal	7.9	12	12	12	12	12	12	−48
3 July 28 57	Mexico	7.9	13	13	13	43	43	−43	−17
4 Sept 24 57	Off Mindanao	7.6	11	11	1	1	1	1	41
5 Nov 29 57	Bolivia	7.8	7	−3	−3	−3	−3	−3	17
6 Dec 4 57	Outer Mongolia	8.3	−8	2	2	2	2	2	−18
7 Dec 17 57	Santa Cruz Is.	7.8	5	−5	−5	2	2	15	−5
8 Jan 19 58	Off Ecuador	7.8	8	8	8	8	8	8	28
9 July 10 58	Alaska	7.9	10	−10	−20	−30	−40	30	20
10 Nov 6 58	Kurile Islands	8.7	19	29	19	19	19	19	19
11 Apr 26 59	Off Formosa	7.75	20	10	0	0	−10	−20	−30
12 May 4 59	Off Kamchatka	8.0	28	18	8	8	2	−12	−22
13 Sept 14 59	Kermadec Island	7.75	1	1	−9	−9	9	−9	−9
14 May 22 60	Off Chile	8.5	12	12	2	−8	8	−18	−28
15 Sept 8 61	Sandwich Island	7.75	36	26	16	16	16	16	6
16 Aug 15 63	Peru-Bolivia	7.75	22	12	2	−58	−58	−58	−68
17 Oct 13 63	Kurile Islands	8.25	21	−39	−49	1	1	1	−9
18 Mar 28 64	Southern Alaska	8.5	−22	38	38	18	18	8	−82
19 May 26 64	So. Sandwhich Island	7.75	−3	−13	−23	−13	77	67	−23
20 Jan 24 65	Ceram Sea	7.6	−20	−20	−20	60	60	50	−30
21 Feb 4 65	Rat-Aleutian Island	7.75	−9	−9	−9	71	71	61	−19
22 Dec 28 66	Off Chile	7.75	13	13	3	43	−117	23	33
$\omega = \pm 10$ days	p		0.29	0.25	0.23	0.20	0.16	0.16	0.15
	No. of correlations		7	6	12	9	8	7	4
	Random Probability RP		0.46	0.48	0.0013	0.02	0.017	0.005	0.42
$\omega = \pm 20$ days	p		0.55	0.48	0.44	0.39	0.32	0.31	0.29
	No. of correlations		15	17	17	16	15	14	10
	Random Probability RP		0.1151	0.00501	0.00159	0.00146	0.00054	0.00161	0.075

breaks corresponding to the May 4, 1959 earthquake are shifted in the forward direction.

An earthquake is considered to correlate with a break if it falls within a correlation window ω. For Table II $\omega = \pm 2\delta$ of a break. For BIH, $2\delta = 20$ days and ILS-IPMS, $2\delta = 0.1$ year. On this basis 15 out of 22 earthquakes correlate with breaks in the BIH and ILS-IPMS data.

An indication of the significance or otherwise of the correlation is the random probability (RP). Assume that there is no relation between pole path breaks and earthquakes; then just by chance a certain number of earthquakes may fall within $\pm 2\delta$ of a break. The probability of obtaining the required number of correlations or better on a chance basis is RP, the random probability. RP is calculated on the basis of well known Bernoulli trials. The conditions for Bernoulli trials are

(i) There must be only two possible outcomes; i.e. success, failure, or hit, miss.

(ii) Successive trials must be independent.

(iii) The outcome of each trial must be entirely by chance and

(iv) The probability for success (or failure) must be constant for all trials.

The RP for obtaining k or more successes in n trials is given by

$$\mathrm{RP} = \sum_{j=k}^{n} \binom{n}{j} p^{j} (1-p)^{n-j} \tag{10}$$

which is the cumulative binomial distribution. The elementary probability p is the probability of a date, drawn at random, falling within ω of a break and is given by the proportion of time axis of the total span of 11 years that is occupied by segments ω wide on each side of a break.

In Table II, $p = 0.31$ for BIH and for the ILS-IPMS, $p = 0.68$. RP is indicated below each magnitude range. For BIH data RP is small but increases with decreasing magnitude. For ILS-IPMS data RP is high, reflecting in part the wider spacing of data.

Table III lists earthquakes and the nearest BIH pole path break for different a. Two values of ω are considered, $\pm 1\delta$ and $\pm 2\delta$. Taking into account the forward movement of a break with increasing a, one sees that a large number of breaks are repeated for values of a from $0\rlap{.}''0125$ to $0\rlap{.}''0225$ (1.25 ft to 2.25 ft.). The correlation window ω determines p, and RP. For $a = 0\rlap{.}''015$, 12 out of 22 earthquakes fall within ± 10 days of a break and 17 out of 22 earthquakes fall within ± 20 days of a break. The corresponding values of RP are 1.3×10^{-3} and 1.6×10^{-3} respectively.

Figure 5 shows fitted arcs and breaks for both BIH and ILS-IPMS data for the years 1957 and 1960. In 1960, the break is preceded and followed by long break free periods. Yet within a few days of the break an earthquake has occurred.

The fitted arcs give rms amplitudes of polar shifts of the order of $0\rlap{.}''1$. Haubrich (1970) calculates the Chandler power from polar shifts and finds it too high as compared to the observed Chandler power. However, this cannot be true, as both estimates of power are derived from the same set of data. The observed Chandler power

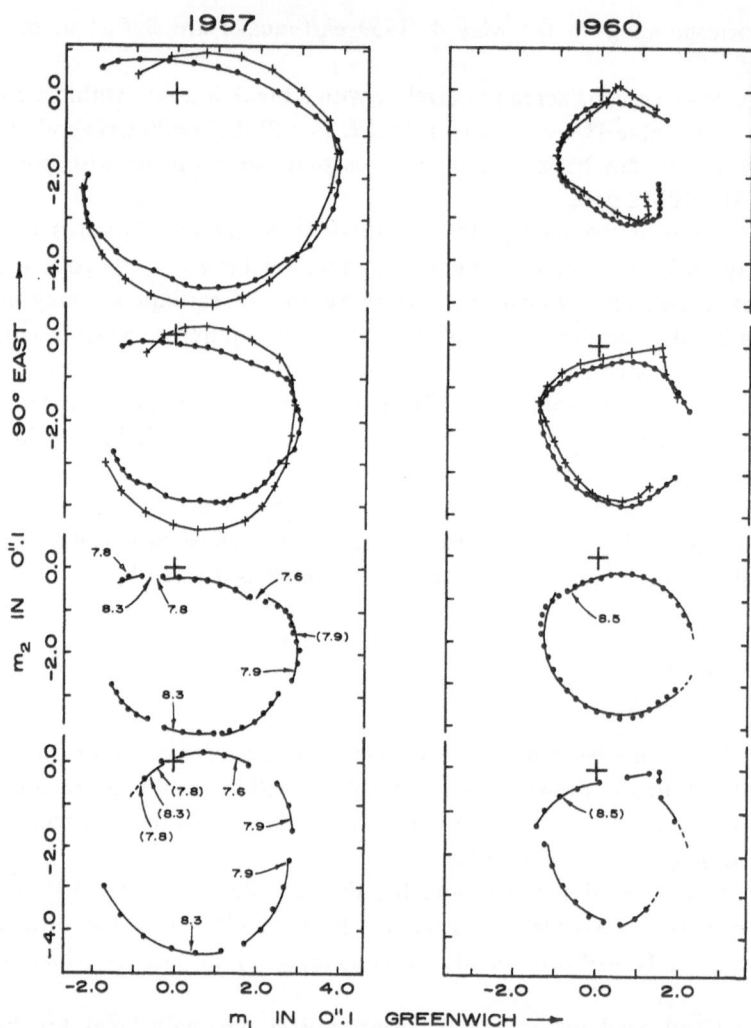

Fig. 5.   Pole paths for the years 1957 and 1960. Pole motion is anti-clockwise. *Top:* overplot of BIH (circles) and ILS-IPMS (crosses) data. Second from top: The same paths with annual component removed. Second from bottom: Arcs fitted to BIH data. *Bottom:* Arcs fitted to ILS-IPMS data. Numbers are magnitudes of earthquakes. Magnitudes of earthquakes that do not fall within $\pm 2\delta$ of a break are given in parentheses.

has been calculated without considering the possibility of sudden shifts in secular pole and change in radius of the pole path. If such shifts are confirmed, then the 'observed' Chandler power should be considered as an underestimate.

5. Conclusion

Calculations using half-space dislocation theory show that large earthquakes can account for roughly one-fifth of the observed Chandler amplitude. There are indications that this is an underestimate. Surface strain observations are at least an order of magnitude greater than predicted by the elasticity theory of dislocations. In addition, the surface displacements due to a dislocation in a homogeneous elastic sphere are generally higher than the corresponding half-space terms (Ben-Menahem *et al*, 1969). Calculation of the contribution to the wobble for a fault in a homogeneous sphere by Ben-Menahem and Israel (1970) as well as by Smylie and Mansinha (1969) for a radialy inhomogeneous self-gravitating sphere with a liquid core, show the contribution of individual earthquakes to Chandler amplitude is several times larger than that given by halfspace theory. Therefore, it is likely that major earthquakes may account for a substantial part of the Chandler wobble.

The number of matches of earthquakes with detected breaks in pole path are far higher than expected on a purely random basis for several combinations of parameters. This would suggest that an optimum combination of parameters can suppress the noise in the data to bring out the correlation with earthquakes. The ultimate observational proof will be matching of earthquakes and pole path breaks in succeeding years.

Acknowledgement

This research was supported by the National Research Council of Canada.

References

Ben-Menahem, A. and Israel, M.: 1970, *Effects of Major Seismic Events on the Rotation of the Earth*, *Geophys. J.* **19**, 367–393.

Ben-Menahem, S., Singh, S. J., and Solomon, F.: 1969, 'Static Deformation of a Spherical Earth Model by Internal Dislocations', *Bull. Seism. Soc. Am.* **59**, 813–854.

Cecchini, G.: 1928, 'Il problema della varizione delle latitudine', *Publicazioni del Reale Observatorio Astronomico di Brera in Milano* **61**, 7–96.

Chinnery, M. A.: 1961, 'The Deformation of the Ground Around Surface Faults', *Bull. Seism. Soc. Am.* **51**, 355–372.

Guinot, B. and Feissel, M.: 1968, 1969, *Annual Reports for 1967 and 1968 of the Bureau International de l'Heure*, Paris.

Gutenberg, B. and Richter, C. F.: 1954, *Seismicity of the Earth and Associated Phenomena*, Hafner Publishing Company, New York.

Haubrich, R. A.: 1970, 'An Examination of the Data Relating Pole Motion to Earthquakes', this volume, p. 149.

Lambert, W. D.: 1926, 'The Variation of Latitude, Tides and Earthquakes, *Proc. Third Pan-Pacific Science Congress, Tokyo*, pp. 1517–1522.

Lambert, W. D., Schlesinger, F., and Brown, E. W.: 1931, 'The variation of Latitude', *Bull. Nat. Res. Council* **78**, 245–277.

Larmor, J.: 1909, 'The Relation of the Earth's Free Precessional Nutation to Its Resistance against Tidal Deformation', *Proc. Roy. Soc., London, Ser. A* **82**, 89–96, and *Monthly Notices Roy. Astron. Soc.* **69**, 480–486.

Larmor, J. and Hills, E. H.: 1906, 'The Irregular Movements of the Earth's Axis of Rotation', *Monthly Notices Roy. Astron. Soc.* **67**, 22–34.

Mansinha, L. and Smylie, D. E.: 1967, 'Effect of Earthquakes on the Chandler Wobble and the Secular Polar Shift', *J. Geophys. Res.* **72**, 4731–4743.

Mansinha, L. and Smylie, D. E.: 1968, 'Earthquakes and the Earth's Wobble', *Science* **161**, 1127–1129.

Maruyama, T.: 1964, 'Statical Elastic Dislocations in an Infinite and Semi-Infinite Medium', *Bull. Earthquake Res. Inst.* **42**, 289–368.

Milne, J.: 1893, 'On the Mitigation of Earthquake Effects and Certain Experiments in Earth Physics', *Seism. J. Japan* **17**, 1–19.

Milne, J.: 1903, *Earthquakes and Other Earth Movements*', Kegan Paul, Trench, Trübner & Co., Ltd., London, p. 267.

Milne, J.: 1906, 'Recent Advances in Seismology', *Proc. Roy. Soc. Lond, Ser. A* **77**, 365–376.

Milne, J.: 1913, *Earthquakes and Other Earth Movements*, 6th ed., Kegan Paul, Trench, Trübner & Co., Ltd., London, p. 376.

Munk, W. H. and MacDonald, G. J. F.: 1960, *The Rotation of the Earth*, Cambridge Univ. Press.

Press, F.: 1965, 'Displacements, Strains, and Tilts at Teleseismic Distances', *J. Geophys. Res.* **70**, 2395–2412.

Press, F. and Brace, W. F.: 1965, 'Earthquake Prediction', *Science* **152**, 1575–1584.

Principal Earthquakes of the World, United States Earthquakes, USCGS, 1957–1965.

Richter, C. F.: 1958, *Elementary Seismology*, W. H. Freeman, San Francisco.

Rochester, M. G.: 1970, 'Polar Drift and Wobble': A Brief History, this volume, p. 3.

Seismological Notes, 1966–1968, *Bull. Seismol. Soc. Am.* **56–58**.

Smylie, D. E. and Mansinha, L.: 1968, 'Earthquakes and the Observed Motion of the Rotation Pole', *J. Geophys. Res.* **73**, 7661–7663.

Smylie, D. E. and Mansinha, L.: 1969, 'The Elasticity Theory of Dislocations in Real Earth Models and Changes in the Inertia Tensor. (Abstract) *EOS* **50**, 645.

Steketee, J. A.: 1958, 'On Volterra's Dislocations in a Semi-Infinite Elastic Medium', *Can. J. Phys.* **36**, 192–205.

Stoyko, N.: 1952, 'Sur les relations entre la variation de la rotation, l'oscillation libre et les tremblements de Terre', *Compte Rend. Acad. Sci.* **234**, 2550–2552.

Yumi, S.: 1968, *Annual Report of the International Polar Motion Service for 1966*, Mizusawa, Japan. See also previous annual reports.

Discussion

Rochester: Do you have any reason to suspect that the earthquakes for which there is no break in the pole path at the time of their occurrence are located and oriented in such a position with respect to the surface of the earth that they are not likely to excite any change in the pole position?

Mansinha: Possibly, but we did not look into this.

Reiter: Could the changes in the pole path act as a precursor of earthquake activity?

Mansinha: There may be enough strain built up prior to the earthquake to cause a shift in the pole. Provided the build up is sudden with respect to the period of the wobble, say over a period of a few days, it would not be counted as a seismic event but it might cause enough displacements of mass to cause a break in the pole path. At the moment, we do not have sufficient results to conclude one way or another.

Runcorn: Is it possible to determine any correlation between the direction of the polar displacement and either the place where the earthquakes occur or the direction of the fault movement?

Mansinha: It is known that the contribution of an earthquake to the displacement of the pole depends upon the fault azimuth. In most cases it is difficult to find the appropriate fault parameters of most earthquakes.

Runcorn: Nevertheless, I think it might be useful if you divided the direction of the displacement into, say, half a dozen groups of 60° sectors and then try to class the earthquakes together in some manner.

Mansinha: This is certainly worth trying but we haven't done it yet.

Bender: What is the rms deviation between the actual points you used and the arc that you calculated?

Mansinha: I haven't calculated the rms deviation but most of the arcs fitted the points to closer than a foot.

Myerson: The size of your breaks is the order of 0″.1 whereas you are predicting breaks on the order of 0″.01. Is this because your theoretical predictions are averaged over all sorts of azimuths?

Mansinha: This may be a fault of the least squares fitting since we have ignored one point of constraint – namely that neighbouring arcs should intersect. We ignored this because it complicates our computer program too much. In general, we are just looking for the event and not the magnitude. It is possible to look at both if one takes the extra constraint into account.

Ulrych: What random probability would you assign to the wobble causing earthquakes?

Mansinha: Very small. This was investigated by Lambert (1926). He compared the periodicities of the wobble with the frequency of earthquakes and dismissed the idea because the nature of the forces that cause deformation due to the wobble are identical to the diurnal tidal forces. Thus if there were any periodicity it would most likely be the tides that caused the wobble or the earthquakes and not the wobble causing the earthquakes, because the tidal deformation is so much larger. The reverse mechanism is also not particularly impressive since, somewhere along the line, one must show that the additional stress caused by the wobble can trigger the earthquake.

Haubrich: Although the tidal forces are larger than the forces due to the wobble, there is no long period tide that produces a force as large as the wobble.

Mansinha: Perhaps I should throw the question back to you Dr. Haubrich. Is it possible for the deformation caused by the wobble to trigger earthquakes?

Haubrich: I don't know.

CORE-MANTLE INTERACTIONS:
GEOPHYSICAL AND ASTRONOMICAL CONSEQUENCES

M. G. ROCHESTER

Dept. of Physics, Memorial University of Newfoundland, St. John's, Newfoundland, Canada

Abstract. The existence of the liquid core, and its coupling to the mantle, give variety and complexity to the possible ways in which the Earth's rotation can change. The interactions are inertial (including what Hide has called topographic coupling), viscous (laminar or turbulent boundary layer friction) and electromagnetic. This paper reviews the roles of the several kinds of coupling in detectably affecting periodic and secular changes both in the rate of rotation of the mantle and crust, and in the geographical and spatial orientation of its rotation axis.

1. Introduction

The problem of core-mantle coupling has roots extending back into the 19th century. Hopkins (1839–42) and Kelvin (1876) both discussed the effect which an interior cavity of spheroidal shape, filled with inviscid homogeneous imcompressible liquid, would have on the rotation of an otherwise rigid Earth. Hopkins' discussion was imperfect, and Kelvin never published the mathematical treatment which led him to his conclusions, correct as they were. The first rigorous analyses were published independently but nearly simultaneously by Hough (1895) and Sludskii (1896). A few more papers on the subject followed, culminating in the particularly elegant and penetrating solution by Poincaré (1910). Strangely, though it was just at that time that seismology was providing decisive evidence for the existence and dimensions of the liquid core, interest in the problem languished for nearly 40 years, until it was revived by Bondi and Lyttleton (1948). During the past two decades the subject of core-mantle interactions has taken a new lease on life, largely because of the increasing precision of astronomical observations and the development of a satisfactory theory of geomagnetism.

I propose in this paper to review the mechanisms which in principle can couple the core to the mantle and thus affect the rotation of the latter, and also to discuss the extent to which each mechanism appears to play a significant role in the various observed changes both in the rate of spin of the mantle and in the celestial and geographical orientation of its pole. Although a substantial number of interesting detectable effects have been attributed to the exchange of angular momentum between the core and mantle, the subject still presents an intriguing array of gaps and uncertainties, and we are far from a satisfactory understanding of core-mantle coupling.

There are essentially three kinds of mechanisms to discuss:

(1) Inertial coupling (including what Hide (1966, 1969) has recently described as 'topographic' coupling);

(2) Viscous boundary layer friction (laminar or turbulent);

(3) Electromagnetic coupling.

L. Mansinha et al. (eds.), Earthquake Displacement Fields and the Rotation of the Earth, 136–148.

The strength of each of these core-mantle interactions can be discussed

(a) locally, in terms of the tensor $\mathscr{S}$ describing the stress on the liquid just below the core-mantle boundary, and

(b) globally, by calculating the net torque **L** exerted by the core on the mantle:

$$\mathbf{L} = - \int_{S_c} \mathbf{r} \times \mathscr{S} \cdot \hat{n} \, dS \tag{1}$$

where $\hat{n}$ is the unit normal to the core-mantle boundary S_c (pointing into the mantle) at the point distant **r** from the Earth's centre of mass. For all but the global inertial coupling due directly to the ellipticity of S_c (considered in Section 2 following), this surface may be regarded as spherical. Then $\hat{n} \simeq \hat{r}$ and

$$\mathbf{L} \simeq \int_{S_c} \mathscr{S}_t \hat{t} \times \mathbf{r} \, dS \tag{2}$$

where $\mathscr{S}_t \hat{t}$ is the vector shear stress applied to the liquid tangential to the core-mantle boundary.

2. Inertial Coupling

The operation of this mechanism, which was the one considered by all the early students of core-mantle interaction, depends directly on the core-mantle boundary departing from a prefectly spherical shape. The radial distribution of density revealed by seismology, together with the assumption of hydrostatic equilibrium, yields a value

$$\varepsilon \simeq \tfrac{1}{400}$$

for the ellipticity of the core-mantle boundary (MacDonald, 1966). Any rotation of the mantle about an axis inclined to the axis of figure (e.g., as in precession) will bring into play geometrical constraint forces imposed by the core-mantle boundary on the contained liquid. These create an asymmetric pressure distribution p over the boundary of the liquid which tends to entrain the core in a motion following that of the mantle, the coupling torque being derived from (1) by setting $\mathscr{S} = -p\mathbf{1}$:

$$\mathbf{L} = \int_{S_c} \mathbf{r} \times p\hat{n} \, dS. \tag{3}$$

In the case of the ideal liquid considered by Poincaré and his predecessors, the steady state is one in which the core follows the rotation of the mantle apart from a relative motion of uniform vorticity (Poincaré flow). Stewartson and Roberts (1963), Roberts and Stewartson (1965) and Busse (1968) have examined the effect of viscosity and find that, provided the precession has gone on for a time $> a^2/\nu$ and the kinematic viscosity

$$\nu \ll \Omega a^2 \simeq 10^{13} \text{ cm}^2 \text{ sec}^{-1},$$

the ultimate pattern of flow set up, apart from a boundary layer of thickness

$$\Delta \simeq (\nu/\Omega)^{1/2},\tag{4}$$

differs from Poincaré flow only by a differential rotation of the interior of the core which is negligible in the case of the Earth. Here $\Omega = 7.3 \times 10^{-5}$ rad sec^{-1} is the axial rate of rotation of the mantle and $a = 3.5 \times 10^3$ km is the radius of the core-mantle boundary.

Viscosity, however, is only one (and possibly not the most important) way in which the real core differs from the ideal model. The flow associated with inertial coupling may be turbulent rather than laminar (Malkus 1968; but see Sections 4 and 5 of this paper). The disturbing effect of the solid inner core on the topology of the flow stirred by precession is as yet unexamined. The presence of the Earth's magnetic field, quite apart from its own direct contribution to coupling (considered in Section 6 below), is likely to seriously modify the hydrodynamics of inertial coupling. Experimental studies of the internal flows in a spinning and precessing liquid contained by a rigid spheroidal boundary are being carried out in Malkus' laboratory, but so far the theoretical studies have not been satisfactorily extended beyond the laminar viscous model.

3. Topographic Coupling

A different kind of inertial coupling has been discussed by Hide (1966, 1969) in connection with the departure of the core-mantle boundary from perfect sphericity on a much smaller scale horizontally, i.e. the topography of the core-mantle boundary. Provided that the thickness of any viscous boundary layer

$$\Delta \ll h,$$

where h is the vertical height of a bump or depression on the core-mantle boundary, the Coriolis force due to the flow of liquid past the mantle permits the topographical feature to entrain a Taylor column of liquid. Quite small scale roughness of the boundary therefore offers the possibility of coupling the core to the mantle by a core-full of Taylor columns. Even the possibility, however, must remain speculative until much more is known about the hydrodynamics of Taylor columns in close proximity.

A rough estimate of the tangential shear stress on the mantle can be made by considering a rectangular bump of height h and sides of length x, y. If the pressure against one side of the bump (of length x) is p_1, and against the opposite side is p_2, then

$$\mathscr{S}_t \simeq (p_2 - p_1)\, xh/xy.$$

Balancing the horizontal pressure gradient against the Coriolis force we have

$$(p_2 - p_1)/y \simeq |\hat{\imath}\cdot\nabla p| \simeq |\hat{\imath}\cdot 2\varrho\boldsymbol{\Omega} \times \mathbf{u}| \simeq \varrho\Omega u$$

where ϱ is the density of the liquid and u is the order of magnitude of the horizontal speed of the core liquid past the mantle.

$$\therefore \mathscr{S}_t \simeq \varrho\Omega uh.\tag{5}$$

Hide and Horai (1968), assuming the topography of the core-mantle interface is responsible for the fairly low order harmonics in the non-hydrostatic part of the Earth's external gravitational field, find $h \simeq 5$–10 km. This is well below the acceptable maximum defined by the ability of present-day seismological techniques to resolve such topography. Putting $u \simeq 0.01$–0.1 cm sec^{-1} (as inferred from the geomagnetic westward drift) and $\varrho \simeq 10$ gm cm^{-3} we find

$$\mathscr{S}_t \simeq 4\text{–}70 \text{ dyne cm}^{-2}, \tag{6}$$

with the latter figure probably an upper limit to the available topographic coupling stress (Hide, 1969). A meaningful calculation of the contribution of this kind of interaction to the net torque on the mantle would require not only a knowledge of the actual topography of the core-mantle interface (which seismology may provide in the not-too-distant future) but an extension of the theory of Taylor column formation to the case of multiple, rather than single, bumps and/or depressions on the interface.

4. Laminar Viscous Friction

Topographic coupling cannot be effective (i.e. Taylor columns cannot be set up) if the bumpiness of the boundary does not greatly exceed the thickness of the laminar viscous boundary layer. Viscous friction itself, of course, can transfer angular momentum from the core to the mantle – in fact this was the mechanism considered by Bondi and Lyttleton (1948), whose work inaugurated the recent history of core-mantle coupling. The strength of the viscous shear stress on the mantle can be estimated by a method analogous to that employed in the last section, by considering the balance of the Coriolis and viscous forces on a rectangular 'box' of liquid with the thickness of the viscous boundary layer, ∇:

$$|2\mathbf{\Omega} \times \mathbf{u}| \simeq |\nu \nabla^2 \mathbf{u}|$$
$$\therefore \Omega u \simeq \nu (u/\Delta^2)$$

i.e. $\nabla \simeq (\nu/\Omega)^{1/2}$, as was given in (4);

$$\therefore \mathscr{S}_t \simeq \left| \varrho \nu \frac{\partial u}{\partial r} \right| \simeq \varrho \nu \frac{u}{\Delta} \simeq \varrho \Omega u \Delta. \tag{7}$$

Unfortunately viscosity is one of the least well-bracketed parameters of the Earth's core. The estimates quoted in the literature over the last two decades range over 13 orders of magnitude (Table I). Bondi and Lyttleton extrapolated over an enormous range of pressure and failed to allow for the effect of increased temperature, whereas Bullard allowed for neither and simply commented that 10^7 cm^2 sec^{-1} is 'highly improbable'. The standard of criticism has not significantly improved since these early estimates, owing to the great difficulty of experimental or theoretical progress in this field. In view of the availability of other mechanisms to couple the core to the mantle, the estimates in Table I based on viscous torques *alone* providing the necessary cou-

TABLE I

Kinematic coefficient of viscosity v in the core

References	v (cm^2 sec^{-1})	Basis of estimate
Bondi and Lyttleton (1948)	10^7	Extrapolation from experimental values obtained by Bridgman for liquid mercury at pressures up to 10 kb. No allowance for temperature
	$\geqslant 1$	Required if viscosity is to couple core to mantle in secular deceleration
Bullard (1949)	10^{-3}	Experimental values for liquid metals at NTP
Miki (1952)	10^{-3}–10^{-2}	Quantum statistical thermodynamics of liquid metal
Jeffreys (1959), corrected in Gutenberg (1951)	$< 10^9$	Absence of significant damping of P waves in the core
Toomre (1966)	$> 6 \times 10^4$	Required if viscosity is to couple core to mantle in steady precession
Sato and Espinosa (1967)	10^8–10^{10}	Required if viscosity in core and mantle is to damp $_0T_2$ mode of free oscillation
Backus (1968)	50	Extrapolation from experimental values at NTP using equation fitting Nachtreib and Petit experiments on liquid mercury up to 10 kb and 400°K

pling may not constitute strong arguments for high values of core viscosity. On the other hand *if* the viscosity *is* as high as 10^6 cm^2 sec^{-1} its effects will be quite comparable in strength to those of electromagnetic coupling, with respect to which it has for long been dismissed as negligible by nearly everyone concerned with the latter mechanism. With this value of v and the estimate of u used in Section 3, the viscous shear stress on the mantle is

$$\mathscr{S}_t \simeq 1\text{–}10 \text{ dyne cm}^{-2}. \tag{8}$$

Such a high viscosity will be associated with a laminar boundary layer about 1 km thick, which is quite reasonable and would probably seriously reduce the effectiveness of topographic coupling.

The point of my discussion is that the arguments so far advanced for very low viscosity in the core are not sufficiently convincing, and therefore that it is still quite uncertain whether or not viscous friction is competitive with other coupling mechanisms. Until there is an improvement in the theoretical basis for extrapolating from experimental data obtained at low pressures and temperatures, or the range of experimental determinations is itself greatly extended in the pressure and temperature domains, the role of viscosity in the core will remain enigmatic.

5. Turbulent Boundary Layer Friction

If the boundary layer is in turbulent rather than laminar shear flow the dissipative mechanism is an 'eddy viscosity', which transfers energy toward the small wavelength end of the eddy spectrum in a fashion analogous to the conversion of ordered mechanical energy to thermal energy by molecular viscosity. If l is the dimension of the largest eddies,

$$v_{eddy} \simeq ul$$

(Batchelor, 1953), but the Reynolds stress arising from eddy friction is typically

$$\mathscr{S}_t \lesssim (0.01-0.1) \varrho u^2 \tag{9}$$

(Toomre, 1966). Roughness of the core-mantle boundary, or instability in the laminar flow, will give rise to turbulence if the Reynolds number

$$R \simeq ul/v \gg 1 .$$

With the estimate of u used in Section 3 and $l \simeq 10^3$ km, $v \ll 10^6 - 10^7$ cm^2 sec^{-1} for turbulence.

Even if the flow in the boundary layer is turbulent, the Reynolds stress given by (9) will be $\lesssim 10^{-2}$ dyne cm^{-2}. This is well below the estimate calculated by Toomre because he sets $u \simeq 2.5$ cm sec^{-1}, which seems at least an order of magnitude too high for tangential velocities at the core-mantle boundary. At any rate it appears that turbulent shear coupling of the mantle to the core can be ignored.

6. Electromagnetic Coupling

The main geomagnetic field and its secular variation originate in the Earth's core and penetrate the core-mantle boundary. Because the lower mantle is an electrical conductor the time taken for a change in the magnetic field just below this interface to manifest itself at the surface of the Earth is of the order

$$t \simeq \pi\sigma d^2$$

where σ is the average electrical conductivity, and d the thickness, of the appreciably conducting part of the mantle. The analyses by Lahiri and Price (1939) and McDonald (1957) show that $d \simeq 2 \times 10^3$ km, and $t \lesssim 4$ yr according to Walker and O'Dea (1952), so

$$\sigma \lesssim 10^{-9} \text{ emu}$$

(Runcorn, 1955). The argument can be made more sophisticated to deal with poloidal and toroidal field harmonics of different degree, spherical geometry, and a conductivity distribution increasing with depth in the mantle, but the upper limit to the average conductivity is increased only by a factor of 2.5 (Smylie, 1965). From a spectral analysis of the secular variation assuming hydromagnetic turbulence in the core Currie (1968) finds $\sigma \simeq 2 \times 10^{-9}$ emu. According to McDonald (1957), who treated the secular variation spectrum differently, the electrical conductivity at the base of the mantle is about 2×10^{-9} emu, within a factor of 3. As Runcorn has pointed out, these values are compatible with what we should expect from semiconductors like the silicates in the lower mantle at the temperatures obtaining there.

Any relative motion at the core-mantle boundary will induce electrical currents in the lower mantle. By Lenz's law these will flow in a pattern such that their Lorentz interaction with the inducing field will produce a torque tending to oppose the slipping of the mantle past the core. On the other hand, changes in magnetic field strength just below the core-mantle boundary will also induce currents in the mantle, giving a Lorentz torque which may either accelerate or decelerate the rotation of the mantle. Thus the mantle is electromagnetically coupled to the core both passively and by direct excitation from the secular variation.

The Maxwell shear stress at the core-mantle boundary is

$$\mathscr{S}_t = H_r H_t / 4\pi \tag{10}$$

and the net torque on the mantle is

$$\mathbf{L} = -\frac{1}{4\pi} \int\limits_{S_c} (\mathbf{r} \times \mathbf{H})\, H_r\, \mathrm{d}S$$

(Rochester, 1962), where H_r and H_t are the radial and tangential components of the magnetic field $\mathbf{H}$.

I have recently reviewed the theoretical studies of the passive response of the electro-magnetically-coupled core-mantle system to both axial and equatorial excitations (Rochester, 1968). The relevant fields in (10) are the poloidal field H_r extrapolated from the Earth's surface to the core-mantle boundary, and the toroidal field H_t induced there by the acceleration of the mantle past the core. For low-frequency (say $f \lesssim 0.1$ cpy) excitations of whatever source, and a rigid-sphere model of the core, the passive response torque can be well represented by

$$\mathbf{L} \simeq -\Lambda \cdot (\mathbf{\Omega}_m - \mathbf{\Omega}_c)$$

where $\mathbf{\Omega}_m - \mathbf{\Omega}_c$ is the angular velocity of the mantle past the core, and the coupling tensor Λ is diagonalized in the geographic frame of reference attached to the mantle. If only the dipole part of the poloidal field is taken into account, the non-vanishing components are

$$\Lambda_{11} \simeq \Lambda_{22} \simeq 3.5 \times 10^{43}\ \sigma\ \text{dyne cm sec}$$

$$\Lambda_{33} \simeq 5.1 \times 10^{43}\ \sigma\ \text{dyne cm sec}$$

and the increase due to the non-dipole harmonics will be at most a factor of 2. The time constant associated with purely electromagnetic coupling about an equatorial axis is

$$\tau_2 \simeq \frac{A}{\Lambda_{11}} \frac{\omega}{\Omega}$$

and about the axis of figure is

$$\tau_3 \simeq \frac{C}{A_{33}} \frac{\alpha}{1 + \alpha}.$$

Here $A \simeq C \simeq 7.2 \times 10^{44}$ gm cm^2 are respectively the equatorial and axial moments of inertia of the mantle, αC is the moment of inertia of the core ($\alpha \simeq 0.1$), and $\omega \simeq 0.84$ cpy is the Chandler wobble frequency. For $\sigma \simeq 10^{-9}$ emu, $\tau_2 \simeq 10^5$ yr and $\tau_3 \simeq 25$ yr. A more sophisticated representation of the electrical conductivity distribution in the lower mantle might enable these relaxation times to be reduced by a factor of 5 or so (Roden, 1963).

A rough numerical estimate of the Maxwell shear stress associated with low-frequency electromagnetic coupling can be obtained by calculating that involved in the passive coupling by the dipole part of the poloidal field: since $H_t \simeq 2\sigma H_r a u$ where $H_r \simeq 4$ gauss, then from (10) we find $\mathscr{S}_t \simeq 0.01$–$0.1$ dyne cm^{-2} for $\sigma \simeq 10^{-9}$ emu and the estimate of u used in Section 3. A more realistic model for the electrical conductivity distribution in the lower mantle could raise this estimate of the available low-frequency electromagnetic coupling stress to

$$\mathscr{S}_t \simeq 0.1\text{–}1 \text{ dyne cm}^{-2}. \tag{11}$$

A simple skin-depth argument shows that the high-frequency (say $f \gtrsim 1$ cpy) part of the secular variation spectrum cannot penetrate the mantle, so we have as yet no way of estimating the intensity of the perturbations in the magnetic field in this frequency range. The researches of Hide (1966), Braginskii (1967), Malkus (1967) and Stewartson (1967) on the free hydromagnetic modes of the core suggest that this part of the secular variation may contribute appreciably to core-mantle coupling, but so far the only mathematical discussion of high-frequency electromagnetic coupling is by Mac-Donald and Ness (1961; corrected by Toomre, 1966). In order to examine the shortening of the periods of the toroidal free oscillations of the mantle due to stiffening of the core-mantle boundary by electromagnetic coupling, these authors used a drastically simplified model in which the boundary is plane and the core liquid is driven hydromagnetically by one-dimensional transverse vibration of the rigid plate representing the mantle. For excitation at frequency f, they find $H_t \simeq H_r u \, (\sigma/f)^{1/2}$. While the effective mantle conductivity in high-frequency coupling is that just above the core-mantle boundary (and therefore higher by perhaps nearly an order of magnitude than the average over the lower 2000 km of the mantle), the coupling strength is less dependent on conductivity than in the low-frequency case, and decreases with increasing frequency. Even setting $\sigma \simeq 10^{-8}$ emu we find that for diurnal excitations $\mathscr{S}_t$ is smaller by a factor of 10^3 than the estimate given by (11).

7. Precession

Because of the different dynamical ellipticities of the core and mantle the lunisolar precessional torque on the core is only 3/4 of that on the mantle. If the core and mantle were decoupled, enormous relative velocities would develop at the core-mantle boundary within a few thousand years. In the absence of any evidence for their existence we can assume that the coupling in the precessional mode is tight. Poincaré (1910), using the core model described in Section 1, found that inertial coupling would suffice to balance the differential precessional torque. Toomre's (1966) review of the viscous, turbulent and electromagnetic coupling mechanisms led him to reject all of them in favour of inertial coupling. He did, however, point out that if the viscosity of the core is of the order 10^5 cm^2 sec^{-1} laminar viscous friction will be adequate to couple the core to the mantle in precession.

The role of electromagnetic coupling in precession is as yet unresolved. The relevant Maxwell shear stress must fluctuate diurnally, and was dismissed as insufficient in strength by Toomre on the basis of a calculation outlined in the concluding part of Section 6. Malkus (1963, 1968) has advanced arguments, based on a somewhat dubious analogy and on order-of-magnitude estimates of the forces involved, to suggest that the core and mantle can be electromagnetically coupled in precession by fields generated in a turbulent hydromagnetic boundary layer. As Toomre points out, further doubt is cast on the validity of Malkus' arguments by the absence of any reference to the electrical conductivity of the mantle. There is as yet no acceptable discussion of the detailed magnetohydrodynamics of such a precess ion-driven dynamo, and its possibility remains a tantalizing problem.

8. Secular Change in Obliquity

Aoki (1967, 1969) and Sekiguchi (1967) have discussed the possibility of attributing to core-mantle coupling the discrepancy of 0″.3 century^{-1} between the observed secular decrease in the obliquity and that calculated from planetary perturbations of the ecliptic. Sekiguchi's work is marred by failure to take into account the dynamical ellipticity of the core. Although Aoki (1969) finds that a core-mantle interaction torque of the strength available from the passive electromagnetic coupling explains the above discrepancy very nicely, his dynamical model at the same time appears to predict a non-tidal secular deceleration of the Earth's axial spin nearly 30 times as rapid as that which is observed and attributed largely to tidal friction!

9. Nutation and Wobble

Poincaré's analysis has been extended to the case of an elastic mantle independently by Jeffreys and Vicente (1957) and by Molodenskii (1961). Besides yielding a satisfactory explanation for the lengthening of the Chandler period, these studies show that the presence of the liquid core enables removal of most (if not all) of the discrepancy between the observed amplitude of the 18.6 year forced nutation and its theoretical value (previously calculated assuming a rigid Earth). Another interesting feature of the inertially-coupled core-mantle system is the existence of a second free wobble mode, with nearly diurnal period (discussed, and further references given, by

Rochester, 1970). Busse (1970) is the first to have taken the solid inner core satisfactorily into account in a discussion of inertial coupling of the wobbling core and mantle.

A series of studies by Elsasser and Takeuchi (1955), Elsasser and Munk (1958), Munk and Hassan (1961), and Rochester and Smylie (1965) have concluded that electromagnetic coupling is inadequate, by several orders of magnitude, to excite wobble at the Chandler frequency. Rochester and Smylie also show that the passive electromagnetic coupling is even less effective than this in damping the wobble. Stacey (1970), however, proposes to revive the core as both source and sink of wobble energy, by means of a nonlinear electromagnetic coupling which he suggests will permit the energy necessary to maintain wobble to be derived from the precessional stirring of the core.

10. Change in the Length of Day

Bondi and Lyttleton (1948) and Elsasser (1949) were the first to discuss the role of core-mantle coupling in the secular slowing down of the axial rotation of the mantle by tidal friction. The former authors considered only viscous coupling and found that the viscosity of the core would need greatly to exceed $1 \text{ cm}^2 \text{ sec}^{-1}$ to effect the necessary tightness of coupling. Dicke (1966) estimated that electromagnetic core-mantle coupling could contribute no more than 5% of the total observed secular deceleration of the mantle, but it is again surprising that the electrical conductivity of the lower mantle does not enter into his discussion. I shall return to this point.

The first consideration of electromagnetic core-mantle coupling was in connection with the problem of accounting for the irregular changes in the length of day (Bullard et al., 1950; Munk and Revelle, 1952; Vestine, 1952; Runcorn, 1954). The explanation of the geomagnetic westward drift offered by Bullard et al. required that there should be an opposite and proportional change in the rate of westward drift whenever there was a change in the length of day, and this was apparently supported by Vestine's matching of a single pulse in the record of each of these phenomena. Recently an even better correlation between such changes has been found by Ball et al. (1968). How strongly such correlations argue for the predominance of electromagnetic coupling over other interaction mechanisms is uncertain now that Hide (1966) has proposed that the westward drift need not necessarily represent a mass motion of the outer core past the mantle.

There is no difficulty in accounting for mantle spin accelerations of several parts in 10^{10} per year (i.e. changes in the length of day of 0.5 msec in a decade) by direct excitation from the core through the agency of electromagnetic coupling, if $\sigma \simeq 10^{-9}$ emu (Rochester, 1960). This result, together with the existence of very long periods in the secular variation, suggests that historical changes in the rate of secular deceleration of the mantle can be similarly explained. At the other extreme the most rapid accelerations, about 30 parts in 10^{10} per year (Markowitz, 1970), are impossible to attribute to electromagnetic coupling unless the electrical conductivity can be increased greatly just above the core-mantle boundary (Roden, 1963). If this is not permissible one must either look to the high-frequency end of the secular variation spectrum to enhance the electromagnetic coupling, or turn to other coupling mechanisms (Hide, 1969).

References

Aoki, S.: 1967, 'On Oort's Constant B', *Publ. Astron. Soc. Japan* **19**, 585.

Aoki, S.: 1969, 'Friction between Mantle and Core of the Earth as a Cause of the Secular Change in Obliquity', *Astron. J.* **74**, 284.

Backus, G.: 1968, 'Kinematics of Geomagnetic Secular Variation in a Perfectly Conducting Core', *Phil. Trans. Roy. Soc. London* **A263**, 239.

Ball, R. H., Kahle, A. B., and Vestine, E. H.: 1968, 'Variations in the Geomagnetic Field and in the Rate of the Earth's Rotation', RAND Corporation Memorandum RM-5717-PR.

Batchelor, G. K.: 1953, *The Theory of Homogeneous Turbulence*, Cambridge University Press.

Bondi, H. and Lyttleton, R. A.: 1948, 'On the Dynamical Theory of the Rotation of the Earth', *Proc. Camb. Phil. Soc.* **44**, 345.

Braginskii, S. I.: 1967, 'Magnetic Waves in the Earth's Core', *Geomag. Aeron.* **7**, 851.

Bullard, E. C.: 1949, 'The Magnetic Field within the Earth', *Proc. Roy. Soc. London* **A197**, 433.

Bullard, E. C., Freedman, C., Gellman, H., and Nixon, J.: 1950, 'The Westward Drift of the Earth's Magnetic Field', *Phil. Trans. Roy. Soc. London* **A243**, 67.

Busse, F. H.: 1968, 'Steady Fluid Flow in a Precessing Spheroidal Shell', *J. Fluid Mech.* **33**, 739.

Busse, F. H.: 1970, 'The Dynamical Coupling between the Inner Core and Mantle of the Earth and the 24-Year Libration of the Pole', this volume, p. 88.

Currie, R. G.: 1968, 'Geomagnetic Spectrum of Internal Origin and Lower Mantle Conductivity', *J. Geophys. Res.* **73**, 2779.

Dicke, R. H.: 1966, 'The Secular Acceleration of the Earth's Rotation and Cosmology', in *The Earth-Moon System* (ed. by B. G. Marsden and A. G. W. Cameron), Plenum Press, New York, p. 98.

Elsasser, W. M.: 1949, 'Non-uniformity of the Earth's Rotation and Geomagnetism', *Nature* **163**, 351.

Elsasser, W. M. and Munk, W. H.: 1958, 'Geomagnetic Drift and the Rotation of the Earth', in *Contributions in Geophysics: In Honour of Beno Gutenberg* (ed. by H. Benioff *et al.*), Pergamon Press, Oxford, p. 228.

Elsasser, W. M. and Takeuchi, H.: 1955, 'Nonuniform Rotation of the Earth and Geomagnetic Drift', *Trans. Amer. Geophys. Un.* **36**, 584.

Gutenberg, B.: 1951, 'Internal Constitution of the Earth', Dover Publications, Inc., New York.

Hide, R.: 1966, 'Free Hydromagnetic Oscillations of the Earth's Core and the Theory of the Geomagnetic Secular Variation', *Phil. Trans. Roy. Soc. London* **A259**, 615.

Hide, R.: 1969, 'Interaction between the Earth's Liquid Core and Solid Mantle', *Nature* **222**, 1055.

Hide, R. and Horai, K.: 1968, 'On the Topography of the Core-Mantle Interface', *Phys. Earth Planetary Int.* **1**, 305.

Hopkins, W.: 1839–42, 'Researches in Physical Geology', *Phil. Trans. Roy. Soc. London* **129**, 381; **130**, 193; **132**, 43.

Hough, S. S.: 1895, 'The Oscillations of a Rotating Ellipsoidal Shell containing Fluid', *Phil. Trans. Roy. Soc. London* **A186**, 469.

Jeffreys, H.: 1959, *The Earth: Its Origin, History and Physical Constitution*, 4th ed., Cambridge University Press.

Jeffreys, H. and Vicente, R. O.: 1957, 'The Theory of Nutation and the Variation of Latitude', *Monthly Notices Roy. Astron. Soc.* **117**, 142.

Kelvin, Lord (W. Thomson): 1876, *Presidential Address, British Association*, reprinted in *Mathematical and Physical Papers*, Vol. 3, Cambridge University Press, 1890, p. 320.

Lahiri, B. N. and Price, A. T.: 1939, 'Electromagnetic Induction in Non-Uniform Conductors, and the Determination of the Conductivity of the Earth from Terrestrial Magnetic Variations', *Phil. Trans. Roy. Soc. London* **A237**, 509.

MacDonald, G. J. F.: 1966, 'The Figure and Long-term Mechanical Properties of the Earth', in *Advances in Earth Science* (ed. by P. M. Hurley), M.I.T. Press, Cambridge, Mass., p. 199.

MacDonald, G. J. F. and Ness, N. F.: 1961, 'A Study of the Free Oscillations of the Earth', *J. Geophys. Res.* **66**, 1865.

Malkus, W. V. R.: 1963, 'Precessional Torques as the Cause of Geomagnetism', *J. Geophys. Res.* **68**, 2871.

Malkus, W. V. R.: 1967, 'Hydromagnetic Planetary Waves', *J. Fluid Mech.* **28**, 793.

Malkus, W. V. R.: 1968, 'Precession of the Earth as the Cause of Geomagnetism', *Science* **160**, 259.

Markowitz, W.: 1970, 'Sudden Changes in Rotational Acceleration of the Earth and Secular Motion of the Pole', this volume, p. 69.

McDonald, K. L.: 1957, 'Penetration of the Geomagnetic Secular Variation through a Mantle with Variable Conductivity', *J. Geophys. Res.* **62**, 117.

Miki, H.: 1952, 'Physical State of the Earth's Core', *J. Phys. Earth* **1**, 67.

Molodenskii, M. S.: 1961, 'The Theory of Nutation and Diurnal Earth Tides', *Comm. Obs. Roy. Belgique* **188**, 25.

Munk, W. H. and Hassan, E. S. M.: 1961, 'Atmospheric Excitation of the Earth's Wobble', *Geophys. J.* **4**, 339.

Munk, W. H. and Revelle, R.: 1952, 'On the Geophysical Interpretation of Irregularities in the Rotation of the Earth', *Monthly Notices Roy. Astron. Soc., Geophys. Suppl.* **6**, 331.

Poincaré, H.: 1910, 'Sur la précession des corps déformables', *Bull. Astron. (Paris)* **27**, 321.

Roberts, P. H. and Stewartson, K.: 1965, 'On the Motion of a Liquid in a Spheroidal Cavity of a Precessing Rigid Body', *Proc. Camb. Phil. Soc.* **61**, 279.

Rochester, M. G.: 1960, 'Geomagnetic Westward Drift and Irregularities in the Earth's Rotation', *Phil. Trans. Roy. Soc. London* **A252**, 531.

Rochester, M. G.: 1962, 'Geomagnetic Core-Mantle Coupling', *J. Geophys. Res.* **67**, 4833.

Rochester, M. G.: 1968, 'Perturbations in the Earth's Rotation and Geomagnetic Core-Mantle Coupling', *J. Geomag. Geoelec.* **20**, 387.

Rochester, M. G.: 1970, 'Polar Wobble and Drift: A Brief History', this volume, p. 3.

Rochester, M. G. and Smylie, D. E.: 1965, 'Geomagnetic Core-Mantle Coupling and the Chandler Wobble', *Geophys. J.* **10**, 289.

Roden, R. B.: 1963, 'Electromagnetic Core-Mantle Coupling', *Geophys. J.* **7**, 361.

Runcorn, S. K.: 1954, 'The Earth's Core', *Trans. Amer. Geophys. Un.* **35**, 49.

Runcorn, S. K.: 1955, 'The Electrical Conductivity of the Earth's Mantle', *Trans. Amer. Geophys. Un.* **36**, 191.

Sato, R. and Espinosa, A. F.: 1967, 'Dissipation Factor of the Torsional Mode $_0T_2$ for a Homogeneous-Mantle Earth with a Soft-Solid or a Viscous-Liquid Core', *J. Geophys. Res.* **72**, 1761.

Sekiguchi, N.: 1967, 'On the Cause of the Unexplained Secular Change in the Obliquity of the Ecliptic', *Publ. Astron. Soc. Japan* **19**, 596.

Sludskii, F.: 1896, 'De la rotation de la terre supposée fluide à son intérieur', *Bull. Soc. Natur. Moscou* **9**, 285.

Smylie, D. E.: 1965, 'Magnetic Diffusion in a Spherically-Symmetric Conducting Mantle', *Geophys. J.* **9**, 169.

Stacey, F. D.: 1970, 'A Re-examination of Core-Mantle Coupling as the Cause of the Wobble', this volume, p. 176.

Stewartson, K.: 1967, 'Slow Oscillations of Fluid in a Rotating Cavity in the Presence of a Toroidal Magnetic Field', *Proc. Roy. Soc. London* **A299**, 173.

Stewartson, K. and Roberts, P. H.: 1963, 'On the Motion of a Liquid in a Spheroidal Cavity of a Precessing Rigid Body', *J. Fluid Mech.* **17**, 1.

Toomre, A.: 1966, 'On the Coupling of the Earth's Core and Mantle during the 26000 Year Precession', in *The Earth-Moon System* (ed. by B. G. Marsden and A. G. W. Cameron), Plenum Press, New York, p. 33.

Vestine, E. H.: 1952, 'On Variations of the Geomagnetic Field, Fluid Motions, and the Rate of the Earth's Rotation', *Proc. Nat. Acad. Sci. Wash.* **38**, 1030.

Walker, G. B. and O'Dea, P. L.: 1952, 'Geomagnetic Secular-Change Impulses', *Trans. Amer. Geophys. Un.* **33**, 797.

Discussion

Kaula: I do not know whether topographic coupling exists but I do not think you can use the gravity field as firm evidence of it. I think the most you can infer about the shape of the core from the gravity field is that possibly the second degree harmonics of the gravity field (the equatorial ellipticity) has something to do with the shape of the core since it does not correlate too well with any near surface features. I read a preprint of Aoki's paper and got the impression that he could only explain the obliquity as due to an unreasonably anisotropic friction between the core and the mantle. Has that been resolved?

Rochester: He was worried about the effect on the change in the length of the day but there are other ways of accounting for this so that you can have anisotropic friction if you wish.

Kaula: You mean that only one component effects the obliquity?

Rochester: Yes.

Markowitz: I understand that you find no way of having core mantle couplings to account for torques of the size called for by astronomical observations.

Rochester: I would not go as far as that. All that I can say is that the kind of electromagnetic excitation torques which can be inferred from the secular variation spectrum are not strong enough to do the job. However, if there is sufficient excitation at high frequencies, which you would not see at the top of the mantle, then there might be no problem at all.

Markowitz: Do you see any way of explaining the rapid changes that occur at intervals of around three or four years?

Rochester: It is on the borderline of the impossible to explain them by changes of the secular variation pattern.

Busse: One must be clear about the role of inertial coupling versus viscous coupling. A distinction must be made between those cases where the core is in turbulent motion and those where it is in laminar motion because for turbulent motion, inertial coupling is much less efficient than for the corresponding laminar case. In addition, viscous coupling is much more efficient for the turbulent case than for the laminar case and this may partly explain why, for instance, Malkus emphasizes turbulent coupling and dismisses viscous inertial coupling. Part of the trouble is that turbulent couplingis still out of the reach of any theory.

Runcorn: I do not quite understand Kaula's remarks. It is quite clear that you cannot hope to explain in a satisfactory way the higher harmonics of the geoid by undulations of the core mantle boundary, but I think that the low harmonics could be interpreted in this way while the higher harmonics, say higher than the eighth degree, are assigned to some other cause nearer the surface. I just do not see how one can rule out the idea that undulations of the core mantle boundary explains the lower harmonics.

Kaula: The main point is that the power spectrum of the gravity field goes down by the $10^{-5} L^2$ rule.

Runcorn: Obviously you get into absurd difficulties in trying to explain the very high harmonics around $N = 10$ but what about the lower harmonics.

Kaula: I think you get into absurd difficulties in $N = 4$.

Runcorn: I do not see this. If you had to have a negative mass down there then, of course, you could say it was absurd.

Kaula: That just makes it impossible. I did not say it was impossible I just said it was extremely implausible. For example outside of the sixth degree, you have a definite correlation of the gravity field with the surface topography.

Bostrom: Recent paleomagnetic results indicate that magnetic variations in the Pacific half of the world have been different from those in the other half of the world. Various other bits of geologic evidence suggests that core formation may still be continuing. Could the speaker tell me whether quite sudden changes in electrical conductivity in the core mantle boundary area, which apparently may be tectonic, would account for the quite sudden changes in the mechanical magnitude of the core mantle coupling.

Rochester: What sort of mechanism do you propose for such a large change?

Bostrom: One might imagine that circuits could suddenly become established if there are bodies of conducting material moving by even a small amount. I notice that in most calculations people take resistivities as being more or less constant in that region, and yet one doubts very much whether they do remain constant through time.

Rochester: I do not know whether I have any very sensible thing to say about it.

Runcorn: If the secular variation in the Pacific has been different from other areas of the world one must find an explanation for it. If one supposes that a part of the lower mantle, say within a few tens of kilometers of the core, is actually solid iron or partially iron, then you could get very considerable shielding of the secular variation and this might change the whole problem of core mantle coupling.

Stacey: I do not think you need go that far since the observatory records at Hawaii over the past 70 years indicate a normal secular variation. The results could be explained by assuming that the lavas that were sampled were ejected much more quickly than has previously been supposed.

AN EXAMINATION OF THE DATA RELATING POLE MOTION
TO EARTHQUAKES

RICHARD A. HAUBRICH

Institute of Geophysics and Planetary Physics, University of California, San Diego, Calif., U.S.A.

Abstract. Although Mansinha and Smylie (1967) conclude that earthquake fault displacements excite the Chandler wobble, it has been found that the excitation from actual historical events is at least an order of magnitude too low to maintain the observed wobble.

Smylie and Mansinha (1968) report evidence of a relationship between observed pole motion and large earthquakes. The pole data have been re-examined; it was found that the significance level of the correlation varies widely with two rather arbitrary parameters. The first of these is the coincidence time span; the second is the acceptance level for picking breaks in the pole path. The data have been reexamined using an optimum least square fit procedure utilizing dynamic programming. The correlation found by this method for the six largest earthquakes was not significant. The results throw doubt on the validity of the proposal that earthquakes move the pole an observable amount.

1. Introduction

The motion of the earth's axis of rotation relative to points fixed on the earth results in changes of latitude which are observed by astronomical measurement. Munk and MacDonald (1960) treat the general problem of earth rotation. Munk and Hassan (1961) found the seasonal term in the pole motion was due to atmospheric excitation. They found, however, that the atmosphere could not account for the 14-month Chandler wobble; the atmospheric spectrum density is one or two orders of magnitude too low.

Mansinha and Smylie in two recent papers suggest that large earthquakes can and do excite the Chandler wobble. Mansinha and Smylie (1967, hereafter referred to as reference (1)) compute the changes in the products of inertia of the earth which results from motion on faults during large earthquakes. They conclude that the resulting changes based on the extensive displacement fields resulting from the elastic theory of dislocations provide sufficient excitation to maintain the wobble.

In the second paper, Smylie and Mansinha (1968, hereafter referred to as reference (2)) examine the latitude observations. Strong correlations are found between the occurrence of earthquakes and 'breaks' in the pole path as inferred from latitude. The breaks represent times at which the pole switches from a previous circular path to a new circular path about a displaced equilibrium pole position.

In the following sections the earthquake excitation is recomputed as in reference (1) except that the actual rate of earthquake occurrence was used in place of that predicted from empirical formulas. Next, some implications of the breaks found in reference (2) are discussed. The following section uses an alternate method to that of reference (2) for finding breaks which utilizes dynamic programming. Finally, the correlations of reference (2) are critically reexamined.

L. Mansinha et al. (eds.), Earthquake Displacement Fields and the Rotation of the Earth, 149–158.

2. Earthquake Excitation

Table I shows the rate of earthquake occurrences for different magnitudes. Column 2 gives the rate in units of events per 64 years computed from the empirical relation used in reference (1),

$$\log_{10} N = A - bM$$

where N is the number of events of magnitude M and larger, $b = 1$ and $A = 8.37$. The empirical relation is not necessarily a good approximation especially for the largest earthquakes. Table I shows this to be the case; column 3 gives the number of observed events for the 64 years 1904–67. The earthquake magnitudes were taken from Gutenberg and Richter (1954) for 1904–45 and from *Principal Earthquakes of the World* for 1946–67.

TABLE I

Magnitude	Predicted events/64 yrs	Observed events/64 yrs	P_1 $(0''01)^2$	P_2 $(00''1)^2$
8.9	4	0	0	0
8.8	5	0	0	0
8.7	6	1	4.8×10^{-2}	8.3×10^{-3}
8.6	8	2	3.8×10^{-2}	4.1×10^{-3}
8.5	10	4	3.0×10^{-2}	2.0×10^{-3}
8.4	12	2	5.9×10^{-3}	2.5×10^{-4}
8.3	16	12	1.4×10^{-2}	3.7×10^{-4}
8.2	19	10	4.5×10^{-3}	7.6×10^{-5}
8.1	25	13	2.3×10^{-3}	2.4×10^{-5}
7.1–8.0			3.3×10^{-3}	1.5×10^{-5}
Total			1.46×10^{-1}	1.52×10^{-2}

Columns 4 and 5 of Table I give the mean square pole motion per year in units of $(0''01)^2$ based on the observed earthquake rates, for the two models relating magnitude to pole shift given in Figure 4 of reference (1). Model 1 is for strike slip faults and Model 2 is for dip slip faults. For each magnitude, M,

$$P_1 = R_M S_1^2; \quad P_2 = R_M S_2^2$$

where

$$\log_{10}(S_1/0.69) = 2.03(M - 8.5)$$

$$\log_{10}(S_2/0.18) = 3.04(M - 8.5).$$

S_1 and S_2 are the polar displacement in units of $0''01$ produced by earthquakes of magnitude M according to models 1 and 2 of reference (1). R_M is the yearly average number of observed events of magnitude M. For magnitudes 7.1 to 8.0 the mean square pole motion has been summed over the magnitudes in that range. The total is the summed mean square amplitude for all magnitudes from 7.1 to 8.9.

Earthquakes smaller than magnitude 8.1 contribute only about 2% of the power for model 1 and less than 0.1% for model 2. The use of the empirical formula to estimate earthquake rates results in an overestimate for the largest events which produce the bulk of the power in the pole motion.

The results from Table I have been converted into Chandler power for values of Q between 10 and 100 where

$$Q = \pi f_0 / \alpha.$$

f_0 and α are the real and imaginary parts of the Chandler frequency. The Fourier transform of a unit step function is $1/2\pi f$ so that the excitation spectrum density resulting from steps with mean square, P, is

$$S(f) = (1/2\pi f)^2 \, P.$$

The total Chandler power is obtained from the excitation spectrum at the Chandler frequency (Munk and Hassan, 1961) and is:

$$2\pi f_0 Q S(f_0) = \frac{Q}{2\pi f_0} \, P.$$

Replacing P with the total power from all earthquakes of magnitude 7.1 to 8.7, the resulting Chandler power was found and is given in Table II for models 1 and 2. The Chandler frequency, f_0, was taken as 0.85 cycles per year.

TABLE II

Chandler power in units of $(0\overset{''}{.}01)^2$

	Model 1	Model 2	Atmospheric
$Q = 10$	0.27	0.028	0.83
$Q = 30$	0.82	0.084	2.5
$Q = 100$	2.7	0.28	8.3

Since the observed Chandler power is about 200 $(0\overset{''}{.}01)^2$, the results in Table II show that the earthquake contribution is too small even for the most favorable case of Model 1 with $Q = 100$.

Richter (1958) revised the magnitudes reported in *Seismicity of the Earth* with the result that M for some of the largest events were increased so that in the 64 year period, two earthquakes are assigned $M = 8.9$ and nine are assigned $M = 8.7$. Using the revised magnitudes for $Q = 100$ the earthquake contribution is 25 $(0\overset{''}{.}01)^2$ and 7 $(0\overset{''}{.}01)^2$ for Models 1 and 2.

The extrapolation of the models to magnitude 8.9 leads to some difficulty. Such an earthquake would have a fault length greater than 2000 km and fault motion extending to a depth of 500 km. There is no evidence that fault motions over such great extents have occurred. Iida (1965) in a study of the largest earthquakes finds no evidence for fault lengths greater than about 1000 km. Furthermore, most large earthquakes show no evidence of motion at depths below 100 km. As an example, Tandon (1954) found for the Assam earthquake of 1950, $M = 8.7$, that the aftershocks extended over a length of about 900 km and that the deepest aftershock was at less than 30 km.

Thus it seems that the Alaska 1964 earthquake was an exceptional event in regard to the extent of its fault motion even when compared to events with larger values of M. For $Q = 100$ and Model 1, more than 20 Alaska size events per year are required to maintain the wobble.

For comparison the Chandler power due to atmospheric excitation (Munk and Hassan, 1961) is given in the last column of Table II. For the revised magnitudes atmospheric excitation is larger than the earthquake contribution for Model 2 and smaller by a factor of 3 than the contribution for Model 1. The highest estimate for earthquake contributed Chandler power ($Q = 100$, Model 1, revised magnitudes) falls short of the observed by about an order of magnitude.

3. Breaks in the Latitude Data

Reference (2) examines two independent sets of latitude data covering the period 1957 to 1967. The data from the International Latitude Service – International Polar Motion Service showed no correlation with earthquakes. Thus only the second data set, that reduced by the Bureau International de l'Heure will be considered. The data consist of 402, 10 day values of the pole position, $\mathbf{m} = m_1 \mathbf{i}_1 + m_2 \mathbf{i}_2$ where $\mathbf{i}_1$ and $\mathbf{i}_2$ are unit vectors in the direction of $0°$ and $90°$ east longitude.

Breaks were found in reference (2) by least square fitting circular arcs to the pole data. A break represents a time at which the mean pole shifts by some amount $\Delta\mathbf{m}$; at the same time the Chandler amplitude vector changes by an almost equal but opposite amount. There are two puzzling things about this model. First, the pole shifts from breaks in the latitude data have root mean square amplitudes on the order of $0\overset{''}{.}1$. In reference (1) the theoretical shift produced by the largest earthquakes is only $0\overset{''}{.}002$. Thus the observed shifts from the latitude data are far too large to be caused by earthquakes. Second, it appears that the observed shifts are too large in comparison to the observed Chandler power. If the shifts corresponding to breaks are due to step changes in the earth's moments of inertia from whatever cause, then one can estimate the total resulting Chandler power by the method outlined above for earthquake step functions. For 32 breaks in 11 years each with root mean square amplitude, $0\overset{''}{.}1$. the mean power per year is:

$$P = \tfrac{32}{11}(10)^2 = 290\,(0\overset{''}{.}01)^2 .$$

For $Q = 30$ the corresponding Chandler power is:

$$\frac{Q}{2f_0}\,P = 1600\,(0\overset{''}{.}01)^2$$

compared to $200\,(0\overset{''}{.}01)^2$ for the observed Chandler power.

Finally there is some evidence that the observed pole spectrum is noisy at frequencies outside the Chandler band. Assuming that the data are noise-free at $f = -1$ cpyr. the computed value of Q is 6, which is somewhat low compared with other estimates (Munk and MacDonald, 1960). The low Q estimate can be explained by assuming that the pole data contain a significant amount of noise outside the Chandler band. The least square fit procedure for finding breaks is essentially a deconvolution. It relies strongly on the broad band portion of the pole data. If earthquakes are correlated with pole data, they are thus not necessarily correlated with pole motion. Rather, it would seem more likely that observed correlations would be between earthquakes and noise in the pole data.

4. Dynamic Programming

In this section results are given for a method of finding pole breaks which is somewhat different from that used in reference (2). For a segment of data containing n data points specify a number of breaks, b. Partition the n data points into $b+1$ intervals such that each interval contains at least two data points. A least square fit to circular arcs is made for each of the $b+1$ intervals; the residual sum of squares from all intervals is then a measure of the goodness of fit for that particular partition. For all partitions of the n data points which yield b breaks pick that with the least residual as the optimum fit of breaks to the data.

The number of partitions can be large being the binomial coefficient $\begin{pmatrix} n-b-2 \\ n \end{pmatrix}$

For example, if $n = 120$ and $b = 15$ the number of possibilities is about 10^{18} and it is impossible to examine them all. One uses the method of dynamic programming which is well suited to this problem (Bellman and Dreyfus, 1962). This method finds the best partition without examining all of them.

The BIH data were analysed by the above method using $n = 120$ and $b = 15$. Four separate analyses were made for data points 1–120, 101–220, 201–320, and 281–400.

The dynamic programming method which finds the optimum solution for b breaks also gives optimum solutions for all numbers of breaks less than b. It was found that the optimum set of breaks differs with the parameter, b. For example, a solution for 7 breaks does not necessarily result in a subset of the solution for 10 breaks.

As a choice of the number of breaks, I took numbers that corresponded to those found in reference (2). Thus for the first section of 120 data points, reference (2) obtained 15 breaks; I therefore chose $b = 15$ for this section and similarly for the other three sections of the data. The result of picking breaks by this method was that only three out of six of the largest earthquakes fell within 20 days of a break. The probability of this occurring at random is 0.27.

5. The Correlation between Earthquakes and Breaks

Reference (2) considers three classes of earthquakes according to magnitude which I will call:

Class A; $M > 8.0$
Class B; $7.75 \leqslant M < 8.0$
Class C; $7.5 < M \leqslant 7.75$.

I now examine the correlation between breaks and events obtained in reference (2). First consider class C events; the random probability is given as RP = 0.29. An earthquake has been missed (Taiwan, March 12, 1966, U.S.C.G.S.) which belongs to this class and does not fall in a break. The revised RP is 0.38. The events of class C appear to be uncorrelated with breaks.

Events of class B have RP = 0.034. I believe that this value is an overestimate. Probability estimates based on the binomial distribution assume that the data samples are independent. Independent samples of events in time follow a Poisson distribution. There is a weak suggestion that the series of breaks is not Poisson distributed since for example 15 out of 32 occur in the first 3 out of 11 years. There is a strong suggestion that the series of class B events is not Poisson since 6 out of 7 occur in a span of slightly over a year and none occur in a span of $8\frac{1}{2}$ years.

If the class B events are Poisson distributed with $\lambda = \frac{7}{11}$ per year as the Poisson parameter, then the probability of obtaining a sample as bunched as the observed one is

$$p_1 \cdot p_2 = 2.7 \times 10^{-6}$$

where

$$p_1 = \sum_{k=5}^{\infty} e^{-\lambda t_1} (\lambda t_1)^k / k! = 6 \times 10^{-4}$$

$$p_2 = e^{-\lambda t_2} = 4.5 \times 10^{-3}.$$

p_1 is the probability that after the first event 5 or more events will occur within $t_1 = 1.04$ years; p_2 is the probability that no events will occur in $t_2 = 8.5$ years. The effect of dependent samples is to increase the RP calculated from the binomial distribution. A Monte Carlo experiment was performed with computer generated random sets of 7 events having a bunched distribution like the class B events. In 5000 trials, 8% of the bunched random events gave 5 or more out of 7 coincidences with the 32 breaks. I conclude that the RP for class B events is about 8%.

The class A events with a given RP of 0.0009 appears to be a strongly significant result. There are, however, two parameters used in the data analysis of reference (2) which do effect the significance level of the correlation. The first of these I call the coincidence parameter, c, measured in days. An earthquake is said to coincide with a break if it is within n days of the break day (taken as the middle day between data points). The probability of coincidence is

$$p = bc/4020$$

and

$$c = 2n + 1$$

where $b =$ the number of breaks (32). Reference (2) takes $n = 19$. Any other value of n gives a lower RP. Figure 1 shows the effect of using various values of c in terms of the odds against randomness, $D = 1/RP - 1$.

The second parameter used in reference (2) I call the acceptance parameter, a. In least square fitting the pole path to data points, a break is found when the next point deviates from the projected least square point by more than a. In reference (2) $a = 0''.02$.

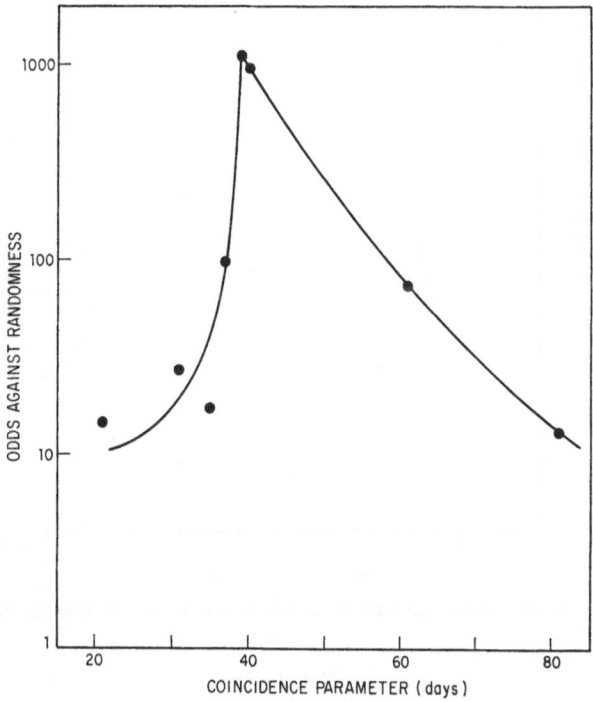

Fig. 1.   Odds against randomness as a function of coincidence parameter a.

Breaks in the BIH latitude data were computed using the method of reference (2) for a range of different values of parameter a. Figure 2 shows the odds against randomness as a function of a. The peak odds lie near to $0''.02$. Most of the other values result in an insignificant RP.

For low values of a the number of breaks is large (59 for $a = 0''.01$). In spite of this the number of coincidences decreases to 4 out of 6. For large values of a the number of breaks is small (14 at $a = 0''.0375$) and the number of coincidences drops to 1 out of 6. Increasing a above $0''.04$ to the point where only 9 breaks occur, results in no coincidences. The largest breaks show no tendency to correlate with earthquakes.

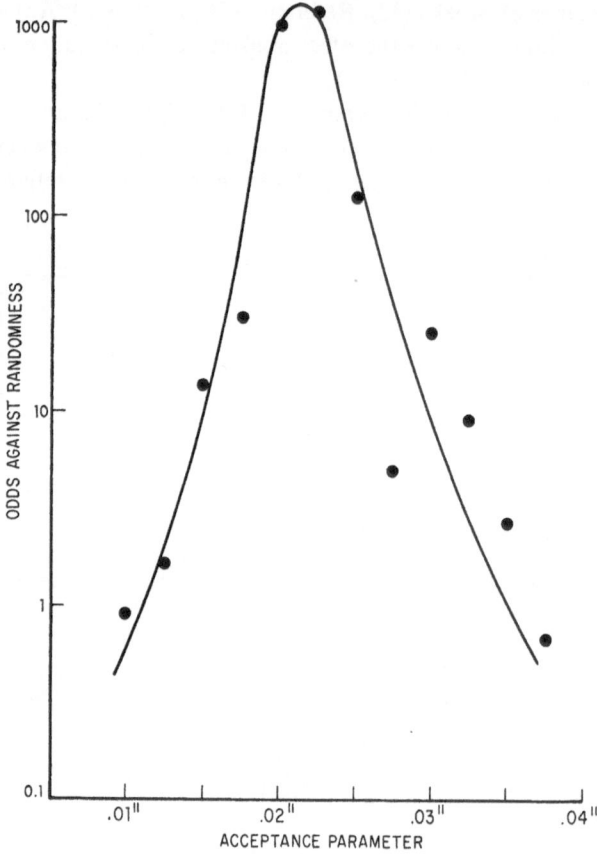

Fig. 2.   Odds against randomness as a function of acceptance parameter c.

6. Conclusions

The accumulative effect of large earthquakes on the Chandler wobble appears to be too small to account for its excitation. At best, earthquakes could account for about 12% of the wobble if one admits the existence of fault dimensions larger than any observed. The dislocation theory used in obtaining these results is based on a semi-infinite, homogeneous half space. The theory for a homogeneous sphere (Ben-Menaham *et al.*, 1969) produces surface displacement fields significantly different from the half space result. But even the homogeneous sphere may be a poor model for the present problem; density and elasticity variations may well give a significantly different answer. At present, however, there seems to be little reason to believe that earthquake excitation is sufficient to maintain the wobble.

I conclude that earthquakes correlate with the pole data but that the correlation is marginally significant. The strength of the correlation lacks robustness; small changes in the analysis procedure produce large changes in results. If correlation does exist, it could very well be between the pole data noise and earthquakes.

References

Bellman, R. E. and Dreyfus, S. E.: 1962, *Applied Dynamic Programming*, Princeton University Press, Princeton.

Ben-Menahem, A., Singh, S. J., and Solomon, F.: 1969, 'Static Deformation of a Spherical Earth Model by Internal Dislocations', *Bull. Seism. Soc. Am.* **59**, 813–853.

Gutenberg, B. and Richter, C. F.: 1954, *Seismicity of the Earth*, Princeton University Press, Princeton.

Iida, K.: 1965, 'Earthquake Magnitude, Earthquake Fault, and Source Dimensions', *J. Earth Sci., Nagoya Univ.* **13**, 115–132.

Mansinha, L. and Smylie, D. E.: 1967, 'Effect of Earthquakes on the Chandler Wobble and the Secular Polar Shift', *J. Geophys. Res.* **72**, 4731–4743.

Mansinha, L. and Smylie, D. E.: 1968, 'Earthquakes and the Earth's Wobble', *Science* **161**, 1127–1129.

Munk, W. H. and Hassan, E. S. M.: 1961, 'Atmospheric Excitation of the Earth's Wobble', *Geophys. J.* **4**, 339–358.

Munk, W. H. and MacDonald, G. J. F.: 1960, *The Rotation of the Earth*, Cambridge University Press, London.

Principal Earthquakes of the World, United States Earthquakes, U.S. Coast and Geodetic Survey, 1946 to 1967.

Richter, C. F.: 1958, *Elementary Seismology*, W. H. Freeman and Co., San Francisco.

Smylie, D. E. and Mansinha, L.: 1968, Earthquakes and the Observed Motion of the Rotation Pole', *J. Geophys. Res.* **73**, 7661–7673.

Tandon, A. N.: 1954, 'Study of the Great Assam Earthquake of August 1950 and its Aftershocks', *Indian J. Met. Geophys.* **5**, 95–137.

Discussion

Smylie: If I understand you correctly, the set of 31 breaks we determined is a subset of more than 10^{19} possible combinations. It is odd that on the first run we could have obtained a high degree of correlation on one of the more than 10^{19} combinations.

Van Flandern: By your method did you find the shifts in location of the secular pole to be of the order of $0''.1$ or significantly less?

Haubrich: A lot of the breaks that I found corresponded with those found by Mansinha and Smylie. Differences between the two methods did not really result in a wild or completely different set of breaks.

Van Flandern: If the shifts in secular pole locations were of the order of $0''.1$, as determined by you, how do you reconcile this with your own statement that shifts of this order result in too much power?

Haubrich: This discrepancy holds only when the number of breaks is large, say 30. If the number of breaks reduces to 10, the discrepancy in power disappears. But then the correlation with earthquakes also disappears.

Major: Although you mentioned that the excitation from historical events is at least an order of magnitude too small to maintain the observed wobble, I am not clear as to how one arrives at the magnitude of the excitation. The main problem is that most of us see strain steps at distances which are embarrassingly large compared to those predicted by either the half-space or spherical dislocation theory. Which is the correct model?

Haubrich: My work is based upon a homogeneous half-space and as far as I know there is no better way of doing this problem at the moment. Nobody has taken even the homogeneous sphere and computed what the changes in the products of inertia are due to an earthquake; but even this may not give you better results because the earth is layered and if this is taken into account, I would expect that things might be changed by a factor of two or three or more.

Major: Does this imply that one should see a step in the pole position associated with each large earthquake, because I think this is extremely improbable?

Haubrich: All I am saying is that the evidence we have now is less convincing than I would want it to be. There is some correlation but, to me, earthquakes as we know them just cannot cause the Chandler wobble. It may well be that when new theoretical work is carried out for a layered earth, or maybe for global tectonics, it will be found that earthquakes can cause the wobble after all.

Whitten: I do not follow why you have altered the data by removing the annual effects before computing centres of the arcs.

Haubrich: I do not see how the annual term can change things because the shifts we are looking for are due to high frequencies in the data, not to the annual term.

Whitten: When the annual term and the Chandler term are out of phase, the polar motions have a very small change throughout the year. When you remove the annual term you introduce a ficticious centre rather than the real centre; the data point represents the real centre.

Haubrich: In my opinion, the annual term remains constant and does not affect the jumps.

EVIDENCE FOR ASSOCIATION OF EARTHQUAKES WITH THE CHANDLER WOBBLE, USING LONG TERM POLAR DATA OF THE ILS-IPMS

ROBERT J. MYERSON

Dept. of Physics, University of California, Berkeley, Calif., U.S.A.

Abstract. A means for comparing yearly earthquake counts with the ILS-IPMS polar data is devised. Using it, one finds strong support for the theory of Mansinha and Smylie (1967, 1968) that earthquakes are associated with the free (Chandler) wobble of the earth's pole. There is some indication, however, that earthquakes, themselves, are not the source of wobble excitation but are, instead, a parallel effect.

1. Introduction

Mansinha and Smylie (1967, 1968) have proposed that earthquakes are the source of excitation for the Chandler wobble of the earth's pole. Using polar data from the Bureau International de l'Heure (BIH) for the years 1957–67, they have found that very large (magnitude $\geqslant 7.5$) earthquakes correlate quite well with sudden, step function, breaks in the center of polar motion. An example of such a break is given in Figure 1. Note that, although the center of wobble changes sharply in this break, there is no discontinuity in the polar motion itself. The break thus produces a change not only in the center of wobble (secular pole) but in the amplitude of wobble as well. Mansinha and Smylie (1967) have shown that mass shifts due to earthquakes may cause enough breaks of this kind to maintain the wobble amplitude in a random-walk fashion.

In their 1968 study, Smylie and Mansinha failed to find any significant association of earthquakes with breaks when they used polar data of the International Latitude Service (ILS) – International Polar Motion Service (IPMS). This was partly because the ILS-IPMS did not publish pole positions as frequently as the BIH. There were

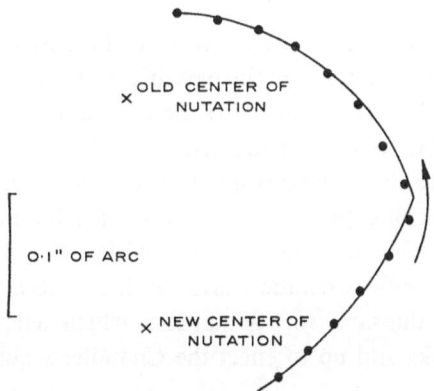

Fig. 1. An observed break in the center of Chandler wobble. (Based on Mansinha and Smylie's (1968) reduction of the BIH data for the last $\frac{1}{3}$ of 1958.)

L. Mansinha et al. (eds.), Earthquake Displacement Fields and the Rotation of the Earth, 159–168.

also, however, discrepancies of as much as 0''.1 of arc between the two services. This is large enough to cast some doubt upon the validity of any claimed break – particularly, as is often the case, when the break is not so pronounced as the one in Figure 1.

The present paper seeks to test the ILS-IPMS data in a different way. Rather than attempt to correlate specific polar breaks with specific earthquakes, yearly means of the Chandler amplitude are compared with yearly earthquake counts. Because the ILS-IPMS has been giving pole positions since the beginning of the century enough years are available.

No such comparison is possible using the BIH data, however. The BIH has only been giving pole positions since 1955.

2. Background

As Mansinha and Smylie pointed out in their 1967 article, the build-up of the Chandler component – as distinguished from the secular and annual components of polar motion – is probably a random walk type of phenomenon. The mathematics of polar motion is given in their 1967 article. The claim of random walk build-up will hold true, so long as excitation is by step-function breaks in the polar motion (such as might be, although not necessarily must be, caused by earthquakes) and, further, that these breaks occur at random times. No other specification of the excitation mechanism is needed.

Given a random walk development, one can readily establish a useful equation for looking at the ILS-IPMS data. To begin, note the case of a simple two-dimensional random walk, a drunk constrained to walk with both feet on the ground, if one wishes. The drunk takes N steps of mean square size $(\Delta R)^2$ apiece. Statistics tells us that, on the average, these N steps (if N is large) will add up to take him a distance:

$$\langle R^2 \rangle = N(\Delta R)^2 \tag{1}$$

from where he started.

One may use an analogous expression to give the development of the Chandler amplitude. Suppose that, at a time τ in the past, breaks were occurring at a rate $N(\tau)$ – that is, in the time interval from τ to $\tau + d\tau$ there were $N(\tau) \, d\tau$ breaks. These breaks are perfectly comparable to steps in the drunk's walk. Let us call the mean square break size $(\Delta M)^2$. Since the wobble is damped, contributions from past breaks will be multiplied by a factor of $\exp(-2\gamma(t-\tau))$, where t is the present time and γ is the Chandler damping coefficient. The presence of damping is the major difference between the kind of Chandler excitation assumed here and a simple two-dimensional random walk. Finally, the same sort of statistics which tells how the drunk's steps add up, tells how breaks add up to effect the Chandler wobble. Hence this integral equation:

$$|M|^2(t) = \int\limits_{-\infty}^{t} e^{-2\gamma(t-\tau)} (\Delta M)^2 \, N(\tau) \, d\tau. \tag{2}$$

Here $|M|^2(t)$ is the square of the absolute value of the amplitude of the Chandler component at time t. For practical purposes, it is determined for a given year by fitting the polar data of that year to a curve of the form

$$M(t + \delta) = |M|(t) \, e^{i(\omega(t)[\delta+t]+\varphi(t))} + M_0(t).$$

Here δ is varied from $-\frac{1}{2}$ year to $+\frac{1}{2}$ year. $\omega(t)$, $\phi(t)$, $M_0(t)$, as well as $|M|(t)$ are determined by least squares fit to the data. One uses data from which the annual component has been removed in doing this.

To put Equation (2) in a useful form, we differentiate it and get the expression:

$$d\,|M|^2/dt = -2\gamma\,|M|^2(t) + (\Delta M)^2\,N(t). \tag{3}$$

Our problem now is to use Equation (3) to test the claim that earthquakes are associated with wobble breaks. To first order, we guess that such a claim implies that, within an additive constant, the break rate is proportional to a yearly earthquake count. In this paper, the count of magnitude $\geqslant 7.0$ earthquakes will be used. To consider a larger earthquake magnitude range would mean sharply reducing the number of years of reliable data. A significantly smaller magnitude range would (sharply, again) cut down the total number of earthquakes used.

In any case, if earthquake rates and break rates are proportional, it would be possible to relate earthquake and polar data by:

$$d\,|M|^2/dt = -2\gamma\,|M|^2(t) + K\eta(t) - K\eta_0. \tag{4}$$

Here $\eta(t)$ denotes the yearly earthquake count and K is the product of ΔM^2 from Equation (3) and the (unknown) factor of proportionality relating earthquake rate and break rate. η_0 is the number of earthquakes per year which are not associated with wobble excitation. η_0, K and γ are to be determined by a least squares fit to the data. The accuracy of this fit will indicate the aptness of Equation (4) – and, hence, the aptness of the hypothesis that earthquakes are associated with (but not necessarily the source of) wobble excitation.

It should be made clear that Equation (4) has been formulated not as truth, but, rather, as a guide for testing data.

3. Observations

The actual data are presented in Figures 2–4. Figure 2 gives yearly values of $|M|^2(t)$, the square of the absolute value of the Chandler amplitude. Except for the last five years, this data is taken from work done by Iijima (1965) with the ILS-IPMS data. The last five years have been extracted by myself from Mansinha and Smylie's (1968) plots, again of the ILS-IPMS data. It should be noted that Iijima, in separating the Chandler from the annual component, uses a six-year running average – and, hence, the data in Figure 2 have undergone a fair amount of smoothing.

In Figure 3 are plotted values of the yearly rate of change in $|M|^2$. These are determined by the formula:

$$d\,|M|^2/dt_i = \tfrac{1}{2}(|M|_{i+1}^2 - |M|_{i-1}^2).$$

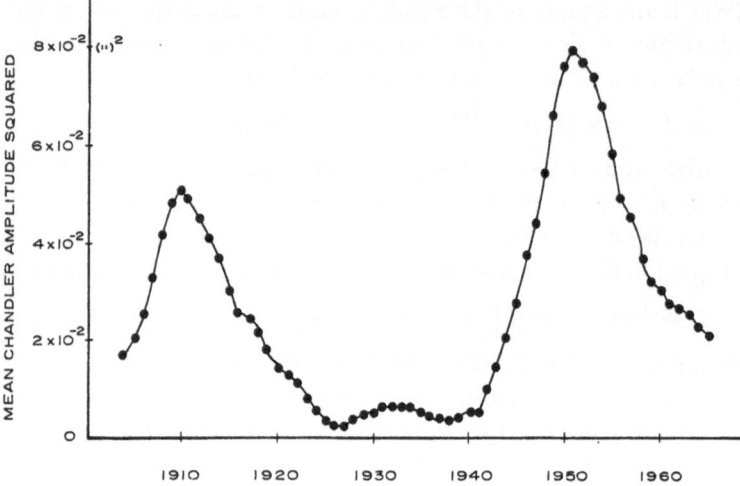

Fig. 2.   Temporal plot of mean Chandler amplitude squared. $(")^2$ represents seconds of arc squared.

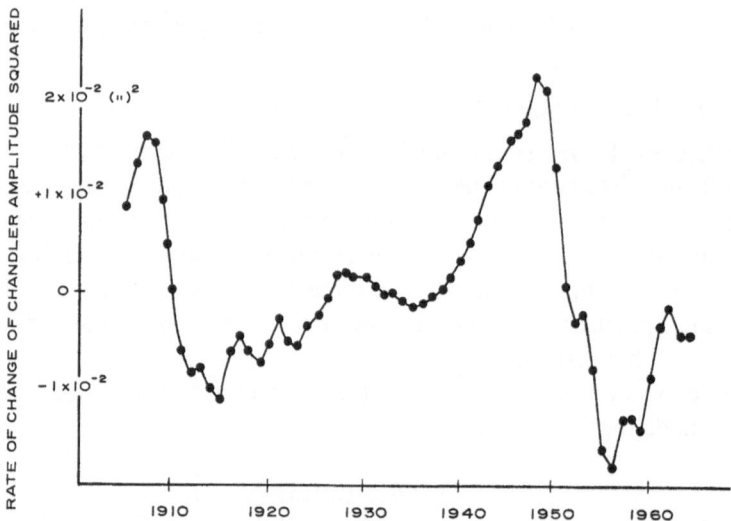

Fig. 3.   Temporal plot of rate of change of Chandler amplitude squared $(")^2$ represents seconds of arc squared.

In Figure 4 are plotted yearly counts of large (magnitude ≥ 7.0) earthquakes. Data here have been smoothed over by the formula:

$$\eta_i' = \tfrac{1}{4}(\eta_{i-1} + 2\eta_i + \eta_{i+1}) \tag{5}$$

to cut down the natural Poisson distribution uncertainty. Considering the amount of smoothing which went into Iijima's polar data, this step is justified.

Earthquake counts up to the year 1951 are from Gutenberg and Richter (1954). For years after 1951, they are taken from *Principal Earthquakes of the World,* by the Coast and Geodetic Survey (1952–65). As Gutenberg and Richter do not consider their data reliable before 1918, counts before that year – while plotted in Figure 4 – are not used in any actual computations for this paper. The years used for computations in this paper are thus 1918–62.

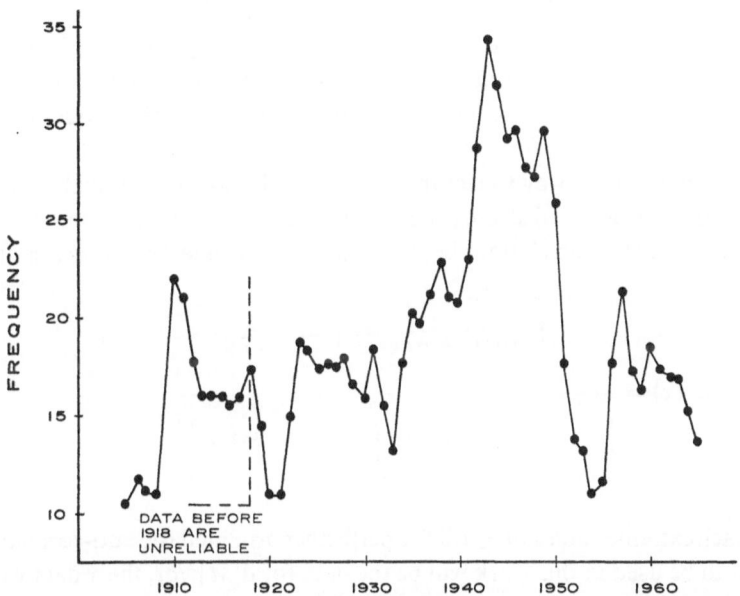

Fig. 4. Number of large earthquakes (mag. ⩾ 7.0) per year. Data has been smoothed by using
$$\eta_i' = (\eta_{i-1} + 2\eta_i + \eta_{i+1})/4.$$

The similarity between Figures 3 and 4 is striking – in itself a strong argument for association between earthquakes and the wobble. It is unfortunate, however, that much of the main feature in the data occurs during World War II, when instruments might be subject to doubt.

In any case, the values of K, γ, and η_0 are determined by least squares fit. In doing the least squares fit, we assume that most of the uncertainty is in the η data – that is, we seek values of K, γ, and η_0 which minimize the sum:

$$\sum (\delta\eta)^2 \equiv \sum_{i=1918}^{1962} \left(\eta_i - \frac{1}{K} \frac{d|M|^2}{dt_i} + \frac{2\gamma}{K} |M|_i^2 - \eta_0 \right)^2. \tag{6}$$

The values thus determined are:

$$K = 1.04 \times 10^{-3} \, (")^2/\text{quake}$$
$$\gamma = (0.05 \pm 0.05) \, \text{yrs}^{-1} \tag{7}$$
$$\eta_0 = 18.5 \, \text{earthquakes/yr}.$$

The value of γ found here is not inconsistent with estimates made by other workers.

Perhaps the most important number, however, is the minimal value of $\sum(\delta\eta)^2$ itself. This gives a good measure of the aptness of Equation (4), which, after all, is what we seek to test. The minimal value of $\sum(\delta\eta)^2$ (for earthquake data *not* smoothed over by Equation (5)) is:

$$\sum(\delta\eta)^2 = 1308. \tag{8}$$

The total number of magnitude $\geqslant 7.0$ earthquakes in the period considered was 908, this number gives one a 'natural' (Poisson distribution) lower limit to $\sum(\delta\eta)^2$ – and one which is comparable to the figure of 1308 obtained by fitting Equation (5) to the data.

Perhaps a more direct way to test the association between earthquakes and Chandler excitation is to take faded cross-correlations between the data shown in Figures 2–4. The 'faded cross-correlation' between two sets of data, x_i and y_i, is defined as:

$$\varrho_{xy}(\tau) = \frac{\displaystyle\int_{-\infty}^{\infty} (x(t) - \bar{x})(y(t+\tau) - \bar{y})\,dt}{\left[\displaystyle\int_{-\infty}^{\infty} (x(t) - \bar{x})^2\,dt \cdot \int_{-\infty}^{\infty} (y(t) - \bar{y})^2\,dt\right]^{1/2}} \tag{9}$$

and will reach extreme values of ± 1.0 for perfectly correlated (or anti-correlated) data. The y data to be used in this work will be the data for $d|M|^2/dt$, the x data will be that for $|M|^2(t)$ in Figure 5 and that for $\eta(t)$ in Figure 6.

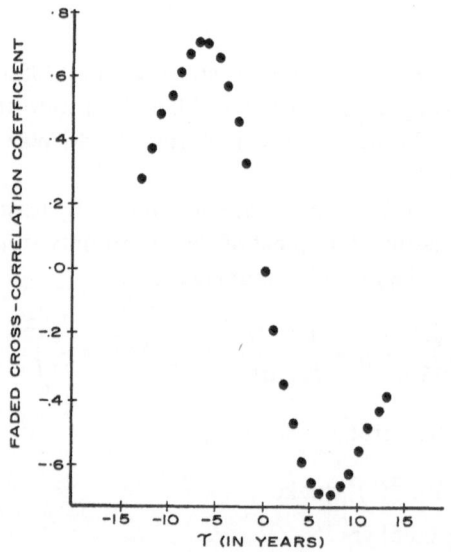

Fig. 5. Faded cross-correlation of $|M|^2$ with $d|M|^2/dt$.

In Figure 5 the cross correlation between $|M|^2$ and $d|M|^2/dt$ is plotted. As might be expected, it has a fairly high peak which is 'faded off' zero by about 6 years – a number which should be roughly $\frac{1}{4}$ the period of the primary Fourier component of the data.

In Figure 6 the cross correlation between η' (yearly large earthquake counts smoothed over by Equation (7)) and $d|M|^2/dt$ is plotted. Its peak value is high: 0.8. Using confidence belts from Dixon and Massey (1957) and assuming that the number of independent parameters is small (this because the data in Figures 2 and 3 are slowly varying), we find that there is roughly a 10% chance data sets randomly correlating as well as or better than they do.

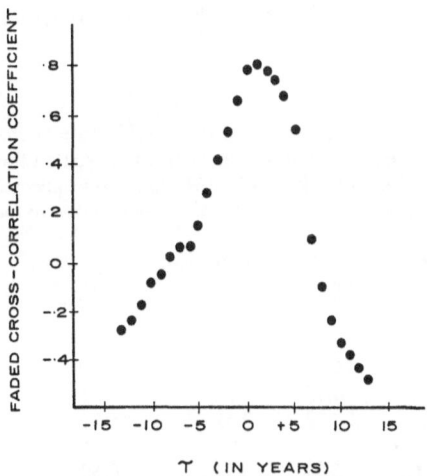

Fig. 6.   Faded cross-correlation of the yearly large earthquake (mag. $\geqslant 7.0$) count with $d|M|^2/dt$.

The fact that the peak value in Figure 6 occurs with virtually zero fading is not surprising: it results from the fact that the primary Fourier component of the data is comparable with the Chandler damping time. To understand this, note that, if:

$$\eta(t) - \eta_0 = \eta_c e^{i\omega t}$$

then, by Equation (4):

$$\frac{d|M|^2}{dt} = \frac{K\eta_c\omega}{(4\gamma^2 + \omega^2)^{1/2}}\, e^{i\omega[t - (\theta/\omega) + (\pi/2\omega)]} \quad \left[\theta = \tan^{-1}\!\left(\frac{\omega}{2\gamma}\right)\right].$$

But, remembering that we are considering only data from after 1918, it may be seen, by inspection of Figures 3, 4, and 6, that $\omega \cong (2\pi/25)$ years^{-1}. The phase difference between $d|M|^2/dt$ and η is $\varDelta \equiv (\pi/2\omega) - (\theta/\omega)$. If we assume that γ, the Chandler damping coefficient is $\frac{1}{30}$ years^{-1} then:

$$\varDelta \equiv (\pi/2\omega) - (\theta/\omega)$$
$$\cong \tfrac{2.5}{4} - (25/2\pi)\cdot\tan^{-1}[2\pi \cdot 30/25] \approx 0.$$

4. Further Investigations

In order to better understand the association between earthquakes and wobble excitation, various subsets of the total large (magnitude $\geqslant 7.0$) earthquake count were
also cross-correlated with $d|M|^2/dt$. In all cases the peak correlation occurred at
virtually zero fading.

The results of this work are listed in Table I. In the first column of Table I are
listed the types of earthquakes considered. Except for the last four these are self-
explanatory. For the last four: 'North-East' earthquakes are those with epicenters
at latitudes north of the equator and longitudes east of Greenwich, 'South-East'
earthquakes are those with epicenters south of the equator and east of Greenwich,
and so on.

TABLE I

Correlations of earthquake rates with $d|M^2|/dt$ (rate of change of the Chandler radius
squared). Expected correlation coefficients are based on the observed correlation of all
magnitude $\geqslant 7.0$ earthquakes with $d|M^2|/dt$. If the expected and observed correlations
differ by more than 0.15, I have noted so in the last column

Type of quake	Mean # per yr.	Expected corr.	Observed corr.	Do exp. and obs. correlations differ by more than 0.15?
Mag. $\geqslant 7.0$	19.32	–	0.80	–
Mag. $\geqslant 7.5$	4.73	0.56	0.36	yes
Depth $\geqslant 70$ km mag. $\geqslant 7.0$	5.82	0.59	0.70	no
$7.2 \leqslant$ mag. $\leqslant 7.3$	5.07	0.57	0.59	no
North-East mag. $\geqslant 7.0$	7.53	0.72	0.67	no
North-West	2.80	0.45	0.52	no
South-East	5.51	0.58	0.76	yes
South-West	4.18	0.52	0.38	no

In the second column of Table I are listed the mean number of earthquakes per
year of the given types. In the third column are listed expected correlation coefficients.
These are estimated by assuming that, for each type, the given average number of
earthquakes per year are drawn arbitrarily from the set of all large earthquakes. The
procedure for making the estimate involved assuming that:

$$\sigma_x^2 = \sigma_{x_e}^2 + \sigma_{x_r}^2$$

where, in the language of Equation (9),

$$\sigma_x^2 \equiv \int\limits_{-\infty}^{\infty} (x(t) - \bar{x})^2 \, dt.$$

$\sigma_{x_r}^2$ is the contribution to σ_x^2 from 'real' fluctuations in the data, $\sigma_{x_e}^2$ is the contribubution from random uncertainty. One further assumes that random uncertainty in the y data (the $d|M|^2/dt$ data) is negligible and that 'real' fluctuations are perfectly correlated with those in the x data. Hence, considering Equation (9):

$$\varrho_{xy} \cong \frac{\sigma_{x_r} \cdot \sigma_{y_r}}{[(\sigma_{x_r}^2 + \sigma_{x_e}^2)(\sigma_{y_r}^2 + \sigma_{y_e}^2)]^{1/2}} \approx \sigma_{x_r}/(\sigma_{x_r}^2 + \sigma_{x_e}^2)^{1/2}. \tag{10}$$

When yearly counts of earthquakes of magnitude $\geqslant 7.0$ are used for the x data, one finds that $\varrho_{xy} = 0.8$. This enables us to use (10) to express σ_{x_e} in terms of σ_{x_r}:

$$\sigma_{x_e} \approx \tfrac{3}{4}\sigma_{x_r}. \tag{11}$$

We now consider an arbitrarily chosen subset of the large earthquake count, containing an average of η' earthquakes per year. To estimate the correlation coefficient between this earthquake data and $d|M|^2 dt$ (let us call it $\varrho_{x'y}$) we assume that $\sigma_{x'_r} = (\eta'/19.32)\,\sigma_{x_r}$ and that (assuming, once again, a Poisson distribution of yearly earthquake counts about their expected values) $\sigma_{x'_e} = (\eta'/19.32)^{1/2} \cdot \sigma_{x_e}$, where 19.32 is the average number of all magnitude $\geqslant 7.0$ earthquakes per year. Using these assumptions and Equation (11), we have:

$$\begin{aligned}
\varrho_{x'y} &= \sigma_{x'_r}/(\sigma_{x'_r}^2 + \sigma_{x'_e}^2)^{1/2} \\
&= \frac{\sigma_{x_r} \cdot (\eta'/19.32)}{[\sigma_{x_e}^2(\eta'/19.32) + \sigma_{x_r}^2 \cdot (\eta')^2/(19.32)^2]^{1/2}} = \frac{1.33}{\sqrt{(1.78 + 19.3/\eta')}}.
\end{aligned} \tag{12}$$

Taking this formula and the various values of η' listed in column 2 of Table I, the expected correlation coefficients given in column 3 of Table I are computed. It should be emphasized that these expected correlations are estimates and not much more. Nonetheless they do seem to be the best way to take into account differences in the mean number of earthquakes per year.

In column 4 of Table I are listed observed correlations between the various types of earthquakes and $d|M|^2/dt$. In the last column, it is noted whether or not the observed and expected correlations differ by more than 0.15.

The most important finding presented in Table I is the poorer than expected correlation of very large earthquakes (magnitude $\geqslant 7.5$) with $d|M|^2/dt$. This seems to conflict with Mansinha and Smylie's theory that earthquakes themselves are the source of wobble excitation. According to their (1967) estimates, almost all of the excitation would be by very large (magnitude $\geqslant 7.5$) earthquakes. Yet the poorer than expected correlation of yearly counts of these earthquakes with $d|M|^2/dt$ argues to the contrary.

Table I also shows that the combined yearly count of deep and intermediate earthquakes correlates significantly *better* than expected with $d|M|^2/dt$. The South-Eastern quadrant, which has a disproportionate number of such earthquakes, also correlates significantly better than expected. One conclusion may be that the wobble is not directly excited by earthquakes, but is, instead, excited by events occurring deeper within the earth – events which, however, are also associated with earthquakes.

Discussion

The primary findings presented in this paper are:

(a) The good correlation between $d|M|^2/dt$ (the rate of change in the Chandler amplitude squared) and $\eta(t)$ (yearly count of magnitude $\geqslant 7.0$ earthquakes) indicates that earthquakes are associated with wobble excitation.

It should be pointed out that, in years to come, the Chandler amplitude may develop in such a manner as to make the $-\gamma(|M|^2(t))$ term dominate the right side of Equation (4). In this case, we would no longer find strong correlation between $d|M|^2/dt$ and $\eta(t)$. Nonetheless, if Equation (4) still continued to fit the data well, we would conclude that the theory of earthquake-wobble association remained supported.

(b) The poor correlation of $d|M|^2/dt$ with yearly counts of very large (magnitude $\geqslant 7.5$) earthquakes and, on the other hand, the good correlation of $d|M|^2dt$ with deep and intermediate earthquakes suggests a deeper mechanism for wobble excitation – one which triggers earthquakes as well as maintaining the wobble.

(c) A value of $(5\pm5)\times10^{-2}$ yrs^{-1} was obtained for the Chandler damping coefficient. While the precision is not exactly overwhelming, this value is consistent with estimates of other workers.

Note. R. W. Tanner of the Dominion Observatory, Ottawa has drawn the author's attention to recent minor changes in Iijima's data for the years 1949.0 to 1955.0. The author has not yet incorporated these into his calculations.

Acknowledgements

The author is grateful to Professors R. H. Dicke and W. J. Morgan of Princeton for their encouragement and advice in the course of this work. The business of working in a new area is a difficult one – and would have been impossible without their guidance.

References

Dixon, W. J. and Massey, F. J.: 1957, *Introduction to Statistical Analysis*, McGraw-Hill, New York, 2nd ed., p. 200.
Gutenberg, B. and Richter, C. F.: 1954, *Seismicity of the Earth*, 2nd ed., Princeton University Press, Princeton, N.J.
Iijma, S.: 1965, *Ann. Tokyo Astron. Observ.* **9**, 155–81.
Mansinha, L. and Smylie, D. E.: 1967, *J. Geophys. Res.* **72**, 4731–4743.
Smylie, D. E. and Mansinha, L.: 1968, *J. Geophys. Res.* **73**, 7661–7673.
Principal Earthquakes of the World, United States Earthquakes, U.S. Coast and Geodetic Survey (1952 to 1965).

AN INVESTIGATION OF PZT OBSERVATIONS FOR EVIDENCE OF THE EXISTENCE OF POLAR DISTURBANCES DUE TO LARGE EARTHQUAKES

N. P. J. O'HORA and D. V. THOMAS

Royal Greenwich Observatory, Hailsham, Sussex, England

Abstract. The latitude and time observations of the Herstmonceux and Ottawa PZT's for the years 1958.0–1969.0 have been jointly analysed to discover whether discontinuities or other systematic changes in the results can be associated with the incidence of major earthquakes.

In a recent paper, Smylie and Mansinha (1968) published results of observational tests of the theory that large earthquakes excite the Chandler Wobble and produce the observed secular polar shift. In a discussion of the published coordinates of the pole of rotation, significant correlation was shown to exist between the epochs of discontinuities in the circular arcs representing the free polar motion and the dates of large earthquakes. The two independent tables of polar coordinates published by the BIH and the IPMS for the years 1957–67 were analysed and evidence was derived from both which showed that the Chandler Wobble is affected by earthquakes. There were, however, notable differences in the results obtained from the two sets of data, and it seemed desirable to examine direct observations of the polar motion to discover whether polar disturbances could be associated with earthquakes. It appeared reasonable to assume that the small changes which were shown to exist in the published polar coordinates should be detectable in the accurate observations obtained with a PZT.

The Herstmonceux PZT has been in use since 1955 (Thomas, 1964). In 1968 the observations made up to the end of 1967 were discussed in a determination of the errors in the adopted positions of the catalogue stars. The errors were evaluated by two independent methods, which gave results in close agreement with each other. The time and latitude observations from 1958.0, when the current observing programme began, were therefore re-reduced using the corrected star positions. In this way an accurate series of observations over the years for which the polar coordinates had been examined by Smylie and Mansinha was made available by the Herstmonceux PZT. However, there is an unfortunate gap in the data from 1962.6 to 1963.6, when the instrument was out of service during a period when the time and latitude service of the Royal Greenwich Observatory was based on the results of the Herstmonceux astrolabe.

A method was first developed for the investigation of the effects of polar disturbances on the Herstmonceux PZT observations. The time observations were expressed as measures of UTO-GA2, with a suitable ephemeris removed (GA2 is the atomic time scale of the Royal Greenwich Observatory). Annual and semi-annual terms were removed from both the time and latitude observations to free them from the effects of the annual forced term in the polar motion, and from any residual periodic errors in the adopted star places. It was assumed that in intervals between polar disturbances

L. Mansinha et al. (eds.), Earthquake Displacement Fields and the Rotation of the Earth, 169–175.
All Rights Reserved. Copyright © 1970 by D. Reidel Publishing Company, Dordrecht-Holland

the free pole of rotation revolved about the secular pole in an anti-clockwise direction with an angular speed of $2\pi/433$ radians per day (433 days being the Chandlerian nutation period recently deduced by Jeffreys, 1968), and that the rate of rotation of the Earth remained constant. It was also assumed that a polar disturbance, such as might be caused by a large earthquake, could change the position of the secular pole, the radius of the free polar motion, and both the phase and the rate of rotation of the Earth. A system of equations was derived, similar to those described below for the case of observations obtained with two instruments, for the determination of the projection of the coordinates of the secular pole on the meridian of the observatory, the radius of the polar motion, and the rate of rotation of the Earth. The use of observations made only by a single instrument made it impossible to distinguish between abrupt phase changes in the Earth's rotation and sudden displacements of the pole of rotation in the direction perpendicular to the meridian of the observatory.

Information on earthquakes was obtained from a table compiled at the Geophysical Laboratories, Edinburgh, listing all earthquakes of magnitude >7.0. It is note-worthy that there are comparatively large differences in the estimated magnitudes of an earthquake published by different reporting agencies. The magnitudes quoted in this paper are the means of the highest and lowest estimates supplied by various agencies. Earthquakes considered as 'large', for the purpose of this paper, are tabu-lated in Table I, which contains all reported earthquakes with a mean magnitude greater than 7.3.

Four separate analyses of the time and latitude data were made as follows:

(1) Using all earthquakes given in Table I.

(2) Using only earthquakes of magnitude $\geqslant 7.7$.

(3) Using the same number of control events as the number of earthquakes in the first analysis.

(4) Using the same number of control events as the number of earthquakes in the second analysis.

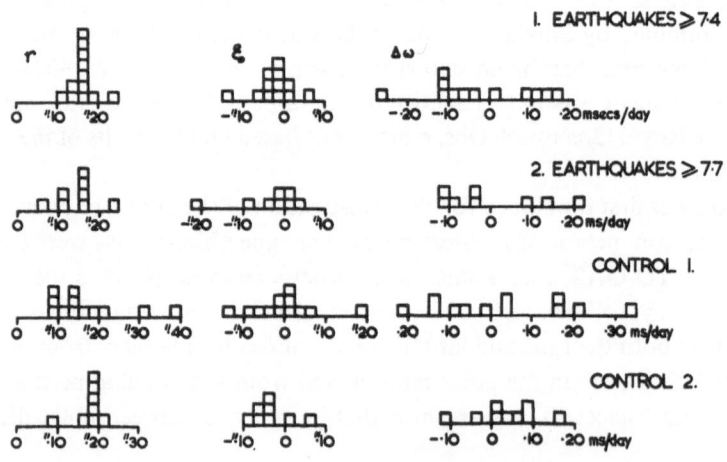

Fig. 1.   Herstmonceux PZT polar disturbances.

TABLE I

Earthquakes, 1958.0–1969.0, with mean magnitudes > 7.3

Civil date		Julian date	Magnitude	Longitude	Latitude
1958, Jan.	1	243 6223	7.8	79W	15N
Jul.	10	6395	7.8	137W	59N
Nov.	6	6514	8.2	148E	44N
1959, Apr.	26	6685	7.6	121E	25N
May	4	6693	8.0	160E	52N
Sep.	14	6826	7.7	177W	28S
1960, Mar.	26	7020	7.5	142E	37N
May	22	7077	7.9	74W	39S
1961, Apr.	13	7403	7.6	78E	39N
June	11	7462	7.6	55E	28N
July	23	7504	7.4	168E	18S
Sep.	8	7551	7.4	27W	56S
1962, May	15	7800	7.3	129E	6S
Sep.	1	7909	7.4	49E	50N
1963, Sep.	17	8290	7.4	165E	10S
Oct.	3	8306	7.7	132E	32N
	13	8316	8.0	150E	44N
Dec.	18	8382	7.4	177W	25S
1964, Mar.	28	8483	8.4	148W	61N
1965, Feb.	26	8818	7.6	152E	36N
Mar.	14	8834	7.6	71E	51N
	30	8850	7.4	178E	51N
Aug.	1	8974	7.4	158E	56S
	23	8996	7.4	96W	16N
1966, Mar.	12	9197	7.5	122E	24N
June	15	9292	7.4	161E	10S
Dec.	28	9488	7.6	71W	26S
	31	9491	7.6	166E	12S
1967, Jan.	5	9496	7.4	103E	48N
1968, Jan.	29	9885	7.4	147E	44N
Apr.	1	9948	7.5	132E	32N
May	16	9993	8.2	143E	41N

The dates of the control events for the third and fourth analyses were obtained by integrating the intervals between earthquakes in a random order. The main results of the investigation are displayed in Figure 1, which shows for each of the four analyses histograms of values of the following variables: r, the radius of the free polar motion, ξ, the displacement of the secular pole along the Greenwich meridian, and $\Delta\omega$, the excess rate of rotation of the Earth.

The notable feature of these diagrams is the high selectivity in the values of r derived from solutions of the data between earthquake dates. If this selectivity is not accidental it suggests that there is a preferential radius in the free polar motion and that this radius remains constant between the occurrence of large earthquakes. There is little significant difference between the displacements of the secular pole, and the changes in the rate of rotation of the Earth, arising from earthquakes and those which could be associated with the control events. The accuracy of the solutions between the control events is not significantly less than that of solutions between earthquakes.

Although this investigation did not yield definite information on the influence of earthquakes on polar disturbances, the results seemed to be sufficiently encouraging to warrant strengthening the analysis by the inclusion of observations from an observatory separated from Herstmonceux by approximately 90° of longitude. It was decided to use the results of the Ottawa PZT in conjunction with those of the Herstmonceux PZT, as these instruments differ by 76° in longitude and only 5° in latitude. The Ottawa observations were prepared for analysis in the same way as the Herstmonceux observations had been prepared in the earlier investigation.

Let the current position of the secular pole be given by (ξ, η) referred to axes through the adopted mean pole with $+\xi$ in the direction of Greenwich and $+\eta$ in the direction 90°W of Greenwich. It is assumed that the secular pole remains fixed between the dates of earthquakes (or control events) and that the motion of the free pole of rotation is an anti-clockwise rotation about the secular pole in a circular arc with radius r, with an angular speed of v radians/day. The position of the pole of rotation with respect to the secular pole is then given by

$$x = r \cos \theta$$
$$y = r \sin \theta$$

where the x, y axes are parallel to the ξ, η axes, respectively, the (x, y) origin is the secular pole, and θ is measured clockwise from the direction $+x$. The effect of the displacement of the pole of rotation from the adopted mean pole on the latitude of an observatory with coordinates (ϕ, λ) is given by

$$\Delta\phi = X \cos \lambda + Y \sin \lambda$$

and the effect on the observed time is given by

$$\Delta t = - \tfrac{1}{15} \tan \phi (X \sin \lambda - Y \cos \lambda),$$

where $X = x + \xi$, $Y = y + \eta$, and $\Delta\phi$ and Δt are increases in the observed values. When X and Y are expressed in seconds of arc Δt is in seconds of time. The values of x and y for any date can be computed in terms of their values at an arbitrary initial epoch. During the interval between two polar disturbances, the following equations of condition are provided by latitude and time observations, respectively, of unit weight:

$$\xi \cos \lambda + \eta \sin \lambda + a \cos (v\tau + \lambda) + b \sin (v\tau + \lambda) = \Delta\phi$$
$$- \xi \sin \phi \sin \lambda + \eta \sin \phi \cos \lambda - a \sin \phi (\sin v\tau + \lambda) + b \sin \phi \cos (v\tau + \lambda)$$
$$+ 15c \tau \cos \phi + e = 15 \cos \phi \Delta t,$$

where the unknowns are ξ, η, a, b, c and e. The difference between the date of an observation and a reference epoch, taken to be the middle of the inter-earthquake interval, is denoted by τ, and a and b are the values of x and y respectively at the reference epoch. The terms involving c and e result from changes in the rate and in the phase of the rotation of the Earth.

In the joint analyses of the results of the two instruments, only those earthquakes in Table I with magnitudes $\geqslant 7.6$ were used. The same method as before was used to determine the dates of control events. In the first joint analysis the latitude and time results of the two PZT's were discussed simultaneously in solutions of the observations between earthquake dates. No solutions were attempted when the intervals between earthquakes were less than 30 days or when the total weight of the observations was less than an arbitrary lower limit. Larger values were obtained for the probable errors of unit weight than had been anticipated. This was a reflection of the fact that the time and latitude values computed for the two observatories from the values of the parameters determined in a solution failed to agree satisfactorily with the observations.

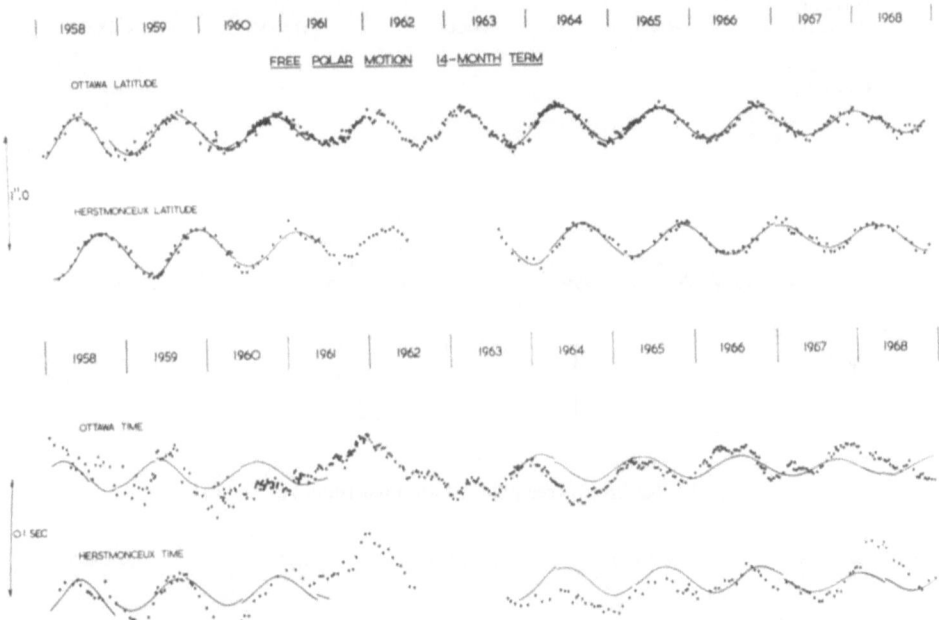

Fig. 2. Comparison of computed and observed variations in latitude and time for the Herstmonceux and Ottawa PZT's.

A second joint analysis was therefore made using only the latitude observations of the two instruments between earthquake dates. This analysis gave results with smaller probable errors. The computed latitudes based on the solutions are in very good agreement with the observed values, as can be seen in Figure 2 where the computed latitudes are represented by curves through the plotted points. Each point represents an observation of weight 100, the result of approximately 100 star transits. The computed times, derived from the latitude solutions, and the observed times are also compared in Figure 2. It has not yet been possible to examine the discrepancies between the observed and computed times to see to what extent they are due to non-uniformity in the rate of rotation of the Earth, and to what extent they are the result of local variations in the time results of either or both observatories.

A third joint analysis was performed using control events, instead of earthquakes. Comparisons of the results of the second and third joint analyses show that the results obtained in solutions of the observations between earthquake dates are not significantly different from those obtained in solutions between random events. The displacements of the secular pole on earthquake dates were not larger than those determined for random dates. The range in the values of the radius of the free polar motion was approximately equal in the two analyses, as is apparent in Figures 3 and 4, in which the values of r determined for each interval are plotted.

The conclusion to be drawn from the investigation of the combined results of the

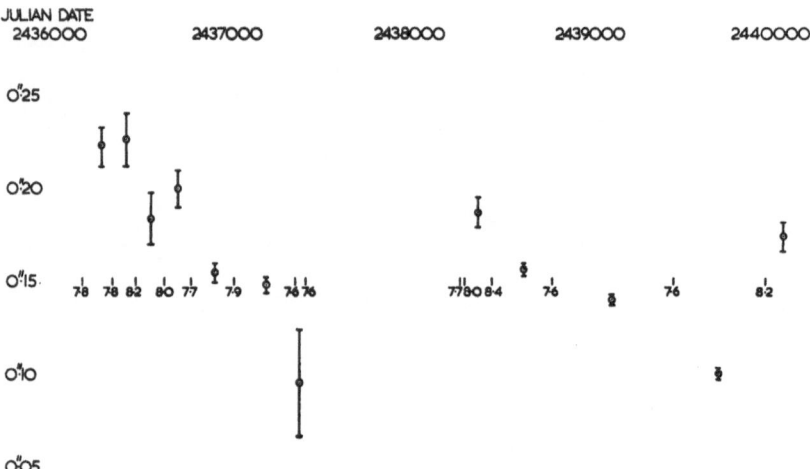

Fig. 3.   Radius of free polar motion (earthquakes ≥ 7.6).

Fig. 4.   Radius of free polar motion (control events).

Herstmonceux and Ottawa PZT's at this stage must be that they offer no support for earthquakes as the mechanism for the excitation of the Chandlerian nutation. However, this negative result may be due to an unfortunate choice of what constitutes a 'large' earthquake. It is therefore intended to repeat the analysis with both higher and lower limiting earthquake magnitudes. Further examination of the time results may also throw some light on the problem.

Acknowledgements

We are indebted to the Dominion Observatory, Ottawa, and to the Geophysical Laboratories, Edinburgh, for supplying data used in the investigation. We should also like to express our gratitude to Mr. J. V. Carey and Miss P. J. Watson for their assistance in the preparation of the data, to Mrs. M. J. Everest, Mrs. A. Strong and Mr. D. A. Calvert for preparing the diagrams and to Mrs. D. E. Hobden for her cooperation in the programming. The computations were carried out on the ICL 1909 computer of the Royal Greenwich Observatory. We are grateful to Mr. H. M. Smith for his encouragement in the work and to the Astronomer Royal for permission to publish the paper.

References

Jeffreys, Sir Harold: 1968, *Monthly Notices Roy. Astron. Soc.* **141**, 255.
Smylie, D. E. and Mansinha, L.: 1968, *J. Geophys. Res.* **73**, 7661.
Thomas, D. V.: 1964, Roy. Observ. Bull. No. 81.

A RE-EXAMINATION OF CORE-MANTLE COUPLING AS THE CAUSE OF THE WOBBLE

F. D. STACEY

Physics Department, University of Queensland, Brisbane, Australia

Abstract. If the coupling of mantle wobble to the core is independent of other motions, then the conclusion of Rochester and Smylie (1965) that core-mantle coupling is inadequate to explain the wobble, must be accepted. However, evidence is accumulating that the geomagnetic dynamo is driven by differential precessional torques between the core and mantle, as proposed by Malkus (1963, 1968). The resulting core motion is turbulent, in the sense that the pattern of motion is not constant in time, and his leads to a cross-coupling between precession and wobble. By virtue of the misalignment of rotational axes of the core and mantle, estimated to be about 40 arc sec, this feeds very much more energy into the wobble than is possible in the absence of precession.

On the other hand, if the mantle is the only source of wobble energy then it must almost certainly be the only sink. For a wobble Q of 30, this demands a mechanical Q of 4 for the mantle and this is implausibly low.

1. Introduction

Simple, order-of-magnitude arguments indicate that core motion and its coupling to the mantle are much more complex than existing theories, which evidently omit an important factor. In this paper it is argued that the missing factor is precession and that the wobble energy is derived from core-mantle coupling by cross-coupling to precession. The approach taken is very simple minded and appeals largely to the difficulties with existing theories. The required pattern of core motion is not specified, but the important feature is an angular deviation of its axis of rotation from that of the mantle, which Aoki (1969) found to be 2×10^{-4} radian.

2. Energy Dissipation by Core-Mantle Coupling – the Linear Coupling Model

Consider first the simple picture of a core and mantle, having slightly different angular velocities, and with a linear coupling between them, that is the mutual torque is proportional to the angular velocity difference ($\Delta\omega$). Assuming the coupling to be electromagnetic, the current flow is proportional to the angular velocity difference and the energy dissipation (dE_R/dt) to the square of current flow. Thus we can define a coupling coefficient K_R by

$$\frac{dE_R}{dt} = - K_R (\Delta\omega)^2 . \tag{1}$$

Then

$$K_R = \frac{1}{\tau_R ({}^1/I_m + {}^1/I_c)} = 1.2 \times 10^{36} \text{ erg secs} , \tag{2}$$

where τ_R is the decay time for ($\Delta\omega$), determined from the spectrum of length of day

L. Mansinha et al. (eds.), Earthquake Displacement Fields and the Rotation of the Earth, 176–180.
All Rights Reserved. Copyright © 1970 by D. Reidel Publishing Company, Dordrecht-Holland

(l.o.d.) fluctuations to be about 1.6 years, and I_m, I_c are moments of inertia of the mantle and core respectively.

Similarly we can suppose that the wobble excitation is due to a linear coupling to the core, with coupling coefficient K_w:

$$\frac{dE_w}{dt} = - K_w (\Delta\omega')^2 , \tag{3}$$

where

$$\Delta\omega' \approx \omega_w \alpha , \tag{4}$$

α being the angular amplitude of wobble and ω_w its angular frequency. But

$$E_w = \tfrac{1}{2} I_m H \omega^2 \alpha^2 , \tag{5}$$

ω being the angular frequency of rotation and H the dynamical ellipticity. Differentiating (5) and comparing with (3) we obtain

$$K_w \approx \frac{I_m H}{\tau_w} \left(\frac{\omega}{\omega_w}\right)^2 = 10^{39} \text{ erg secs} , \tag{6}$$

where τ_w is the wobble decay time, 12.4 years. Comparison of (2) and (6) shows that if both l.o.d. fluctuations and wobble are to be explained in terms of the coupling, $K_w > 800 \, K_R$. This is a simple way of looking at the result of Rochester and Smylie (1965), who concluded that wobble was not explained by core-mantle coupling. It is still assumed that the coupling explains the l.o.d. fluctuations. This method of presenting the problem follows that given in greater detail elsewhere by the author (Stacey, 1969).

3. The Mantle as a Sink of Wobble Energy

If the mantle (and crust) are postulated as a source of wobble energy, because core-mantle coupling is inadequate, then the mantle must also be the only effective sink of wobble energy. The required mechanical Q of the mantle is easily calculated.

$$\frac{2\pi}{Q} = \frac{\Delta E}{E_s} = \frac{\Delta E}{E_w} \cdot \frac{E_w}{E_s} = \frac{2\pi}{Q_w} \cdot \frac{E_w}{E_s} , \tag{7}$$

where ΔE is the energy loss per cycle, E_s is the strain energy associated with the wobble motion, E_w is the total energy of the motion (Equation (5)) and Q_w is the observed Q for the wobble (about 30). E_w/E_s can be expressed in terms of the observed wobble period, $\tau_0 = 430$ days, compared with the period for a hypothetical rigid earth, $\tau_R = 305$ days:

$$\frac{E_w}{E_s} = \frac{I_m \omega^2}{\mu H V \left(1 - \dfrac{\tau_R}{\tau_0}\right)^2} = 7.5 , \tag{8}$$

where μ is the effective rigidity of the mantle, volume V. From Equation (8) the required mantle Q at the wobble frequency is $Q=4$. From the variation of Q over seismic and free oscillation frequencies there is no evidence to suggest that Q decreases so rapidly at lower frequencies as to make this value plausible. There must therefore be another sink and also another source of wobble energy.

4. Cross-Coupling of Wobble and Precession

By virtue of the different dynamical ellipticities of the core and mantle, they tend to precess at different rates, the core slower than the mantle. The observed rate of precession is due to the ellipticity of the earth as a whole and is dominated by the mantle; the core, being coupled electromagnetically to the mantle, follows the precession, but with a slight lag, which it is of interest to calculate. The rotation of core and mantle about slightly different axes causes a stirring of the core, which Malkus (1963, 1968) proposed as the driving mechanism for the geomagnetic dynamo. We are here interested in the coupling to the mantle of irregularities in the core motion. The details are unclear, but an order-of-magnitude calculation shows what is possible.

We first take from the work of Aoki (1969) the value of the misalignment of core and mantle axes ($\varepsilon=2\times10^{-4}$ radian). What Aoki showed was that this misalignment explained both the slower rotation of the core (westward drift) and the disparity between the observed and calculated rates of change of the obliquity (at present 23.5°). Then the angular velocity difference between the core and mantle due to the precessional misalignment is

$$\Delta\omega_p = \omega\varepsilon. \tag{9}$$

It is important to the argument that the coupling be slightly non-linear, i.e., the energy dissipation by precession be not quite proportional to $(\Delta\omega_p)^2$, due perhaps to deformation within the core being a function of $\Delta\omega_p$. However, to calculate an order-of-magnitude coupling, let us assume first that the coupling is linear; then

$$\left(\frac{dE}{dt}\right)_p = -K_p(\Delta\omega_p)^2 \tag{10}$$

and if we equate $(dE/dt)_p$ to the 'missing' 10^{19} ergs/sec, which is indicated by the disparity between dissipation by tidal friction, as deduced from astronomical observations (2.74×10^{19} ergs/sec – Munk and MacDonald, 1960) and the best marine tidal estimate (1.5×10^{19} ergs/sec – Miller, 1966), we obtain $K_p=5\times10^{34}$ erg sec. This is an upper limit because the marine tidal estimate is sufficiently uncertain that it may not actually be in disagreement with the astronomical one. By taking this value we are assigning 10^{19} ergs/sec to the precessional dynamo, as in Malkus's theory of the origin of the geomagnetic field. It is of interest to note that K_p is necessarily smaller than K, the coupling for rotational differences about a common axis. This is expected because the field penetration of the mantle oscillates diurnally due to the precessional misalignment but varies only very slowly due to angular velocity differences, thus penetrating much further in the latter case.

Now we must consider the effect of superimposing the wobble motion upon the misaligned axial rotations. If the coupling were perfectly linear then the two energy dissipations would be independent, but an essential feature of the precessional dynamo is that the core motion is turbulent (i.e., time dependent) and this implies immediately that it is driven by a non-linear coupling, in which the mutual torque between core and mantle is not linear with angular velocity difference but depends more strongly upon it. We can suppose that in addition to the linear term in the instantaneous energy dissipation we can add a non-linear term:

$$\frac{dE}{dt} = - K_p(\Delta\omega_p + \Delta\omega_w)^2 - A(\Delta\omega_p + \Delta\omega_w)^3, \tag{12}$$

where A is a new constant. It is of interest to compare the magnitudes of the two terms in Equation (12), with the assumption that the second (non-linear) term provides cross-coupling adequate to excite (and damp) the wobble. The first (linear) term contributes nothing to cross-coupling, because $\Delta\omega_p$ and $\Delta\omega_w$ are both cyclic, with different periods.

By virtue of the superposition of the motions they dissipate energies proportional to their angular velocities, so that the energy dissipation by the wobble alone is (neglecting the linear contribution which was already shown to be negligible)

$$\left(\frac{dE}{dt}\right)_w = - \frac{\Delta\omega_w}{\Delta\omega_p + \Delta\omega_w} \cdot A(\Delta\omega_p + \Delta\omega_w)^3$$

$$\approx - \frac{\Delta\omega_w}{\Delta\omega_p} \cdot A(\Delta\omega_p^3). \tag{13}$$

It is thus equal to $(\Delta\omega_w/\Delta\omega_p) = 2 \times 10^{-5}$ of the energy dissipation by precession due to the non-linear terms in (9). But the required value of $(dE/dt)_w = 1.2 \times 10^{13}$, as may be verified by substitution of values in Equation (3). Thus we require the non-linear contribution to precessional dissipation to be

$$A(\Delta\omega_p + \Delta\omega_w)^3 = \frac{1.2 \times 10^{13}}{2 \times 10^{-5}} = 6 \times 10^{17} \text{ ergs/sec}. \tag{14}$$

But we assigned 10^{19} ergs/sec to the precessional dissipation and so the non-linear term in Equation (12) is required to account for 6% of the total dissipation. This is entirely reasonable.

5. Conclusion

Although cross-coupling of precession and wobble is not here rigorously demonstrated, it must follow from the expected non-linearity in core-mantle coupling. The energy dissipation by precession may be as much as 10^6 times the energy required to sustain the wobble against a damping with a Q of 30. The opportunity of a spill-over of precessional energy is thus so great that this must be regarded as the most plausible of the wobble excitation mechanisms.

Acknowledgement

Constructive criticism of a draft of this paper by M. G. Rochester is gratefully acknowledged.

References

Aoki, S.: 1969, 'Friction between Mantle and Core of the Earth as a Cause of the Secular Change in Obliquity', *Astron. J.* **74**, 284.

Malkus, W. V. R.: 1963, 'Precessional Torques as the Cause of Geomagnetism', *J. Geophys. Res.* **68**, 2871.

Malkus, W. V. R.: 1968, 'Precession of the Earth as the Cause of Geomagnetism', *Science* **160**, 259.

Miller, G. R.: 1966, 'The Flux of Tidal Energy out of the Deep Oceans', *J. Geophys. Res.* **71**, 2485.

Munk, W. H. and MacDonald, G. J. F.: 1960, *The Rotation of the Earth. A Geophysical Discussion,* Cambridge University Press, Cambridge.

Rochester, M. G. and Smylie, D. E.: 1965, 'Geomagnetic Core-Mantle Coupling and the Chandler Wobble', *Geophys. J. R. Astr. Soc.* **10**, 289.

Stacey, F. D.: 1969, *Physics of the Earth,* Wiley, New York.

Discussion

Rochester: I would like to point out that, unless somebody has been working on this recently and not published anything, there is as yet no way that we can rule out the oceans as a sink of the wobble. I also think that the discrepancy in the obliquity cannot really be a true secular term because if it is, the obliquity would disappear in about 20 m.y.

Stacey: I would like to ask one of the astronomers what happens. You have an obliquity of $23\frac{1}{2}°$ dissipated at 50 sec of arc per century at the present time. It could not possibly have lasted for geological time on that basis; so what is preserving it?

Rochester: Presumably there must be some sort of long period source connected with the planetary perturbations.

Stacey: That seems unlikely.

Rochester: Anyway, it is an interesting question. I would like to remark that I think arguments like this are very attractive but also very seductive. I realize that it is an extraordinarily difficult problem but I do not think we can really be satisfied until somebody has produced a truly dynamical argument.

Stacey: I agree.

A POSSIBLE CAUSE OF THE CORRELATION BETWEEN EARTHQUAKES AND POLAR MOTIONS

S. K. RUNCORN

Dept. of Geophysics and Planetary Physics, School of Physics,
University of Newcastle upon Tyne, Newcastle upon Tyne NE1 7RU, U.K.

Abstract. A common geophysical cause of the excitation of the Chandlerian nutation and the irregular changes in the length of the day in impulsive torques resulting in core-mantle coupling is discussed. The 'local' stress distribution in the lower mantle is suggested as the trigger for earthquakes. Possible correlation between earthquakes and polar shifts is explainable in this way.

1. Introduction

The attempt to show a correlation between earthquakes and discontinuities in the polar path of the Chandlerian nutation has caused much interest. The mechanism of excitation of the Chandler wobble is one of the most puzzling unsolved problems in the geophysics of the solid earth. The departures of the polar path from a circle can be well represented as sudden changes in the centre of the circular path and of the radius of curvature (Mansinha and Smylie, 1967). These discontinuities are naturally identified with the random 'impulses' which maintain the Chandler wobble in spite of the natural expected decay in its amplitude. The interesting theory of Mansinha and Smylie that these 'impulses' are due to mass displacements of the crust during earthquakes must be difficult to settle conclusively by the data. The latitude and longitude of the epicentre and the direction of movement of the crust is of as much consequence in determining the effectiveness of the earthquake in displacing the pole of figure as its magnitude is. Thus even if we suppose this theory of the excitation of the Chandlerian nutation to be true, a complete correlation between discontinuities in the polar motion and earthquakes greater than a certain magnitude cannot be expected. Similarly to press an investigation of a correlation between two such sets of events to the limit, i.e. to include smaller and therefore more numerous events, must inevitably, given errors in their timings, lead to the conclusion that the correlation found between the less numerous events of greater magnitude is fortuitous.

The interesting question now arises whether, if the correlation is accepted as a working hypothesis, discontinuities in polar paths and earthquakes might both be caused by a third phenomenon: that the latter cause the former is not the only conclusion to be drawn. The mechanism by which stresses build up in the rigid crust, which thus stores the strain energy released in an earthquake, has recently been considered in terms of plate tectonics. But it is clear in any case that the slow build up of stress over, say, 100 years in one part of the crust is the result of geological or tectonic processes which operate gradually. These movements in the mantle with

L. Mansinha et al. (eds.), Earthquake Displacement Fields and the Rotation of the Earth, 181–187.
All Rights Reserved. Copyright © 1970 by D. Reidel Publishing Company, Dordrecht-Holland

a constant rate over times of the order of tens of years, bring the crust to the breaking point. The nature of these movements is still obscure – whether they are due to convection or to a chemical mechanism need not be discussed – the essential point is that they do not seem to be impulses. Thus the idea that a trigger of some kind sets off an earthquake at a locality, where the slowly accumulated stress has reached the breaking stress, is physically reasonable.

Sudden changes in the angular rotation vector of the mantle gives rise to a stress difference field in the mantle and crust, resulting from the consequent sudden change in the centrifugal force field arising from the change in the axis of rotation. But the change in the pole position does not arise from outside the earth. It is therefore more likely that the change in the distribution of mantle stress difference is a consequence of the mechanism which gives a jolt to the pole, rather than a consequence of the displacement itself.

2. Nature of the Astronomical Data

First suppose the earth to be an elastic body without a fluid core. As it is a nearly axially symmetrical body, the nutation can be considered in terms of three colinear poles in the following order: the pole of figure fixed in the body, the pole of angular momentum fixed in space and the instantaneous rotation pole – for a small ellipticity the latter two poles are very close. In the presence of slight damping the poles will eventually coincide and there will be no nutation. A sudden relative displacement of the poles results in a nutation. This can be brought about by a sudden displacement of the pole of figure relative to the earth – this is Smylie and Mansinha's suggestion; or the displacement of the angular momentum vector relative to the earth – this is the suggestion of Runcorn (1968). In both theories the source is internal to the earth.

Runcorn (1968) has argued that the irregular changes in the length of the day which occur suddenly, i.e. within a year or so, arise from the same physical mechanism as the polar displacements. The classical astronomical data, from which Spencer Jones (1939) proved that irregular changes in the earth's rotation rate occur, are the discrepancies between the observed longitudes of the Moon, Sun, Venus and Mercury and the 'theoretical' ones calculated from gravitational theory. This data is not accurate enough to determine how sudden the changes really are, but they become effective in less than a few years. They are clearly 'sudden' on a geological time scale too (or rather on the scale of the internal movements in the earth's mantle which cause tectonic phenomena). The data suggest that the irregular changes occur at random intervals. The much more accurate quartz and atomic clock data show that smaller changes in the length of the day occur more often and become effective in a few months. Thus only in the last ten years is there possibility of examining the relation of changes in polar path and in the length of the day.

It was recognised early (De Sitter, 1927) that possible changes in mass distribution at the surface of the earth failed by many orders of magnitude to explain the changes in the length of the day. Finally the changes had to be explained by postulating an interchange of angular momentum between the core and mantle, and magnetohydrodynamic forces rather than viscous forces seemed a likely agency. Runcorn (1954) suggested that the change in the relative rotation rate of the core and mantle, measured by the westward drift of the non-axial parts of the geomagnetic field, would correlate simply with the irregular changes in the length of the day. Though the correlation between the westward drift and the changes in the length of the day found by Vestine (1953) has since been shown to be an excellent one (Kahle *et al.*, 1969) with a phase lag (to be expected), it is found that the amplitude ratio is not correct, unless only an arbitrarily chosen part of the outer core exchanges angular momentum with the mantle. There are in fact quantitative difficulties in the theory of the electromagnetic coupling of core and mantle (Rochester, 1960; Roden, 1963). Rochester and Smylie (1965) consider that the electromagnetic torques on the lower mantle are not great enough (by a factor of 20) to cause the observed angular accelerations. Nevertheless there seems no other source than electromagnetic processes for the necessary torques. Therefore despite these quantitative difficulties I suppose the coupling of the mantle to a turbulent core by magnetohydrodynamics to be the *vera causa* of the irregular changes in the length of the day. Hide (1969) has recently suggested a possible way out of these difficulties by increasing the coupling by boundary irregularities.

3. Theory of Chandler Nutation

The question therefore arises whether the core-mantle electromagnetic coupling explains the maintenance of the Chandler wobble. I argued (Runcorn, 1968) that the magnitude and time scale of the torques required to change the length of the day in the way observed and those required for the continual excitation of the Chandlerian nutation are so similar, that it is reasonable to suppose they are the same physical process. I stated the argument too briefly perhaps and failed to relate it to the opposite conclusion reached by Rochester and Smylie (1965). Of the larger changes in the length of the day, the increase of $\frac{1}{300}$ sec which occurred around 1897, becoming effective over 2–4 years, has been much discussed. It involved a reduction in the angular momentum of the mantle of 1 part in 2.6×10^7. This was the result of an impulsive torque equal to $H_M/2.6 \times 10^7$, where H_M is the angular momentum of the mantle. Its cause is a hydromagnetic turbulence or disturbance in the core stronger and more short lived than those normally postulated to explain the geomagnetic secular variation. The natural damping of the Chandlerian wobble confuses the question of the magnitude and frequency of the kicks necessary to maintain it but these seem to be of the order of $0''.01$–$0''.1$ of arc, see, for example, Lambert *et al.* (1931). These represent changes in the angular momentum of the earth's mantle of 1 part in 2×10^7 and 2×10^6 respectively. Thus the impulsive torques on the mantle parallel to the axis of rotation necessary to explain the 'sudden' changes in the length of the day and impulsive torques on the mantle perpendicular to the axis, i.e. in the equatorial plane, required to excite the Chandler wobble are within an order of magnitude of each other. It is a reasonable first hypothesis that they originate in the same geophysical process.

It may be asked whether it is correct to talk in terms of impulsive torques rather than 'sudden' changes in the magnitude of the torques. The earlier analyses of the observed minus the theoretical longitudes of the moon by De Sitter (1927) were fitted by a number of straight lines, the intersections representing instantaneous or 'sudden' changes in the rate of the earth's rotation. Brouwer (1952) introduced a new method of treating the errors in the observations of the moon's longitude by fitting them to a number of parabolic arcs. Their intersections represent instantaneous changes in the magnitude of the torques on the mantle. The latter representation of the data was held to be a better representation as the points around the sudden changes of slope, being scattered, are better fitted by a curve. At first sight it also seems more physically reasonable; sudden changes in the earth's angular velocity imply infinite torques. However fitting by parabolic arcs does in fact smooth out any sharp changes of slope and sudden changes in the magnitude of the torques are not inherently more realistic than sudden changes in the angular velocity of the mantle. The astronomical data still awaits fitting to a physical theory of the phenomena. I think that no other way of dealing with its errors is statistically sound: arbitrary smoothing procedures seem unlikely to separate observational uncertainties from real geophysical effects. As the nature of the hydromagnetic processes in the core which give rise to the changes in the rotation rate have not been developed theoretically, this desirable procedure cannot be followed. However disturbances in the core propagate with the velocities of magnetohydrodynamic (Alfvén) waves of the order of 1 cm/sec: thus a disturbance of the size of an eddy in the core (perhaps 100 km across) will 'surface' in the order of a year. As it is known that the electromagnetic time constant of the lower mantle is of the same order, it seems possible that short lived, but intense, current systems could be established in a year or so in a part of the lower mantle. The torques resulting from these current systems could change the rotation of the earth, the mantle being stressed locally until the torques decayed. Such a non uniform stress pattern in the mantle would seem a possible trigger for an earthquake. The theory of such a process has not been developed, but the 'local' origin of the torques required is not unreasonable.

Indeed Moore (1955) drew attention to the relevance, for the study of the 'sudden' changes in the length of the day, of the similarly sudden changes in the geomagnetic secular variation rates. Moore, too, tried to establish a correlation between these and the irregular changes in the length of the day. In particular he drew attention to the interesting coincidence in time, around 1897, of the largest observed 'sudden' increase in the length of the day, $\frac{1}{300}$ sec, and possibly the most remarkable establishment of a major secular variation focus ever observed, that off South Africa. The secular variation there has a strength as great as ever recorded (160γ/year). This rate trebled within a year or two about 1897.

However further study revealed large numbers of 'smaller' sudden changes in both the secular variation rates and the length of the day and the attempt at a correlation became insuperable, for the same reason as that given above. But the significance of these secular variation impulses (as they have been termed by Deel (1945)) for the theory of the electromagnetic coupling of the core and mantle remain.

I conclude therefore that the representation of both the changes in the length of the day and the changes in pole position by impulsive torques occurring every few years is a satisfactory approximation to the truth and has geophysical importance. The process is not a random one and so there is no satisfactory method of analysing the international latitude service data which only covers 60 years to determine the free decay time of the Chandler wobble. The two phenomena could have very different time scales. The values given by various stochastic calculations, 7–30 years, seem unlikely to be any guide to the nature of non-elastic processes in the mantle.

4. Impulsive or Step Function Torques

From a physical point of view Euler's equation is best expressed in vector form

$$\mathbf{L} = \dot{\mathbf{H}} + \boldsymbol{\omega} \times \mathbf{H},\tag{1}$$

where $\mathbf{L}$ is the external torque applied to the body, $\mathbf{H}$ is its angular momentum measured with reference to coordinates rotating in space with angular velocity $\boldsymbol{\omega}$. Let C be the moment of inertia about the axis of symmetry and A the moment of inertia about a line in the equatorial plane, assuming axial symmetry, then:

$$\mathbf{H} = C\omega_3\mathbf{k} + A\omega_2\mathbf{j} + A\omega_1\mathbf{i}.$$

The notation of Munk and MacDonald (1960) is useful: $\omega_3 = \Omega(1+m_3)$, $\omega_2 = m_2\Omega$ and $\omega_1 = m_1\Omega$ and $\mathbf{m} = m_1 + im_2$, where for small nutation, the squares of m_1, m_2, m_3, $\mathbf{m}$ can be neglected.

Thus $L_3 = C\dot{m}_3\Omega$, the torque required to cause the irregular changes in the length of the day and

$$L_2 = A\dot{m}_2\Omega - m_1\Omega^2(C-A)$$
$$L_1 = A\dot{m}_1\Omega + m_2\Omega^2(C-A),$$

or

$$\dot{\mathbf{m}} + (1/i)\mathbf{m}\sigma_r = \mathbf{L}_w/A\Omega,\tag{2}$$

where $\mathbf{L}_w = L_1 + iL_2$, the torque required for maintaining the wobble and $\sigma_r = \Omega(C-A)/A$.

Solving (2) by Laplace transform method, the subsidiary equation is

$$[p + (\sigma_r/i)]\bar{\mathbf{m}} - \mathbf{m}_0 = \bar{\mathbf{L}}_w/A\Omega\tag{3}$$

$$\therefore \mathbf{m} = [\mathbf{m}_0 + \mathbf{H}_w/A\Omega]\exp(i\sigma_t t),\tag{4}$$

where $\mathbf{H}_w$ is the impulsive torque at $t=0$ discussed above, $\mathbf{m}_0$ is the pole position at $t=0$.

This result is that given by Runcorn (1968) that is, the application of the impulsive torque $\mathbf{H}_w$ in the equatorial plane displaces the pole of rotation on the earth's surface in the direction of vector $\mathbf{H}_w$ by an amount $H_w/A\Omega$ and thereafter the pole moves over the surface in a circle, but the centre of this remains the pole of figure as before, neglecting the damping. It is incorrect to state, as has sometimes been stated, that the simplified discussion in Runcorn (1968) is erroneous: the impulsive torques required to excite the Chandler wobble are within an order of magnitude of those inferred from the irregular changes in the length of the day to exist.

However, a different conclusion is reached if it is supposed that a unit function torque in the equatorial plane is applied to the mantle: the physics of the situation is then different.

If $\mathbf{L}_w = 0$ for $t < 0$
          $= 1$ for $t > 0$,

Equation (3) gives $\mathbf{m} = i\mathbf{L}_w [1 - \exp(i\sigma_r t)]/A\sigma_r \Omega$ for $t > 0$. Then $L_w/L_3 = \sigma_r |m|/\dot{m}_3$.

Taking the amplitude of the Chandler wobble as $0\overset{\prime\prime}{.}14$ and $\dot{m}_3$ as $8 \times 10^{-9}/y$, $L_w/L_3 \simeq 400$. By such an argument Munk and Hassan (1961) reached their conclusion

Thus if the core were responsible for both the wobble and the changes in diurnal rotation, the equatorial torque components would have to be several hundred times larger than the axial torque, whereas one expects the equatorial components to be, if anything, less than the axial component. The conclusion is that the core is a negligible source of the Chandler wobble.

For Rochester and Smylie (1965) in their study of the electromagnetic coupling of the core and mantle find that $L_w/L_3 \simeq 0.2$. Thus the conclusion was reached, mentioned again by Smylie and Mansinha (1968) that the core-mantle coupling is "at least several hundred times too weak to explain the observed wobble amplitude".

The application of a step function torque does not displace the pole suddenly but causes its path to alter suddenly so that it now moves in a circle of radius $L_w/A\sigma_r\Omega$ about a mean point displaced (as long as the torque is applied) an angular distance $L_w/A\sigma_r\Omega$ from the pole of figure in a direction $90°$ to the vector $\mathbf{L}_w$. It is clear that a considerable torque is now required to hold the mean pole of rotation away from the pole of figure by this amount because of the gyroscopic effect.

Of course, no mechanical work is done by the applied torque components over whole numbers of nutations. For the rate of work done is the sum of

$$L_1 m_1 \Omega = -[L_1^2 \sin \sigma_r t + L_1 L_2 (1 + \cos \sigma_r t)]/A\sigma_r,$$

and

$$L_2 m_2 \Omega = [L_1 L_2 (1 + \cos \sigma_r t) - L_2^2 \sin \sigma_r t]/A\sigma_r,$$

that is

$$[2 L_1 L_2 \cos \sigma_r t - (L_1^2 + L_2^2) \sin \sigma_r t]/A\sigma_r.$$

But I do not think the evidence supports the interpretation that the Chandler wobble is excited by step function torques: the changes in the circular polar path and the variation in the length of the day are sudden, compared to the period of the damping of the Chandler nutation which is variously estimated as between 7 and 30 years.

Thus the discussion assumes that the time constant of free decay of the Chandler wobble is long compared to the duration of an impulse. Bearing in mind the observed shortness of the latter, best seen in the data on the length of the day, and the considerable uncertainty about the former, this is reasonable.

However a very different conclusion is reached if the time constant of decay of the Chandler wobble is taken to be short. Then obviously it can only be maintained by a torque continually applied to the mantle – with some variation with time to account for the non-uniform non-circular motion of the pole.

Of course relative rotation of the core and mantle would be brought to zero in the absence of the turbulent effect, referred to above, by the eddy currents induced in the mantle by the 'steady' magnetic field generated in the core. However the westward drift can be traced back for the last 300 years, so that the damping time constant in the core and mantle oscillations seem to be large.

References

Brouwer, D.: 1952, *Astron. J.* **57**, 125.
Deel, S. A.: 1945, Paper U.S. Coast Geod. Survey, No. 664.
De Sitter, W.: 1927, *Bull. Astron. Inst. Neth.* **4**, 21.
Hide, R.: 1969, *Nature* **222**, 1055.
Kahle, A. B., Ball, R. H., and Cain, J. C.: 1969, *Nature* **223**, 165.
Lambert, W. D., Schlessinger, F., and Brown, E. W.: 1931, *The Figure of the Earth*, Bulletin 78, National Research Council, Washington D.C., p. 245.
Mansinha, L. and Smylie, D. E.: 1967, *J. Geophys. Res.* **72**, 4731.
Moore, A. F.: 1955, Ph. D. theses Cambridge University.
Munk, W. H. and Hassan, E. S. M.: 1961, *Geophys. J.* **4**, 339.
Munk, W. H. and MacDonald, G. J. F.: 1960, *The Rotation of the Earth – A Geophysical Discussion*, Cambridge University Press, London.
Rochester, M. G.: 1960, *Phil. Trans. Roy. Soc. London* **A252**, 531.
Rochester, M. G.: 1962, *J. Geophys. Res.* **67**, 4833.
Rochester, M. G. and Smylie, D. E.: 1965, *Geophys. J.* **10**, 289.
Roden, R. B.: 1963, *Geophys. J.* **7**, 361.
Runcorn, S. K.: 1954, *Trans. Amer. Geophys. Union* **35**, 49.
Runcorn, S. K.: 1968, in *Continental Drift, Secular Motion of the Pole, and Rotation of the Earth, IAU Symposium No. 32* (ed. by Wm. Markowitz and B. Guinot), D. Reidel, Dordrecht, The Netherlands, p. 80.
Smylie, D. E. and Mansinha, L.: 1968, 'Earthquakes and the Observed Motion of the Rotation Pole', *J. Geophys. Res.* **73**, 7661–7673.
Spencer Jones, H.: 1939, *Monthly Notices Roy. Astron. Soc.* **99**, 541.
Vestine, E. H.: 1953, *J. Geophys. Res.* **58**, 127.

DEFORMATION FIELDS: OBSERVATION

THE MEASUREMENT OF SMALL EARTH STRAINS

R. C. BOSTROM

Dept. of Geologic Sciences, University of Washington, Seattle, Wash., U.S.A.

Abstract. Fossil evidence indicates that parts of the earth are subject to secular deformation at rates less than 10^{-17} per second. Evidence is presented to suggest that non-reversing strain steps associated with seismicity are concentrated in the tectonic belts. It is unknown whether mass displacement in the earth's outer shells is accomplished principally as strain steps or by means of processes of flow and glide so far undetectable. Strain observations are required in the earth's old stable cratons as well as in the tectonic belts. To progress in the observation of secular strain imposition a stable length standard is required. No information exists that changes in the velocity of light are less than changes in the strain state of earth materials. For this and other reasons it is concluded that a stable etalon is more suitable as reference than a length unit defined by a frequency standard such as an atomic clock. The design is described employing strain-imaging of a stress-free coupling to the crust; and of an etalon having stability in respect to geological rates of deformation.

Stresses of internal and external origin affect the body of the earth and result in displacement, permanent or oscillatory, of its material. Our awareness of earth strains is commonly indirect, being based upon the observation of ancient faults or the geophysical changes in seismic wave trains in traversing the deep interior.

The purpose of what follows is to examine means of directly measuring deformation of the planetary surface.

1. Earth Strains

A. RATES AND PERIODICITIES

Our problems arise from the fact that earth strains have periodicities ranging from a fraction of a second per cycle to intervals of time familiar only to the astronomer and geologist; and that the rate at which they are imposed is small, namely a few parts in 10^6 per month (in tectonically active areas) to much less than a millionth of this rate elsewhere. The noise background due to such effects as man-made disturbances and solar heating of the observation site may be greater than the signal it is desired to retain, so that retrieval has to be made on a basis of frequency-discrimination.

Internally generated strain waves resulting from seismic disturbances contain frequencies ranging from hundreds of Hertz to 3×10^{-3} Hertz, the lower limit representing the fundamental tones of the earth in free vibration. Data forthcoming from recent interferometric measurements suggest that earthquakes generate in addition components which have still lower frequencies, correspondingly unable to excite resonance.

Externally generated strains are associated with the tidal attraction of the sun and moon and presumably with the effect on the earth of relativistic gravitational waves emitted by objects subjected to acceleration in distant parts of the universe. The period of the stresses responsible for the earth tides ranges from eight hours (M_3, the ter-diurnal component of the tidal waves) to 19 years (N_1, the period of revolution of the lunar nodes), and strictly speaking must be extended to include those associated with

L. Mansinha et al. (eds.), Earthquake Displacement Fields and the Rotation of the Earth, 191–205.
All Rights Reserved. Copyright © 1970 by D. Reidel Publishing Company, Dordrecht-Holland

the equinoctial precession, in excess of 20000 years. The periodicity of the free nutation of the earth has been shown by Chandler to lie toward the short-period end of this scale, 14 months. The rate of strain imposed at the earth's surface in the form of extensions by the higher-frequency tidal components (the only ones so far directly observed) is a few parts in 10^{13} per second.

B. ELASTIC AND INELASTIC STRAINS

The position is complicated by the fact that in addition to deformation components having a periodicity of many years, deformation observed at the earth's surface contains an inelastic component resulting in displacements of a permanent nature. Having in mind the time intervals involved, it may be seen to be impossible to separate non-reversing, tectonically imposed earth strains from low-frequency tidal deformations solely on the basis of frequency filtering.

The effect of the inelastic component of earth deformation in response to tidal stresses is to produce a phase lag of a few degrees of strain in respect to stress. In the case of the Eulerian free nutation period, expected on the basis of rigid-body theory to be 10 months, the effect of elastic processes in the earth body is to lengthen the period to that observed, represented in the Chandler wobble.

A change to a new static strain state, or strain step, is expected to occur at the time of an earthquake and has been observed (Wideman and Major, 1967). A theoretical exploration of step changes has been conducted by Press (1965) and by Ben-Menahem and Singh (1968). The deformation characteristics of the earth's outer shells in response to stress vary from those of cold granites, experimentally only slightly susceptible to inelastic strain, to those of 'magma chambers' and the material of the low velocity zone. In consequence it is not surprising that observed changes in the strain state seem already to depart from those to be expected in a uniformly layered earth. Instrumental manifestations of steps have relied principally for their validation upon their association with known seismicity. Manifestations similar in all respects but not thus verified have been recorded at times of seismicity ranging in magnitude down to the undetectable (Figure 1). For reasons of instrumental instability, see below, it has not been possible to record with confidence continuous slow changes in the strain state at instrument sites.

It follows that we are ignorant of whether displacement of the material of the earth's surface is accomplished in the main as an accumulation of steps marked by seismicity; or undetected, by continuously operating processes of glide and flow.

2. Strain Meters

We here exclude consideration of such devices as tilt meters and accelerometers, which may be used to detect strain indirectly.

A. THE EXTENSOMETERS OF MILNE, BENIOFF, AND SASSA

The first direct-reading earth strain meter was conceived and built by Milne (1888),

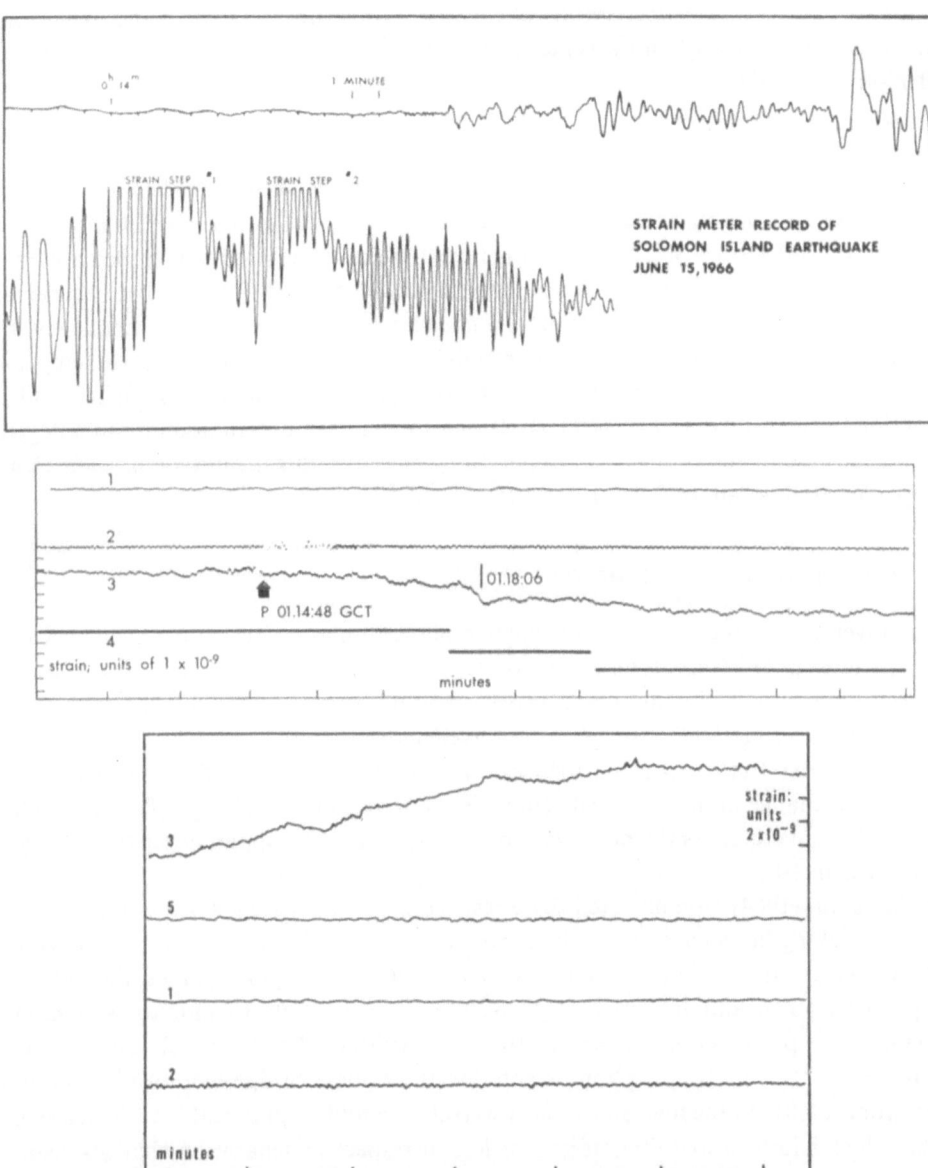

Fig. 1. Strain seismograms showing events having an irregular or imperceptible connection with seismicity. In (a), reproduced from Wideman and Major (1967), two strain events separated by an interval of several minutes were recorded subsequent to the arrival of body waves from a teleseismic event. Similar events have been recorded (Bostrom and Vali, 1968b) before the occurrence of shocks. Strain events of the type illustrated in (b), reproduced by courtesy of *Nature, ibid.*, have been recorded at the time of tremors ranging in magnitude down to the undetectable and (c), possibly non-existent; strain amplitude is apparently unrelated to that of the seismicity. In (b) and (c) traces 3 are those of the interferometric strain seismograph operated by the University of Washington in the Cascade Range at 47°47′N, 120°50′W. The remaining traces are those of high-magnification inertial seismographs as follows: (1) horizontal pendulum, period 20 sec, aligned parallel to interferometer gap; (2) vertical pendulum, period 3 sec; (4) fringe-counting strain seismograph; (5) vertical pendulum, period 20 sec.

but was insensitive to most strains of interest. Direct observation of the extensions associated with tidally and seismically imposed earth strains was first achieved by Benioff (1935), by continuously comparing the distance between benchmarks with the length of a fused-quartz rod; and by Sassa *et al.* (1951) utilising variations in the sag of an invar wire stretched between piers some meters apart.

The extensometers of Sassa and Benioff have been responsible for the detection of phenomena previously unseen, such as the earth's free oscillations. Their limitation in respect to long-period strains lies in the fact that the material of which they are constructed is not known to be dimensionally invariant with age; hence that it is necessary to discard as 'instrumental drift' signals having a very low or zero frequency. In addition, as such materials are not unresponsive to changes in atmospheric temperature, humidity and pressure, isolated observation sites must be constructed and a continuous record kept of thermal variations greater than one-thousandth of a degree C. In the case of fused quartz a temperature change of this magnitude introduces an apparent earth strain of 0.4 parts in 10^9.

B. INTERFEROMETRIC STRAIN METERS

The invention of the laser as a coherent monochromatic light source brought early suggestions as to its use in an interferometer operative as strain meter. Nevertheless, to count changes in the number of waves standing between reflectors at fixed points, if e.g. the latter are 50 m apart and the wavelength is 6000 Å, is to limit precision to scarcely better than a part in 10^8, less than that of the Benioff extensometer. In addition a change in the air density equivalent to a barometric change of a thousandth of an atmosphere alters the propagation velocity, hence the apparent separation, by a few parts in 10^7.

Vali *et al.* (1964) circumvented these restrictions by providing a vacuum light path and by locking the readout of the interferometer to a fixed phase position vis-à-vis the standing wave train. If the width of the phase-lock is then $\pi/100$ radians, the reflector separation 50 m and the wavelength 6328 Å, it is possible to monitor separation changes to a precision of 0.6 parts in 10^{10}. The width of the phase-lock is a statistical function of the number of photons entering the optical bridge employed (Vali and Bostrom, 1968). As the light source is powerful, it would be practicable to decrease the phase-lock width to $\pi/10000$ radians or less in respect to separation changes lasting for the duration of a high-frequency seismic wave (one-hundredth of a second); and correspondingly to decrease this quantity, increasing the precision, in respect to seismic waves of normal, longer, period.

The limitation at present of this form of extensometer is presented by changes in the emission frequency of the laser. Whereas the extensions it is wished to measure take place at rates less than 10^{-9} per hour and at periodicities ranging to infinity, investigations of the wavelength changes of lasers stabilised on the Lamb dip of the gain curve have shown these to exceed two parts in 10^8 over periods of a few tens of hours and to be progressive with time (Mielenz *et al.*, 1968). To progress, it is necessary to stabilise the output of the laser with reference to the time scale of the extensions in nature.

Vali and colleagues (Vali *et al.*, 1968) have used a similar read-out to measure length changes in crossed interferometer arms, detecting a component of shear strain.

C. AIR-PATH EXTENSOMETERS

To achieve the precision required to observe the extensions associated with earth tides and low-level tectonism, and furthermore to obtain fringe patterns so well formed as to permit phase-locking, it will be apparent that the light path of an interferometric extensometer must be in vacuum. Conversely, however, to observe strain imposition in a ground sample large enough to suppress the effect of rock inhomogeneities and to be sensitive to deep crustal strains, the observation field must have a dimension of tens of kilometers.

In practical terms this means that we must employ as strain-translation element an above-ground beam, dispense with the use of a light interference pattern, and account as well as possible for the effects of an atmosphere on ray geometry and propagation velocity.

A choice of light or microwave frequencies is available. Employment of microwaves permits observation free of interruption by cloud; on the other hand, light velocities are less sensitive to changes in the atmospheric moisture content; furthermore, microwave reflectors have to be larger than those employed for experiments using light in the ratio of their wavelengths. Finally, less available in experiments employing microwaves are analogs of such control devices as the Kerr cell. The compromise is available of employing light as the signal carrier, but modulating this at microwave frequencies.

Two successful devices, the tellurometer and the geodimeter, employing respectively radio and light carriers, have made possible the measurement of an electromagnetic path-length to a precision of somewhat better than a part in a million. The correction for curvature and propagation velocity has had to be made indirectly, by means of a computation based on line-end observations of atmospheric temperature, pressure, and humidity. This process downgrades overall measurement accuracy by an order of magnitude, and is oblivious to secular changes in atmospheric composition.

Independently of its value at the source and reflectors, the refractive index of the atmosphere varies continuously along the ray path due to meteorological inhomogeneities and to the vertical gradient in density under gravity. To take account of path changes in the velocity, the approach of Bender and Owens (1965), Thompson and Wood (1967) and Fowler (1968) has been to observe differences in the electromagnetic distance measured at several wavelengths. This technique has the advantage of providing an integrated value of the refractive index for the entire path. Its limitation has been that path-length differences for various frequencies are so small as to be difficult to measure. Thayer (1969) has concluded that for ranges not over 25 km the simultaneous use of 2 optical and a radio frequency permits accuracies of about 0.02–0.03 parts in 10^6 under common refractive conditions. Thompson and Janes

(1967) have shown that in respect to strain frequencies of less than 10^{-6} Hz, tectonically induced changes in the path-length corrected for path-length measured at one frequency are likely to stand well above residual noise.

Air-path techniques employing continuous recording (hence able to employ frequency-filtering to separate the signal) are then in a position to observe distance changes of parts in 10^8 and greater for long-period and secular strains. Precautions are necessary in separation of extensions due to earth tides, by reason of the existence of synchronous tides in the atmosphere. In respect to secular strains, refractivity corrections should be made by direct observation of velocities (Bostrom, 1969) to take account of permanent changes in atmospheric composition.

3. The Sampling Problem

In the techniques discussed, the sampling problem is predominant. No matter how sensitive, a strain meter fixed to the planetary surface observes changes in respect to only one sample of ground. How representative of crustal strain changes are those observed in a sample of limited size?

With respect to an ancient stable craton, the geologist may be able to state that the rock mass containing a cavern observation station is coupled to the regional rock mass, and not separated from it by faults or zones suffering relaxation by flow. In seismic areas, where strain measurements have most commonly been made, this circumstance seldom applies. An obstacle in setting up instrumentation in the San Francisco area, for instance, is finding an observation site demonstrably coupled to the crustal mass in which the energy of the large local seisms must accumulate.

The anomalously large and early strain steps observed at a station 4000 km from small-magnitude seismicity in central America (Bostrom and Vali, 1968b) could be explained by hypothesizing that strain imposition is concentrated along tectonic belts, rather than more uniformly distributed. This hypothesis accords with the large viscosity range from point to point of the earth's surface, but calls for a strain-propagation formulation not based upon an earth composed of layers uniform laterally. It is most desirable that simultaneous strain data be obtained at observation sites located within the earth's stable rigid cratons and within mobile belts.

Techniques of sampling large areas for strain are urgently required. The position is that for separation of the harmonic components of strain such as earth tides continuous observation is desirable. On the other hand in respect to the displacement of masses, for the purpose of seismologic prediction or estimation of shift in the axes of moment as a result of seismicity, it would be valuable to densely sample the deformation of regions.

Repeated geodetic surveys can scarcely sample more often than once every 6 months and their accuracy, >1 part in 10^6, is less than is desirable.

It has been claimed in respect to air photography that satellite-mounted cameras are capable of recording the white lines on runways. Earth-surface deformation in California is distortional and amounts to several cm/yr across the San Andreas rift.

The question might be considered of whether distortion is visible on superimposed air photographs taken from the same point at successive seasons. Unfortunately, ray curvature due to atmospheric inhomogeneities can produce an apparent displacement of many centimeters of points on the ground seen from an aircraft; the amount varies within the area covered by a single photograph.

4. Strain Imaging

Photographic techniques are uniquely capable of storing information of the kind desired. In this respect it has been demonstrated that holography is not entirely confined to the realm of coherent light (Mertz and Young, 1961; Goodman *et al.*, 1966). Furthermore it has been proposed by Horman (1965) and demonstrated (Alexandrov and Bonch-Bruevich, 1967) that superimposed holograms of an object before and after deformation (or the double-exposure of a single plate) permit the construction of an interferogram of extraordinary detail, or strain image.

A. LUNAR STRAIN IMAGING

A coherent light source, for example, a laser setup on a mountain crest on the airless moon, will receive back from the illuminated landscape wave-fronts whose interference pattern could be recorded on a photographic plate without lens. Later plates superimposed, or repeated exposures of the same plate, would result in the recording of interferograms displaying distortions suffered by the illuminated surface between exposures. The beam could be expanded or the laser rotated to cover any angle. Using a motion-picture camera film transport and pulsed laser the sampling interval could be made as short as desired.

B. TERRESTRIAL STRAIN IMAGING

An extension of this process to the surface of the earth introduces difficulties, the chief of which is imposed by the atmosphere. Present evidence is that so much loss of coherence would be introduced by atmospheric inhomogeneities that a useable interferogram of a landscape could not be constructed.

It would seem however that interferometric imaging could be used to observe the imposition of strain at a cavern site. A limitation in the use of extensometers has been the crudeness of the devices by which their sensors are coupled to the earth. The point of attachment of a reflector to bedrock is liable to be the site of strain imposition greater than the long-term earth-strains it is wished to measure. In Figure 2 is shown the design of strain-imaging devices observing extensions and distortions at a cavern site. For such short distances, to stir the air in the cavern by fans and to stabilise the laser source to a fixed air-path wavelength would be preferable to arranging for a vacuum ray-path.

5. Stable Length Standards

Geologic strains are imposed at rates varying down to the very small. Fossil specimens

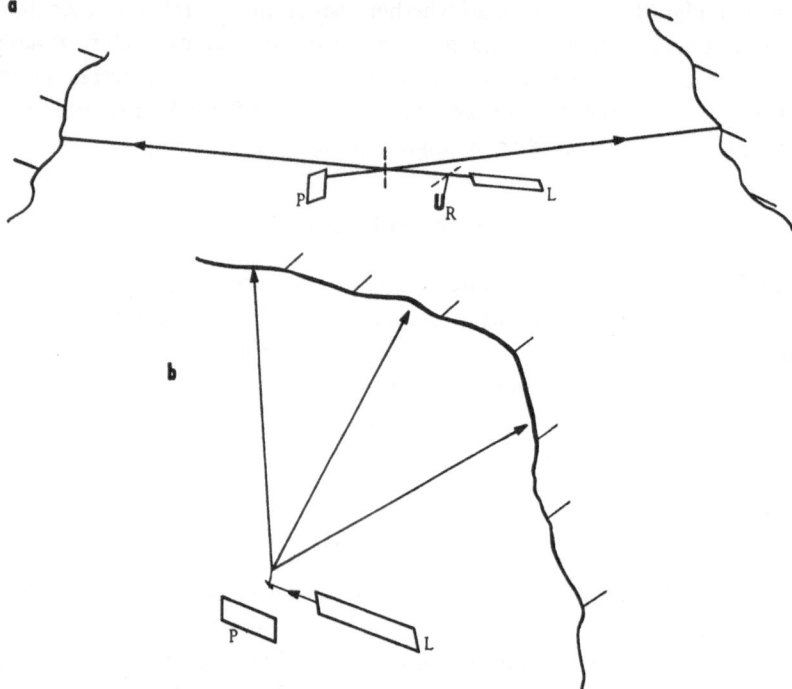

Fig. 2.   Potential use of strain-imaging in a cavern observation station. Both arrangements provide stress-free coupling to the bed-rock. In each, L represents a laser source, P a photosensitive recording device. In (a), changes in the distance between the points at which the laser beam split by a partial reflector impinges on the wall are represented by changes in the interference pattern recorded at P. R is an air-path refractometer, and laser-tune servo maintaining constant the wavelength in the contemporary atmosphere. In (b) a beam-expanding lens permits illumination of a large portion of the cavern walls and roof. Strain changes appear in interferograms recorded at P as fringes. Their direction describes the direction of strains in the rock.

of curled Palaeozoic trilobites demonstrate less than 10% deviatoric strain suffered in an interval of 500 million years. The rocks of which these form part have been strained during this period at an average rate not exceeding one part in 10^{17} per year; probably much less.

We have no such certain knowledge that the physical 'constants' used in mensuration, for instance, the vacuum velocity of light, are unchanging. The Hubble red shift might conceivably be interpreted as contrary evidence, and Dirac has drawn attention to the possibility of age-dependent change in the value of the universal coupling constants.

A difficulty therefore arises, if we are attempting to measure the rate of strain suffered by parts of the earth's crust, in selecting a standard of comparison itself known to be stable.

When a force affects part of a solid medium such as the earth two sorts of deformation generally result, a change of volume and a change of shape. Aftershock studies of strain release have shown that these sorts of strain may to a large extent behave independently. In what follows we consider the problem of standards in measuring firstly the deviatoric and secondly the volumetric strain.

A. THE KELVIN SEISMOGRAPH

In an address to the Royal Society of Edinburgh in 1902 Lord Kelvin proposed the introduction of "a new specifying method for stress and strain in an elastic solid". In this Kelvin (1902) specified the forces acting through the face centres of an elementary tetrahedron and the resulting changes in edge length. His specification avoided the need to suppose that the strains considered are infinitesimally small and also avoided an asymmetry, the consideration of two kinds of strain, 'elongations' and 'shearings'. In a sense this asymmetry persists today; the dilatational wave referred to by seismologists does not represent purely volumetric strain, but consists of a change in dimension in one direction only. Perhaps as a result of the introduction a year earlier of the language of tensors, the tetrahedral specification of strain seems to have been little used.

Kelvin's concept has nevertheless much to recommend it, and forms the basis of an interferometric seismograph (Bostrom and Vali, 1968a; see also Cook *et al.*, 1965) intended to observe deviatoric strains taking place at an arbitrarily small rate and volumetric strains imposed at a rate greater than the known change in the emission frequency of the light source used for mensuration. In this instrument (Figure 3) the beam from a continuous laser is split into six components made to form the arms of

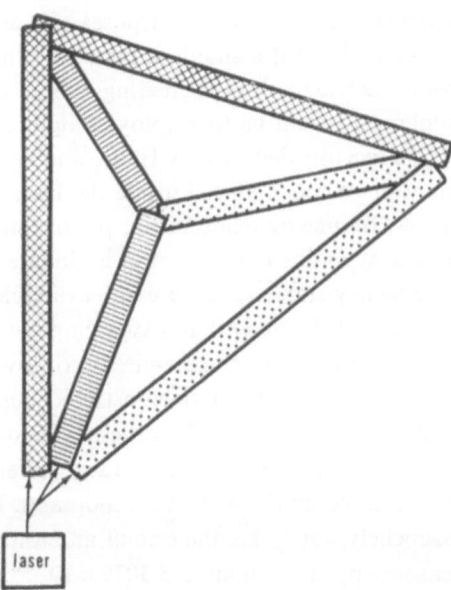

laser

Fig. 3. Kelvin's tetrahedron formed by dividing the emission of a laser into 3 pairs of two beams. Each of the beams so formed is used as input of a Fabry-Pérot interferometer arm observing length changes in the tetrahedron edge it outlines. The response of this instrument is described in Bostrom and Vali (1968a).

an interferometer in the form of a tetrahedron. The reflectors outlining the unit figure will be fixed to the walls of a cavern. Alternatively the stress-free coupling system of Figure 2a could be used, partitioning the emission of the source laser to form beams having the six directions necessary to measure changes in the value of the strain tensor. As deviatoric strains are then defined by changes in the relative lengths of the tetra-hedron edges, changes in the emission frequency of the laser are eliminated as a variable. Dilatational waves of any period are detectable and may be separated from those of other modes. By the same token, secular shear strains taking place at any rate are detectable and may be mode-separated.

The Kelvin seismograph will, however, be unable to distinguish between change in the emission frequency of the source laser and the imposition of pure volumetric strain (uniform change in dimension with reference to all three space axes), if the rate of the latter is less and frequency-discrimination cannot be used. The purely volume-tric element in earthquake waves and eigenvibrations of any natural frequency is thus detectable; volumetric strain taking place at geologic rates, for instance as a result of changes in the gravitational constant, is not.

To detect pure volumetric strain imposed at an arbitrarily small rate it is therefore necessary to preserve constant the vacuum wavelength of the laser emission; or to compare this continually with a length standard known to be invariant.

B. ATOMIC LENGTH STANDARD

It might at first be thought that to create a standard length it is necessary only to generate a standard frequency. For everyday purposes this assumption is sufficient. The wavelength of the emission lines of a standard lamp, for instance, is conveniently taken to be constant. An attractive method of creating a short-term length standard of use in the Kelvin seismograph would be to employ as light source an I_2-controlled continuous laser of the type recently decribed by Hanes and Baird (1969). In this an ab-sorption cell containing iodine vapour placed inside the laser cavity produces Lamb dip in the power-output vs. frequency relationship, permitting servo stabilisation of the laser with reference to a hyperfine component of an iodine absorption line.

The most constant frequency-standard at present available is the emission line from an atomic clock, e.g. one of the Essen-Parry type employing a caesium resonator (Essen and Parry, 1957). A comparison of the emission of several of these operative during the past decade (Guinot, 1967) has demonstrated constancy of output fre-quency in this interval within a few parts in 10^{13} or less. This quantity is not certainly insignificant in comparison with strain imposition rates suffered by samples of the crust, *supra*, and those detectable by the Kelvin seismograph. Furthermore it can be shown (Clemence and Szebehely, 1967) that the rate of an atomic clock varies season-ally for relativistic reasons by an amount, 3.3079×10^{-10}, not insignificant for present purposes. For similar reasons, its rate varies by an unknown amount through time with changes in the Newtonian constant and as such arcane events take place as changes in the sun's mass and changes in the configuration of the galaxy. Finally, if the Hubble red shift were to be explicable by a reduction in the velocity of light through geologic time, its rate exceeds that of the deviatoric strain rate observed on the basis of the fossil record.

For these various reasons, apart from the instrumental difficulties involved in stabilising a light frequency using a microwave standard, it seems unsafe to assume that frequency-generated length standards are time-invariant for purposes of interferometrically observing the imposition of the smallest strains.

C. A MATERIAL LENGTH STANDARD

An additional reason for obviating dependence upon a frequency standard is the interruptibility of observation this arrangement entails. It has been pointed out by Shamsi and Stacey (1967) that once continuity of observation has been broken in the case of an interferometer, as during replacement or failure of a laser, because waves are individually indistinguishable an ambiguity is introduced in the data array. This disadvantage, as well as those previously noted, would be avoided if the distance between reflectors under observation could be referred to the dimension of an invariant etalon. It would then be possible after interruption to employ the method of exact fractions to establish the datum of discrete observation series.

The difficulty in constructing an etalon for this purpose is absence of information on the dimensional stability of materials with time. Fused-quartz may be considered suspect from the long-term viewpoint by reason of its vitreous condition. Not only is a lower energy state available in the direction of its crystallisation, but evidence exists in the form of the drift of gravimeter springs of its ability to inelastically deform. The devitrification with age in the geologic column of glasses demonstrates the instability of materials in this condition.

These objections would not apply to a crystalline etalon and in particular to such an object aggregated in the geologic past rather than one recently wrought. The occurrence of euhedral mineral crystals which have been at surface conditions for extended periods has prompted a review of their availability as a local length standard.

A few minerals, such as Iceland spar and beryl, not only occur as ancient stable optically strain-free crystals, but possess axial thermal-expansion coefficients of opposite sign. This would make it possible simultaneously to control their temperature by interferometric means and to establish a path offering cancellation in length change as temperature changes.

An arrangement using beryl is illustrated in Figure 4. The reflectors shown, conveniently constructed of the same mineral axially oriented, could rest on the natural surface of a large crystal. However, it would be feasible to grind a beryl plaque to a convenient plane surface without introducing residual strain using the techniques used to grind large telescope mirrors. Beryl, $Al_2Be_3(SiO_3)_6$, occurs in optically continuous crystals up to 10 m long. Its magnetic susceptibility is insignificant. Its hardness is 7.5–8 and its response to the hydrostatic stress of being placed in vacuum is almost certainly a step strain without tail.

In Figure 4 the reflectors in Michelson interferometer configuration, having equal arm length, produce at the read-out a fringe pattern insensitive to changes in light

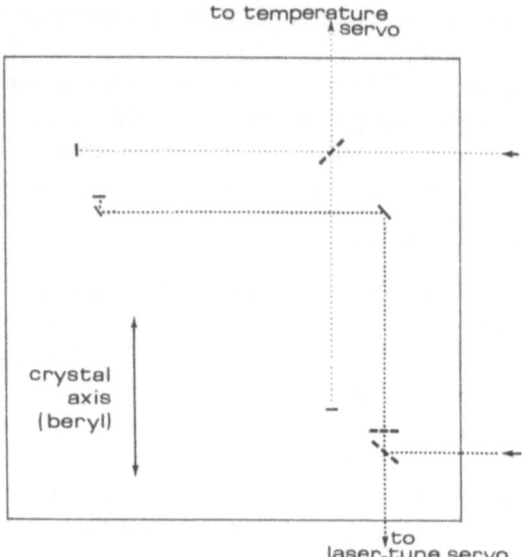

Fig. 4. A reference etalon having long-term stability. The arms of the Michelson interferometer controlling temperature are equal in length. The arm of the Fabry-Pérot interferometer follows a path proportioned in the ratio of the thermal expansion coefficients. The beryl plaque, on which the reflectors are mounted in V-notches parallel to the crystal axes, and all light paths, are in vacuum.

wavelength but sensitive to temperature changes. The fringe pattern produced is observed by an optical bridge sensitive to changes in phase position, signalling a temperature servo. The thermal expansion coefficients, α_+ and α_-, of beryl as determined by Fizeau and others (Tutton, 1922) are as follows where $\alpha = a + 2bT$; α_- and α_+ are the coefficients respectively parallel and normal to the principal crystal axis, having the values:

	a	$2b$
α_-	-0.000001430	0.00000000973
α_+	0.000001417	$0.0000000113\,0.$

It l is the length of each arm, λ the wavelength of the light, T the temperature, and n^{-1} the fraction of a fringe shift detectable by the optical bridge,

$$\Delta Tl(\alpha_- - \alpha_+) = \frac{\lambda}{2n},$$

so that the system maintains constant the temperature to within

$$\Delta T = \frac{\lambda}{2nl(\alpha_- - \alpha_+)}.$$

With respect to the reflectors in Fabry-Pérot configuration, for small temperature changes the increase in length of the short part of the path, proportioned in the ratio of the expansion coefficients, cancels the decrease in length of the long part. But the temperature is maintained stable by the Michelson system within the smallest fringe shift detectable by the optical bridge; so that the path-length remains constant in the Fabry-Pérot system much below the limits of detectability. Apparent changes in the path-length of the Fabry-Pérot system are therefore attributable to changes in the emission frequency of the laser, and are used to signal a tuning servo employing a piezo-electric transducer to move one of the laser mirrors, after the fashion of Lipsett and Lee (1966).

The thermal stabilisation required to reduce path-length changes in the Fabry-Pérot system to less than one-thousandth of a wavelength using the output of a helium-neon laser is several thousandths of a degree Centigrade; this degree of thermal stabilisation could be achieved using an independent thermostat. Doing so would make it possible to maintain continuity of dimensional stability independently of change of laser, or laser breakdown. The vacuum-chamber containing the beryl etalon should be connected to the chamber of the strain interferometer to make the system independent of changes in the propagation velocity.

The ability to control vacuum wavelength to less than a part in 10^{10} for periods without limit, restricting drift to less than 3 parts in 10^{17} per year, would make it possible to observe the imposition of very long period volumetric strain by using a laser so stabilised in the Kelvin seismograph. The removal of site effects such as a slow inflow of the cavern is another matter.

The stable etalon described could be employed in conjunction with an atomic frequency standard to investigate changes in the propagation velocity of light.

Acknowledgements

The seismograms shown in Figures 1b and 1c were made in collaboration with Dr. V. Vali at the Cascades Geophysical Site operated jointly by the University of Washington and Boeing Scientific Research Laboratories. The interpretation reached is solely the responsibility of the author. Our research has been assisted by grants from the National Science Foundation and by equipment and facilities made available by the U.S. Air Force, Project VELA Uniform.

References

Alexandrov, E. B. and Bonch-Bruevich, A. M.: 1967, 'Investigations of Surface-Strains by the Hologram Technique', *Zh. Tekhn. Fiz.* 37, 360–369.
Baird, K. M.: 1968, 'The Role of Interferometry in Long-Distance Measurement', *Metrologia* 4, 135–144.
Bender, P. L. and Owens, J. C.: 1965, 'Correction of Optical Distance Measurements for the Fluctuating Atmospheric Index of Refraction', *J. Geophys. Res.* 70, 2461–2462.
Benioff, H.: 1935, 'A Linear Strain Seismograph', *Bull. Seism. Soc. Am.* 25, 283–309.
Ben-Menahem, A. and Singh, S. J.: 1968, 'Multi-Polar Elastic Fields in a Layered Half-Space', *Bull. Seism. Soc. Am.* 58, 1519–1572.
Bostrom, R. C.: 1969, Reflectors Returning Refractivity and Range Information. A.G.U./E.S.S.A. Symposium on Electromagnetic Distance Measurement, Boulder, Colo., June 1969,

Bostrom, R. C. and Vali, V.: 1968a, 'A Seismograph to Observe Deviatoric Strains', *Trend Eng.* **20**, 7–12.

Bostrom, R. C. and Vali, V.: 1968b, 'Strains Recorded on a High-Magnification Interferometric Seismograph', *Nature* **220**, 1018–1020.

Clemence, G. M. and Szebehely, V.: 1967, 'Annual Variation of an Atomic Clock', *Astron. J.* **72**, 1324–1326.

Cook, A. H., Marussi, A., and Rowley, W. R. C.: 1965, 'A Laser Strain Gauge for Seismology', *Geophys. J. Roy. Astron. Soc.* **9**, 281–282.

Essen, L. and Parry, J. V. L.: 1957, 'The Caesium Resonator as a Standard of Frequency and Time', *Phil. Trans. Roy. Soc. London* **250**, 45–69.

Fowler, R. A.: 1968, 'Earthquake Prediction from Laser Surveying', Nat. Aeron. and Space Admin., SP-5042, 32 pages.

Gaskill, J. D.: 1968, 'Imaging through a Randomly Homogeneous Medium by Wavefront Reconstruction', *J. Opt. Soc. Am.* **58**, 600–608.

Goodman, J. W., Huntley, W. H., Jackson, D. W., and Lehmann, M.: 1966, 'Wavefront-Reconstruction through Random Media', *Appl. Phys. Letters* **8**, 311–313.

Guinot, M. B.: 1967, 'Formation d'une échelle moyenne de temps atomique', *Bull. Astron.* **2**, 449–464.

Hanes, G. R. and Baird, K. M.: 1969, 'I_2-Controlled He-Ne Laser at 633 nm', *Metrologia* **5**, 32–33.

Horman, M. H.: 1965, *J. Appl. Opt.* **4**, 333–336.

Kelvin, Lord: 1902, 'A New Specifying Method for Stress and Strain in an Elastic Solid', *Proc. Roy. Soc. Edinburgh* **24**, 97–101.

Lipsett, M. S. and Lee, P. H.: 1966, 'Laser Wavelength Stabilization with a Passive Interferometer', *Appl. Opt.* **5**, 823–826.

Mertz, L. and Young, N. O.: 1961, *Proc. Internat. Conf. on Optical Instruments, London 1961*, p. 305.

Mielenz, K. D., Nefflen, K. F., Rowley, W. R. C., Wilson, D. C., and Engelhard, E.: 1968, 'Reproducibility of Helium-Neon Laser Wavelengths at 633 nm', *Appl. Opt.* **7**, 289–293.

Milne, J.: 1888, The Relative Motion of Neighbouring Points of Ground', *Trans. Seism. Soc. Japan* **12**, 63.

Press, F.: 1965, 'Displacements, Strains and Tilts at Teleseismic Distances', *J. Geophys. Res.* **70**, 2395–2412.

Sassa, K., Ozawa, I., and Yoshikawa, S.: 1951, 'Observation of Tidal Strain of the Earth', *Intern. Assoc. Geod. Brussels Assembly 1951*.

Shamsi, S. K. and Stacey, F. D.: 1967, 'Michelson Interferometer as an Earth Strain Sensor', *Earth Planet. Sci. Letters* **3**, 466–468.

Thayer, G. D.: 1969, 'Atmospheric Effects on Multiple Frequency Range Measurements', A.G.U./ E.S.S.A. Symposium on Electromagnetic Distance Measurement; Boulder, Colo., June 1969.

Thompson, M. C. Jr. and Janes, H. B.: 1967, 'Correction of Atmospheric Errors in Electronic Measurements of Earth Crust Movements', *Bull. Seism. Soc. Am.* **57**, 641–655.

Thompson, M. C. Jr. and Wood, L. E.: 1967, 'The Use of Atmospheric Dispersion for Refractive Correction of Optical Distance Measurements', in *Electromagnetic Distance Measurements*, I.A.G. and Hilger and Watts, London, pp. 165–172.

Tutton, A. E. H.: 1922, *Crystallography and Practical Crystal Measurement*, II, Macmillan & Co., Ltd., London, pp. 1321–1330.

Vali, V. and Bostrom, R. C.: 1968, 'Use of a Laser Extensometer to Observe Strain in a Large Ground Sample', *Bull. Geod.* **88**, 151–157.

Vali, V., Krogstad, R. S., and Vali, W.: 1964, 'Measurement of Earth Tides and Continental Drift using a Laser Interferometer', *Proc. IEEE* **52**, 7.

Vali, V., Krogstad, R. S., Moss, R. W., and Engel, R.: 1968, 'Some Observations of Strains across the Kern River Fault Using Laser Interferometer', *J. Geophys. Res.* **73**, 6143–6147.

Wideman, C. J. and Major, M. W.: 1967, 'Strain Steps Associated with Earthquakes', *Bull. Seism. Soc. Am.* **57**, 1429–1444.

Discussion

Bender: We have an instrument that has operated intermittently for some time now. The laser is locked onto the fused quartz cavity that is thermally and acoustically isolated. Although it has only run for a total of perhaps 10 days we have not yet seen any evidence of strain steps, except possibly at the time of earthquake activity. The instrument seems to have a stability of something like one part in 10^{12} with reference to how well the laser is locked to the reference interferometer. However, the temperature of the reference interferometer is only constant to a few millidegrees centigrade, so there are possibilities of drift of the order of 1 part in 10^9 in the system. We plan to install a 3.39 μ laser stabilized against the methane absorption band so that we hope to get a very accurate long term reproducible reference for measuring strain.

Markowitz: I must say that I believe that it is extremely difficult to make a material object which will give you stabilities of the order of 1 part of 10^{11} over a year. It is not so much the problems that you have mentioned but one such as insuring that the surface layer of the mirrors which you use does not change by much less than one layer of atoms during the course of time. Even if you use a vacuum of the order of 10^{-6} or 10^{-7} torr, it is still possible to have an accumulation of products on the mirrors which will change the effective position of the reflection. The only way to avoid this kind of thing is to arrange to make changes in the length of the interferometer in such a way that the distance you are using is not related to the surface of the mirror but just to the changes that you make. Then you have to have changes in mechanical positioning arrangements and this also is an extremely difficult thing to do.

Bostrom: I agree that it is extraordinarily difficult. I think the principal difficulties in this area are no longer going to be in the stabilization problem but in site difficulties. Such things as the correct way of fastening mirrors to the walls of the cavities. We can use all sorts of finesse in this kind of thing but then we go and stick a mirror on the wall with epoxy.

Bender: I agree with that. Our instrument is located in Colorado and I think there are some advantages to this site. On the other hand, the instrument is only 30 m long so that our end effect problems are much more serious than yours. Another problem is that one has to balance out the amount of light getting reflected back into the laser so that it is down a part in 10^4 of the light going through the other way, otherwise you can get apparent step changes and so forth. These adjustments can be done but they really require a great deal of care.

Vali: We misalign the instrument a little bit so that the beam does not go back into the laser at all.

Bender: If the laser beams overlap I do not see how you avoid any of the light getting back into the laser by the misalignment.

Vali: If you align it exactly you can see the beam going back into the laser through the telescope, or whatever you have in front of it. If you turn both mirrors slightly you can set it so that no laser beam is going back into the laser at all.

SOME EARTH STRAIN MEASUREMENTS WITH LASER INTERFEROMETER

VICTOR VALI

University of Washington, Seattle, Wash., U.S.A.

Abstract. The use and limitations of laser strainmeters for geophysical strain measurements are discussed. Some observations with the Kern River Fault interferometer and the 1000 m interferometer at Stevens Pass, Washington, are presented.

The development of the laser made it clear that it is practical to build long armed interferometers for accurate displacement measurements. There are two reasons for this, the first being the long coherence length of the laser light and second, the very small divergence (diffraction limit) of the light. The long coherence length, defined as $c/\Delta f$, where c is the velocity of light and Δf the bandwidth of the laser light, means that one can make the optical path difference between two light paths as unequal as one desires (presently all terrestrial distances) without losing the interference pattern. In principle, it is possible to produce highly coherent light by selecting a very narrow spectral range of ordinary light; however, the intensity of such light is so low as to be unusable. One of the most suitable light sources at present is a He-Ne laser, which emits visible red light at 6328 Å units.

Since the atmospheric density fluctuations (caused by turbulence, pressure, temperature and composition changes) would change the optical path (which consists of the geometrical path times the index of refraction of air) of the light beam by a few parts in one million, almost all of the light path between the interferometer mirrors must be inside vacuum. Therefore, a long evacuated pipe is placed between the interferometer mirrors. The reduction of the influence of air density fluctuation is proportional to the ratio of the beam path in vacuum to the path in air.

At present, two different types of interferometer systems are being used. The first is a Fabry-Pérot instrument. It has the advantage of simplicity and ease of operation. In addition, it is possible to mount the interferometer mirrors in air as close to the windows on the end of the vacuum pipe as one desires. The other one is a Michelson interferometer. This arrangement enables one to make the air gaps of the two arms equal, thus eliminating the effect of pressure fluctuations. However, it is still turbulence sensitive; depending on the desired accuracy of measurement, this effect can be important.

To estimate the residual influences of environmental changes on the results, one proceeds as follows. The optical path is defined as

$$l_0 = nl,$$

where n is the index of refraction and l is the geometrical path length. Hence

$$\Delta l_0 = n\Delta l + l\Delta n.$$

L. Mansinha et al. (eds.), Earthquake Displacement Fields and the Rotation of the Earth, 206–216.
All Rights Reserved. Copyright © 1970 by D. Reidel Publishing Company, Dordrecht-Holland

The first term in the sum contains the displacement to be measured while the second term is the error due to index of refraction change in the beam path. For pressures less than 10^{-5} mm Hg the $\Delta n < 3 \times 10^{-12}$. The influence of the residual air in the vacuum pipe is proportional to its length. For a 1000 m path length the error is $< 3 \times 10^{-7}$ cm or 30 Å.

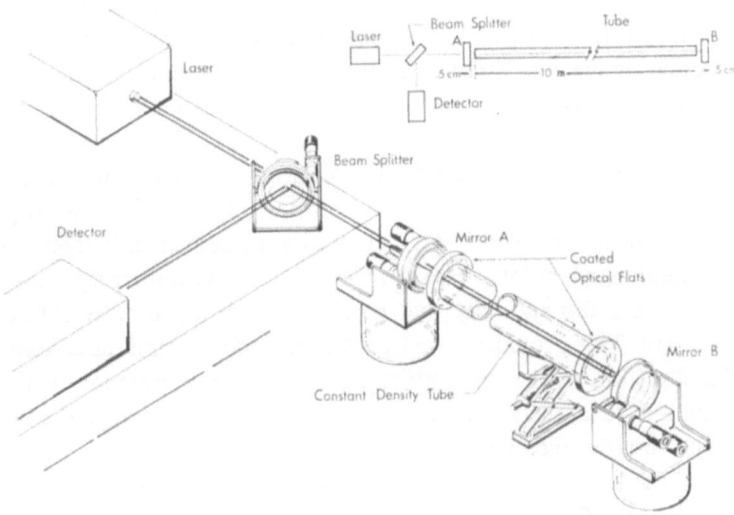

Fig. 1. Fabry-Pérot interferometer configuration.

The change in length of the evacuated tube due to temperature changes also changes the optical path length, because one substitutes air for vacuum over the increment in length resulting from expansion. The influence of $10^{-3}\,^\circ$C temperature change for the aluminum tube is $\Delta l = \Delta l_0 \, (n - n_0)$, where Δl is the change in length of the tube $(\Delta l = \alpha l \Delta T = 2 \times 10^{-5} \times 10^5 \times 10^{-3} = 2 \times 10^{-3}$ cm) and $(n - n_0)$ is the index of refraction difference between air and vacuum. This means that the change of optical path is $\Delta l_0 = 2 \times 10^{-3} \times 3 \times 10^{-4} = 6 \times 10^{-7}$ cm or 60 Å. This is about 2×10^{-2} of a fringe for the 1000 m instrument.

The change in length of the Al tube due to atmospheric pressure changes is

$$\Delta l = \Delta F l / A Y,$$

where ΔF is the change of force on the end of the tube due to atmospheric pressure change, l is the length of the tube, A is the cross section area of the Al making up the tube, and Y is Young's Modulus for this material. Numerically, $\Delta F = \pi r^2 \Delta p$ (r is the radius of the aluminum pipe, $\cong 7.8$ cm, and Δp is the atmospheric pressure variation), since $\Delta F \cong 150 \times 3 \times 10^4 \cong 4.5 \times 10^6$ dynes, and $l = 10^5$ cm, $A = 27$ cm^2, $Y = 7 \times 10^{12}$, dyne cm^{-2}, i.e., $\Delta l = 2 \times 10^{-3}$ cm for the 1000 m instrument.

To measure the distance change between the interferometer mirrors one can count the fringes that pass a point on the screen. In addition, one has to determine the direction of the fringe motion and the corresponding ground motion to obtain unambiguous results. Another and more sensitive method is to build a fringe following servo mechanism. If one uses a speaker cone or a galvanometer for this purpose, the fringe position is indicated by the voltage necessary to keep the speaker cone or the galvanometer mirror locked to an interference fringe. By this means the position of a fringe has been determined with accuracy of 1% of the distance between fringes. Since the distance between two consecutive fringes is equal to half of the wavelength of the light used, it is in this case equal to about 30 Å. For comparison, this is equivalent to about 15 atomic diameters. Noise level, largely due to statistical fluctuations of the number of photons detected by the photomultipliers, in lasers is so low that measurements to 10^{-5} of a fringe separation are practical.

To make the readout system capable of following displacements more than a few fringes, it is provided with a limit switch, the purpose of which is to interrupt the servo mechanism after it has traveled a distance of about one fringe. It then returns to its mechanical equilibrium point where it picks up the adjacent fringe, so that in the recording there is a discontinuity that corresponds to a ground displacement of a half a wavelength. This provides a very convenient and continuous calibration of the system. The direction of the discontinuity determines the direction of the displacement. This kind of readout system makes the dynamic range (that is, the earth motion amplitude the interferometer can record) of the instrument, in principle, infinite. Otherwise (in addition to other problems) if one wants to measure strain with accuracy equal to one part in 10^{12} and still include the earth tides (which are about 5 parts in 10^8) without the use of the limit switch, the width of the recorder paper would have to be 20 m. Similar considerations apply to the magnetic tape recording: one can accommodate only about four orders of magnitude of signal change.

The instrument measures only one component of the earth motion, namely, the distance change between the interferometer mirrors. The mirrors have to be mounted so that the reflecting surfaces are, in effect, fastened to the earth. If, for some reason, there is a slight misalignment of the interferometer and the fringes get narrower or wider, the size of the discontinuity on the recording also changes, but it will still correspond to the earth motion of one half of the wavelength of light.

Another readout system can be used for determining absolute displacements. If a laser wavelength is locked to a Fabry-Pérot cavity (meaning that the number of waves within the cavity stays constant) that consists of two mirrors fastened to the bedrock its frequency change is proportional to the displacement of the ground. When this changing frequency light is mixed with a constant frequency light, the resultant beat frequency change is proportional to the ground displacement. The beat frequency is measured by a frequency counter and recorded. To make sure that all the interesting ground oscillations are observed the time interval of measurement of the beat frequency has to be one fifth or less than the shortest period of ground motion. In this case the laser light source has to be kept outside of the distance over which the displacement is to be measured. The influence of the laser discharge tube would otherwise be too large for meaningful measurements.

The accuracy of locking a laser wavelength to an outside Fabry-Pérot cavity is about the same as the accuracy of a fringe following servo mechanism. However, this system has the great advantage of having no mechanically moving parts. The strain sensitivity of the beat frequency type strainmeter is independent of its length, whereas for the interferometer it is proportional to it. The choice between these two readout systems is determined largely by the availability of the components.

One of the major advantages of laser strainmeters is the lineaɪity of the response, that is, it is independent of the frequency of the ground motion. Further, it does not have any instrumental resonances to complicate the interpretation of results. Since the amount of information recorded in a single day, for example, is very large, it is recorded on magnetic tape to make it easily accessible for computer processing.

When strains are measured on the surface of the earth one detects, in addition to mechanical strains, thermal strains. These are caused by daily and yearly temperature variations and, ordinarily, are not very interesting things to investigate. To minimize the temperature effects, the earth strain measurements should be conducted in mines or abandoned railroad tunnels. Another way is to put the strainmeter mirrors on long pillars that are set deep into the ground such that the thermal strains become negligible.

One of the interferometers is located at the Kern River Fault close to Lake Isabella in California (Vali *et al.*, 1968). Since the main purpose of this instrument is to look

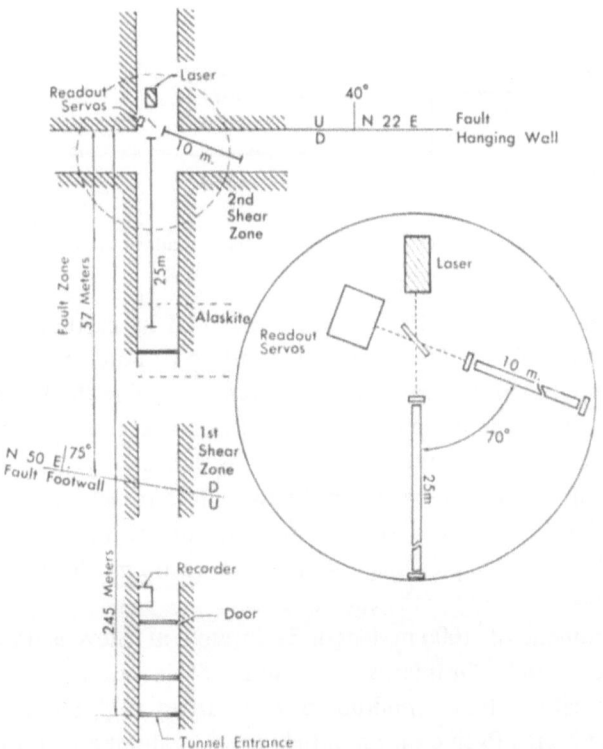

Fig. 2.   The Kern River Fault tunnel and instrument configuration.

at the shear motion across the fault, it is a two armed instrument with relatively short arms of 10 and 25 m.

At the Kern River Fault site the observed amplitude of earth tides is about 10 times larger than normal, indicating, as expected, a magnification of strain across a fault. Another interesting thing observed at that site is the fine structure of the fault motion. A typical recording shows a cluster of short duration back and forth motions which usually continue for a few minutes. The earth normally returns to its original strain state after the event ends. Sometimes, however, a new strain state is established. These events could be caused by small strains that occurred some distance away from the recording site. When such a strain is relieved, the original strain state at the instrument is re-established.

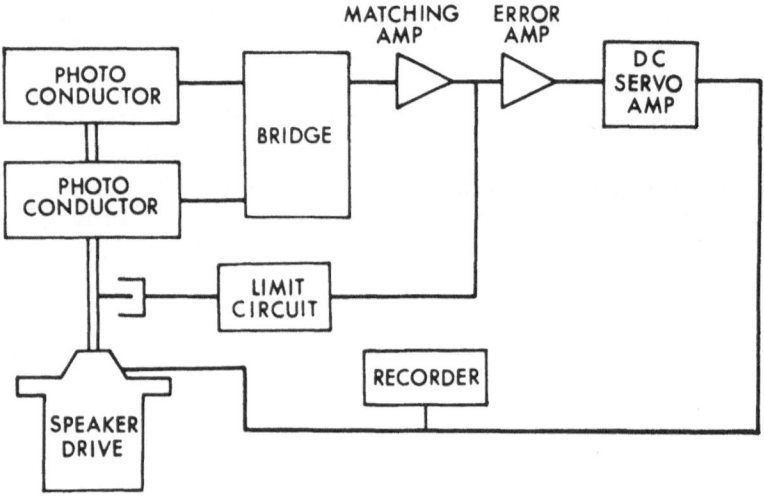

Fig. 3.   Block diagram for the servo readout system.

Some of the events recorded at the site are caused by earthquakes and some by underground explosions in Nevada. It is interesting to note that using only one instrument, a two-component strainmeter, one can determine the location and size of the event. The distance can be determined from the arrival time difference between the pressure waves and shear waves, while the direction can be determined from the amplitude ratio of the pressure waves in the two interferometer arms. The size of the event is determined from the absolute amplitude of the pressure or shear waves. Usually for underground explosions, the shear waves are much smaller (as compared to the pressure waves) than for the earthquakes.

Another instrument of 1000 m length is located at Stevens Pass, Washington (Vali and Bostrom, 1968). Ordinarily a recording of this interferometer usually shows microseismic activity with an amplitude of one part in 10^{10}. Sometimes when there are storms in the North Pacific the amplitude goes up about a factor of 5. The power spectrum obtained from the recording of this instrument shows a pronounced peak at

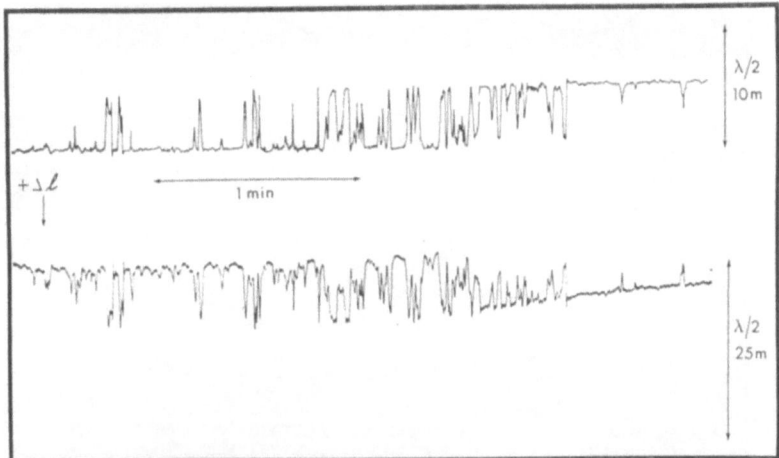

Fig. 4. Fine structure of fault motion across the Kern River Fault.

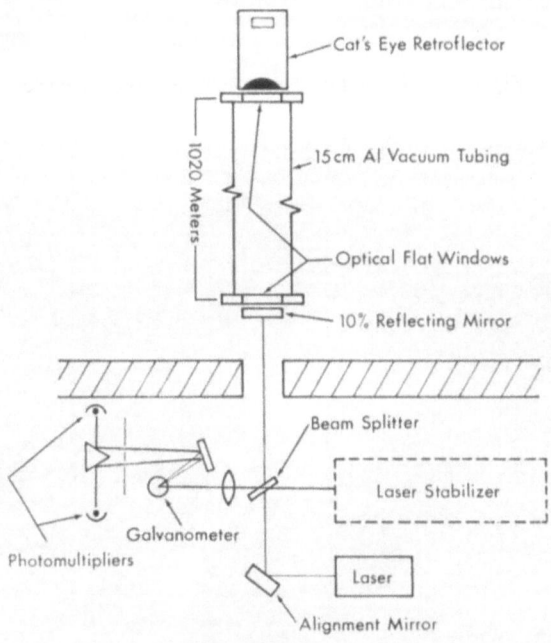

Fig. 5. Diagram of the Stevens Pass interferometer.

microseismic period of 7 sec per cycle. At about 7 min per cycle there is also a strong peak. This is probably due to microbarographic air pressure fluctuations. The earth tides form a strong third reference point in the power spectrum.

The only difficulty with these systems has been the laser wavelength stability: the laser wavelength change would change the output of a strainmeter that would be indistinguishable from the strain caused output change. For especially built lasers that

Fig. 6.   Active end of the Stevens Pass interferometer.

Fig. 7.   1000 m long vacuum pipe at Stevens Pass.

are placed in thermally and acoustically quiet environment, the drift of the wavelength can be made less than one part in 10^{10} per hour. However, this is not sufficient for secular strain measurements. Two different systems for stabilization have been developed and are in operation. The first system (developed by R. H. Lovberg and J. Berger at UCSD) uses a standard length, about 30 cm, made of fused quartz which is well insulated acoustically and placed in a temperature controlled environment. This length, in effect, forms the cavity of a Fabry-Pérot interferometer. Part of the light emitted by a laser is deflected into this cavity. If, for some reason, the wavelength of the laser light changes, the output of this system changes and is fed back to the laser and the original wavelength is restored. The system is as good as the length standard that forms the stabilizing cavity.

Another method developed by the JILA group (J. L. Hall, 1968) utilizes an extremely narrow molecular absorption or emission line and locks the wavelength of the laser to it. They have found that a rotation-vibration line of methane at 3.39 μ wavelength is suitable for this purpose. At millitorr pressures this line has very small shift due to pressure, temperature and other external conditions. By this means one circumvents the requirement of mechanical stability of a length standard and, in effect, refers only to the atomic constants. A reproducibility of better than one part in 10^{11} has already been attained. The ultimate half width of this line is about one part per 10^{12}, so a corresponding laser stability can hopefully be achieved. This would correspond to the present readout sensitivity of the 1 km interferometer.

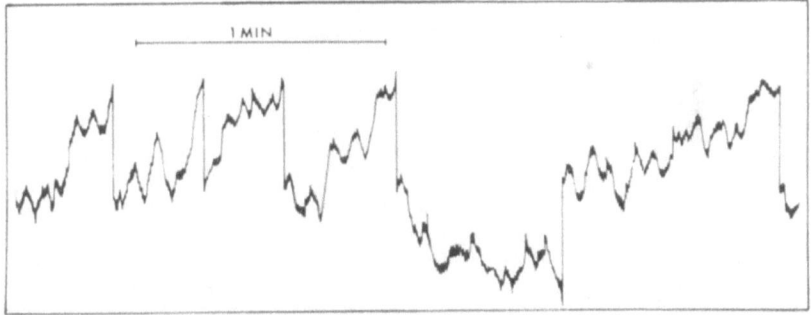

Fig. 8. A few minutes recording of the Stevens Pass instrument. There are six discontinuities. These occur when the limit switch is actuated and the servo travels to the next fringe.

The laser strainmeters now in operation go a long way in increasing the accuracy, reliability, simplicity and span length of the earth over which strain may be measured. However, to specify the strain completely, all six components of the strain tensor would have to be measured. The simplest form of this kind of an instrument is a tetrahedral interferometer where strains on all six sides are measured independently. To obtain all strain components, including the uniform expansion and contraction of the ground spanned by the tetrahedron, the laser wavelength has to be stabilized to within the measurement accuracy.

There are some measurements that indicate that the American continent moves away from Europe and Africa at the rate of a few centimeters per year. This motion is concentrated in the mid-Atlantic rift, a region of seismic activity on the ocean floor. The rift passes through Iceland, however, where the region could be spanned with a laser strainmeter to determine the structure of the drift.

The Barbados Ridge is a zone of compression where the Atlantic basin is moving with respect to the Caribbean block at the rate of a few millimeters per year. It is another interesting location for placing the strainmeter to determine the fine structure of crustal movement.

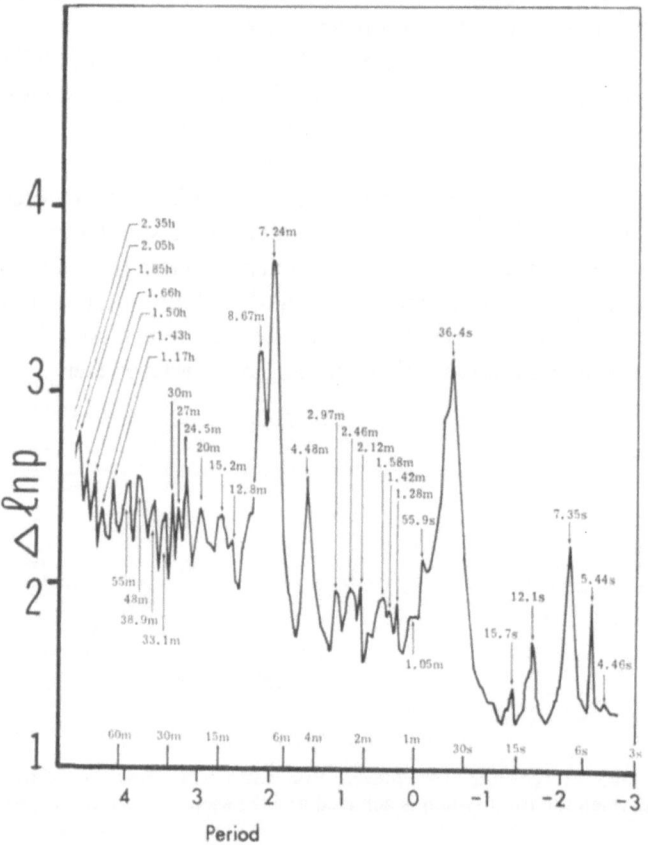

Fig. 9. Power spectrum of earth strains at Stevens Pass (May 28–June 2, 1968).

An interesting application of laser strainmeters would be the use of the earth or the moon as a receiving antenna for detecting gravitational radiation. According to general relativity, a rotating or oscillating mass quadrupole, such as a binary star system, a rotating rod, etc., should emit gravitational waves. The forces of these waves that interact with a detector are transverse to the direction of propagation, analogous to electromagnetic radiation. However, they are not vector forces and therefore do not cause any motion of the center of the mass but produce only expansions and compressions. In 1961 R. L. Forward *et al.* published a note in *Nature* about their attempt to detect interstellar gravitational radiation with the earth as the receiving antenna. The data were obtained during a seismically quiet period with a Benioff gauge at the Lake Isabella site in California. The power spectrum, however, did not show any peaks at the normal mode frequencies of the earth. Therefore, they were only able to set an approximate upper limit for the effects of gravitational radiation. The increased span length and sensitivity of geophysical laser strainmeters make it desirable to continue the search for gravitational radiation by this method. With long recording times one can quite routinely see periodic signals whose amplitudes are 10^4 times smaller than random noise.

Because of the lack of atmosphere, the lunar strain measurements do not require any long evacuated pipes. One of the interesting locations for placing a laser strainmeter would be across the crater Copernicus. It is about 100 km in diameter and the crater rims are high enough to be above the horizon. Furthermore, it is one of the areas that may be explored by future lunar flights. The lunar tides are of 27-day period, caused mainly by the eccentricity of the orbit. On the basis of similarity between the moon and the earth it had been estimated that the lunar tides have an amplitude of about 2 m or more. However, there is increasing evidence that this value may be far too high.

The lunar satellite discovery of mass concentrations (mascons) under some lunar maria indicates, according to Z. Kopal (private communication), Manchester, that the rigidity of the moon has to be at least 1000 times higher than previously estimated. This means that the lunar tide amplitude is less than 1 cm. Hence, the change in the diameter of the crater Copernicus over a lunar cycle is less than 0.5 mm, corresponding to about 2000 fringes in the interferometer. Kopal has estimated that a detectable meteor impact caused moonquake will take place on the average less than once in a century. Small meteors are not effective in causing seismic waves on the moon because their energy is spent in the granular surface layer (regolith). Internally caused lunar seismic activity is probably very small and infrequent. These considerations indicate that the moon might be a very quiet body and therefore a good gravitational radiation receiving antenna. The lowest normal mode of lunar oscillation is about 15 min. This is about 3000 times shorter period than the lunar tides, and offers no difficulty in discrimination even when the amplitude ratio is 10^5 or more. The observation of lunar oscillations would be of interest in itself, no matter what the excitation mechanism is (meteor impact, seismicity, gravitational radiation) because from the damping coefficient, one can learn about the internal constitution of the moon.

References

Forward, R. L. *et al.*: 1961, 'Upper Limit for Interstellar Millicycle Gravitational Radiation', *Nature* Feb. 11, 1961, p. 473.

Hall. J. L.: 1968, 'The Laser Wavelength Stabilization Problem', 5th International Conference on Quantum Electronics, Miami, Fla.

Vali, V. and Bostrom, R. C.: 1968, 'One Thousand Meter Laser Interferometer', *Rev. Sci. Instr.* **39**, 1304.
Vali, V., Krogstad, R. S., Moss, R. W., and Engel, R.: 1968, 'Some Observations of Strains across the Kern River Fault using Laser Interferometer', *J. Geophys. Res.* **73**, 6143.

Discussion

Major: I do not think the peak at 7 or 8 min is a characteristic of the laser. Most strain meter measurements show garbage somewhere in this part of the band; it may be barometric.

Vali: I think so too and we are putting in microbarographic instruments to check this.

Major: One does not see this peak on ordinary seismographs because their response at 7 min, for a traditional 130 sec pendulum system, is practically zero.

Vali: Once you get this instrument stabilized, its frequency response is from zero to whatever your readout frequency is.

ERROR ANALYSIS OF A LASER STRAINMETER

ANTHONY F. GANGI

Texas A & M University, College Station, Tex., U.S.A.

Abstract. A servo-controlled LASER strainmeter has been designed based on an optical interferometer. The design goal is a strain sensitivity of 10^{-10} using a small interferometer – of the order of
1 m. To achieve this sensitivity with such a small instrument, it is necessary to pay careful attention to
possible noise or error sources.

The major error source has been found to be due to temperature fluctuations. The next most important error source is the electrical noise in the instrument circuits. The possible error sources due to
frequency and amplitude instabilities of the LASER and index of refraction changes can be made
small and negligible by proper design. To make these effects negligible, the arm lengths of the interferometer are made equal, the interference or fringe pattern is modulated at an audio frequency and
the light paths of the interferometer are kept in a partial vacuum.

To achieve the necessary temperature stability, the instrument must be placed in a constant temperature environment such as found in deep mines or tunnels. However, it is shown that the stringent
temperature requirements are not unique to small strainmeters.

1. Introduction

A broadband, servo-controlled LASER strainmeter has been designed based on the
Michelson-Morely interferometer (Salisbury and Gangi, 1967). Two configurations
of the servo-control system have been used. One is called the DC servo-control
system and its block diagram is shown in Figure 1a. The other is called the AC servo-
control system and its block diagram is shown in Figure 1b. A complete dynamic
analysis of the two servo-controlled strainmeter configurations has been performed
(Gangi, 1969).

The basic design principle of the instrument is as follows: one arm, the 'free arm'
of the interferometer (defined by the distance from the beam splitter to the 'free
mirror', and denoted by X_2 in Figure 1) is attached to the Earth's surface and its
length changes as the Earth's surface is strained. The other arm of the interferometer
(denoted by X_1 in Figure 1) is servo-controlled to keep its length equal to the length
of the 'free' arm. This maintains a constant fringe pattern at the output of the interferometer; that is, at the photo-detector.

The amount of motion required to maintain the servo-controlled arm equal in
length to the 'free' arm is monitored to provide an output reading of the strains induced in the earth's surface by seismic activity. The motion of the servo-controlled arm
is accomplished by using a voltage-to-mechanical motion (piezoelectric) transducer. It
is the voltage to this transducer that is monitored.

The design goal for the instrument is to achieve a threshold strain sensitivity of
10^{-10} and a flat frequency response from DC to about 1 cps with an instrument that
would fit in a 1 m^2. The Benioff quartz strainmeters (Benioff, 1935) have higher
sensitivity than 10^{-10}, but this is due to the fact that they measure strains over larger
distances (from 10 to 100 m). The sensitivity of the present instrument can be in-

L. Mansinha et al. (eds.), Earthquake Displacement Fields and the Rotation of the Earth, 217–229.

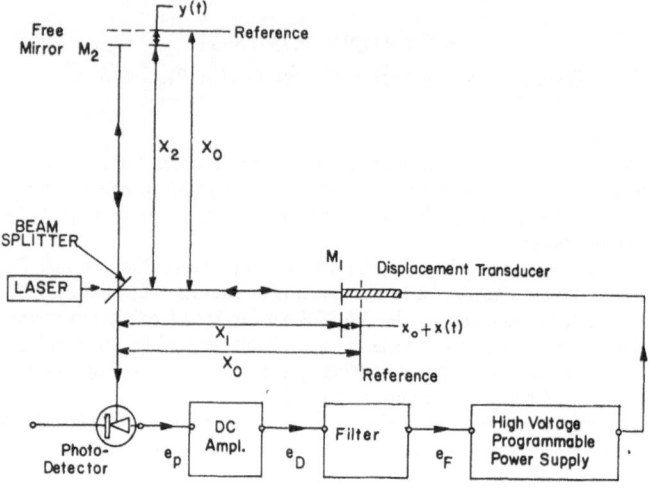

a. DC SERVO CONTROL SYSTEM

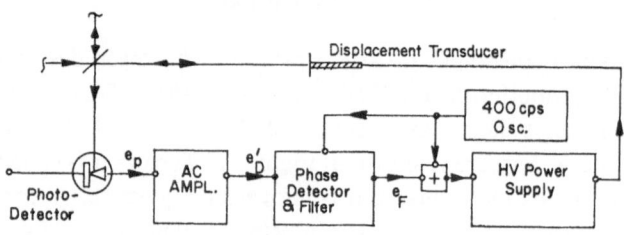

b. AC SERVO CONTROL SYSTEM

Fig. 1.   Servo-controlled LASER strainmeter diagram.

creased by a factor of ten by increasing the arm lengths to 10 m. However, the approach taken here has been to obtain as high a sensitivity as possible with a one meter arm length before attempting to increase the size of the instrument.

An instrument with a strain sensitivity of 10^{-10} is capable of measuring the Earth strains due to Earth tides (Earth tide strains are about 5×10^{-8}); the free oscillations of the Earth excited by large earthquakes; the long period mantle, surface and body waves from earthquakes of about magnitude 6 or greater; and the DC strains from large earthquakes.

To achieve this strain sensitivity with a small instrument, it is necessary to pay careful attention to the noise and error sources present. The major noise or error sources are in the construction of strainmeters, their readout transducers and associated electronic circuits. These latter two noise or error sources remain constant as the size of the instrument increases. The possible error sources, their expected magnitudes and means to minimize them are investigated here.

The instrument discussed here is intrinsically broadband. It can sense strains from zero frequency (DC) to well above 1000 cps. The frequency response of the system is designed to be flat from DC to about 1 cps. The instrument bandwidth is varied by changing the characteristics of the filter in the servo-loop. In its operating characteristics, the instrument is similar to the quartz strainmeters designed by Benioff. The major advantage of LASER strainmeters over the quartz strainmeter is that the ageing characteristics are less severe. The quartz strainmeters have a large long term drift as the quartz rod ages. This long term drift is believed to be associated with the relieving of internal stresses and strains in the quartz rod. In LASER strainmeters, the LASER light beam (usually from a helium-neon gas LASER) serves the same function as the quartz rod and it has no equivalent ageing characteristics. However short term and long term perturbations of the LASER light beam 'length' can be introduced by LASER frequency variations and by turbulence in the air. The turbulence causes changes in the index of refraction along the beam path. To eliminate this latter problem, the LASER light beam is propagated through a low-pressure air path.

One of the major problems limiting the instrument sensitivity is the temperature fluctuation in the instrument environment. The coefficients of thermal expansion of most rock or Earth materials are of the order of $10^{-5}/°C$. This indicates that the temperature must remain constant to better than $10^{-4}°C$ in order to achieve strain sensitivities of 10^{-10}. Temperature control or stability of this order is achieved generally only at depth in the Earth. For this reason most strainmeters are located in deep mines, tunnels or caves.

2. Error Analysis

The error analysis of the servo-controlled LASER strainmeter depends upon the configuration used for the feedback control loop. The expressions describing the operation of the DC and AC servo-loops are derived in detail elsewhere (Gangi, 1969). We will begin the error analysis with the DC servo-loop configuration since it is the simpler of the two configurations. There are a number of advantages to the AC configuration over the DC configuration of the servo-loop and these are best demonstrated by analyzing the DC configuration.

The equation describing the operation of the DC servo-controlled strainmeter is given by Gangi (1969):

$$x(t) = K_K K_T \int_{-\infty}^{t} [e_0(T) + \alpha e_{D_0} \cos(4\pi/\lambda)(x_0 + x(T) - y(T))]$$

$$\times f(t-T)\,dT$$

$$= K_T K_K e_F(t) = K_T K_K e_D(t) * f(t) \tag{1}$$

where $e_0(t) = K_p(e_{p_0} + K_D I_0) - e_B \ll \alpha e_{D_0}$; $K_p = $ DC amplifier gain (volts/volt); $e_{p_0} = $ 'Dark' voltage from photo-detector (volts); $K_D = $ photo-detector conversion coefficient (volts/watt); $I_0 = $ LASER light intensity (watts); $e_B = $ bucking bias from DC amplifier (volts); $K_K = $ high voltage amplifier gain (volts/volt); $K_T = $ displacement transducer conversion coefficient (microns/volt); $\alpha = 2A_1A_2/(A_1^2+A_2^2) = A_1A_2/I_0$; A_1, A_2 = light amplitudes from arms 1 and 2 respectively; $e_{D_0} = K_pK_DI_0$; $e_F(t)$ = voltage output from the filter; $e_D(t)$ = voltage output from the DC amplifier; $f(t)$ = the impulse response of the filter; λ = LASER light wavelength = 0.6328 microns.

The above expression is obtained under the following conditions: (1) we have assumed, for the moment, that there are no variations in the index of refraction along the light paths in the two arms; (2) the only element in the servo-control loop which limits the frequency response is the filter – all other elements are broadband with flat frequency responses beyond 100 cps; (3) the displacement of the transducer is proportional to the output voltage of the filter (the proportionality constant being K_KK_T, the product of the high voltage amplifier gain and the transducer conversion coefficient); (4) the output voltage of the filter is equal to the convolution of its input voltage (i.e., the output voltage from the DC amplifier) with the impulse response of the filter, $f(t)$; and (5) the output voltage of the preamplifier is determined by the fringe pattern at the photo-detector; namely

$$e_D(t) = e_0(t) + \alpha e_{D_0} \cos 4\pi \left(X_1(t) - X_2(t)\right)/\lambda. \tag{2}$$

The variation of the DC amplifier output voltage, e_D (and the photodetector output voltage, e_p), with relative arm length changes is shown in Figure 2.

In the above expression, α is a measure of how well the light beam splitter forms equal intensity beams in the two arms of the interferometer. For equal intensity

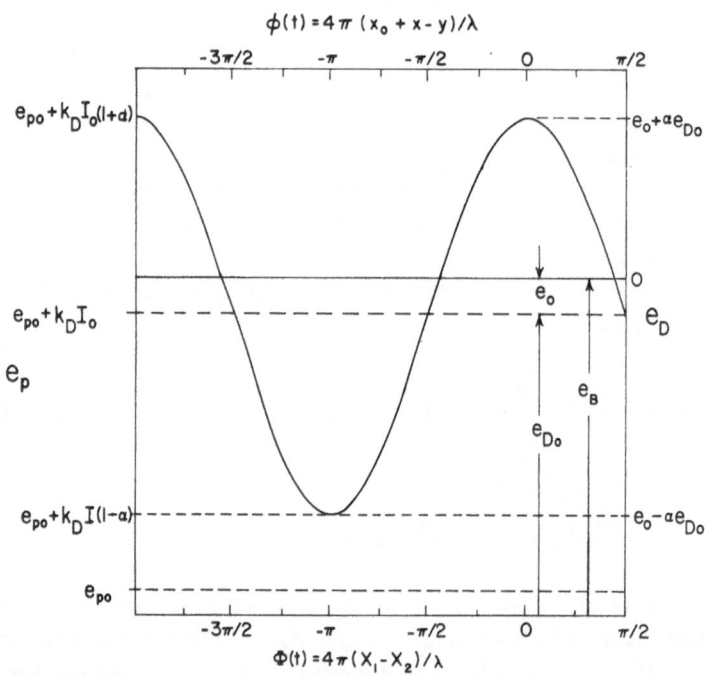

Fig. 2. Photo-detector and preamplifier output voltage vs. $\Phi(t)$ and $\phi(t)$.

beams in each arm, $\alpha=1$. Its value will always be close to one since, for example, for a 3 to 1 ratio in the amplitudes of the light beams in the paths, $\alpha=0.6$.

The transducer displacement $x(t)$ can be considered to be composed of two parts

$$x(t) = x'(t) + d(t) \tag{3}$$

where $x'(t)$ is the part of the $x(t)$ that has been considered in detail elsewhere (Gangi, 1969) and is that part of the transducer displacement which follows the displacements introduced in the 'free' arm by the strains in the Earth. The term $d(t)$ represents the extraneous displacements that are introduced by the various noise and error sources. It is this part of the displacement that we wish to concentrate on here. In the following we will assume that $x'(t)=y(t)$ or that both $x'(t)$ and $y(t)$ are zero.

A. LASER FREQUENCY VARIATIONS

To determine the effects of the LASER frequency variations we find the value of x_0 that must be obtained at equilibrium. To do this most simply, we assume the strains in the 'free' arm are zero (i.e., $y(t)=0$ and, consequently, $x'(t)=0$) and the error terms are zero also (i.e., $e_0=0$ and, consequently, $d(t)=0$). This amounts to saying that when the instrument is put into operation, the lengths of the interferometer arms are adjusted to give a zero voltage from the DC amplifier and the bias is so adjusted that the maximum positive and negative variations from the DC amplifier are equal as the arm X_1 is varied (i.e., $e_0=0$).

The frequency of the light from the LASER is assumed to be constant during the set up time and represented by λ. Then we have

$$0 = K_T K_k \alpha e_{D_0} \int_{-\infty}^{t} \cos(4\pi x_0/\lambda) \, f(t-T) \, \mathrm{d}T \tag{4}$$

and this expression can be satisfied, in general, only if

$$\cos(4\pi x_0/\lambda) = 0.$$

This leads to the condition

$$x_0 = (N \pm \tfrac{1}{2}) \lambda/4$$

where N is some positive or negative integer. From this expression we see that there are a number of equilibrium values for x_0. It also can be shown that the values $x_0=(N-\tfrac{1}{2})\lambda/4$ are positions of unstable equilibrium if the loop gain is positive. If the loop gain is negative, the above values are the stable equilibrium values. In the following we will assume the loop gain is positive and that $x_0=(N+\tfrac{1}{2})\lambda/4$. The value of the integer N is a measure of the difference in the lengths of the two arms of the interferometer. When $N=0$, the two arms lengths are the same to within $\lambda/8$.

If we now allow the LASER frequency to vary slowly, then the impulse response of the filter can be approximated by a delta function and we will have a displacement of the transducer given by

$$d(t)/K_T K_K \alpha e_{D_0} F(0) = \cos\left(4\pi(x_0 - d(t))/\lambda'\right).$$

Where $F(0)$ is the DC response of the Filter (assumed to be a low-pass filter here) and λ' is the new wavelength associated with the new, slowly varying LASER frequency. The term on the left hand side of the above equation is very small (e.g., $K_T = 3$ micron/ 1000 volts, $K_K = 200$; $\alpha e_{D_0} = 10$ volts; $F(0) = 1$; hence $K_T K_K \alpha e_{D_0} F(0) = 6$ microns) since we wish to have $d(t) \ll \lambda/2 = 0.316$ microns. Consequently, the new equilibrium condition will occur for, to a good approximation,

$$x_0 - d(t) = (N + \tfrac{1}{2})\,\lambda'/4$$

from which we have

$$d(t) = (N + \tfrac{1}{2})(\lambda - \lambda')/4 \cong N\Delta\lambda/4.$$

This displacement corresponds to an apparent strain of

$$S = d(t)/X \cong N(\Delta\lambda/\lambda)(\lambda/4X).$$

For this apparent strain to be less than 10^{-10} we have the condition

$$N < 10^{-10}(\lambda/\Delta\lambda)(4X/\lambda).$$

For an arm length X of 1 m, a wavelength of 0.63 microns and LASER frequency stability $(\Delta f/f = -\Delta\lambda/\lambda)$ of 10^{-6}, must satisfy $N < 600$. This means the two arm lengths must be equal to within 150 λ or approximately 0.1 mm to keep the error strains below 10^{-10} for a frequency stability of 10^{-6}. The frequency stability of commercially available helium-neon gas LASERS is better than 10^{-6} (high quality helium-neon gas LASERS are advertised to have frequency stabilities better than 1 part in 10^9. See also Siegman et al. (1967).

Making the two arm lengths equal to within 0.1 mm is not difficult, especially since the instrument itself can be used to determine the equality of the arm lengths. This is most readily accomplished by using a broadband light source (e.g., an incandescent lamp) as the interferometer light source. Because of the broad band of light frequencies in the source, sharp interference fringes will be obtained only when the two arm lengths are equal to better than 0.1 mm. In actuality the broadband source can be used to adjust the arm lengths to within a few tens of wavelengths since the maximum fluctuation in the fringe pattern intensity is obtained when the path lengths are equal.

The above analysis shows that it would be possible to have the path lengths of the two arms differ by as much as a meter if the LASER frequency is stable to 1 part in 10^{10} and a strain threshold of 10^{-10} is satisfactory. In this case, the instrument would become essentially a one meter Fabry-Perot interferometer strainmeter. This analysis also shows that a LASER source need not be used if the two arm lengths are equal. However, the LASER has the two advantages of high light intensity and monochromaticity. The latter means that less noise would be generated in the photo-detector by mixing or beating between the different frequency components in the light source.

In the following we will assume that the arm lengths are equal to within a wavelength (i.e., $N=0$) and/or that there is no frequency variation in the LASER light source.

While the above analysis was performed for the DC servo-loop, the results obtained are equally valid for the AC servo-loop.

B. DC DRIFT ERRORS

Amplifiers, photo-detectors and LASER light sources have large, very low frequency output variations. These variations generally have a $1/f$ frequency spectrum. Consequently, these very low frequency noise or error sources are of major importance. The more deleterious noise sources are those closer to the input to the servo-loop (the input to the servo-loop is the fringe pattern at the photo-detector), since the later elements of the servo-loop will amplify these noise or error signals.

To determine the effects of these slow drift variations, we assume the impulse response of the filter can be adequately represented as a delta function for these variations. This means we are assuming the filter is a low-pass filter with a cut-off frequency much higher than any variation in the signal. Equation (1) then becomes (with $N=0$, $x_0 = +\lambda/8$, $y(t)=x'(t)=0$)

$$d(t) = K_T K_K \left(e_0(t) - \alpha e_{D_0} \sin 4\pi d/\lambda\right) F(0) \tag{5}$$

where $F(0)$ is the DC response of the filter. In general, $K_T K_K F(0)$ is so large that in order to satisfy Equation (5), the term in parentheses on the right must be zero (or quite close to zero). Thus we have

$$e_0(t) = \alpha e_{D_0} \sin 4\pi d/\lambda$$

as the condition to be satisfied. This is what we would expect since the equation indicates that the fringe pattern will be displaced so that it gives a voltage which cancels the 'bias' error voltage $e_0(t)$. We also note that if $e_0(t)$ is greater than αe_{D_0}, it will not be possible to satisfy the above equation and the system will not come to equilibrium. The instrument is designed so the $e_0(t)$ is very much less than αe_{D_0}, consequently the sine function will be much less than one and we can approximate it by its argument. This gives

$$d(t) = \lambda e_0(t)/4\pi\alpha e_{D_0} .$$

In terms of the system parameters, this expression is

$$d(t) = \left(K_p(e_{po} + K_D I_0) - e_B\right) \lambda/4\pi\alpha K_p K_K I_0 . \tag{6}$$

We see that significant drift will occur as the DC preamplifier gain K_p varies, as the dark voltage of the photo-detector e_{po} varies, as the LASER intensity I_0 varies and as the preamplifier bias voltage e_B varies. Less significant drift occurs for variations in the high voltage amplifier gain K_K and in the beam splitter characteristic α. The LASER frequency variation is negligible since $d < \lambda/4\pi$ and the wavelength variation is at most 1 part in 10^6. With a 1 m path length and light wavelength of 0.6328 microns, frequency variations of 1 part in 10^6 introduce apparent strains of the order of 10^{-12} which are well below the instrument sensitivity. However, the other effects can bring about changes in $d(t)$ of the order of a quarter of a wavelength. These changes would correspond to strain fluctuations of the order of 10^{-7} which are considerably larger than the desired strain sensitivity of 10^{-10}.

The problems associated with most of these variations can be eliminated by using the synchronous detector (AC) configuration of the servo system. In this case, the DC voltage changes due to the above effects will not be passed through the AC amplifier. It is for this reason that the AC configuration is utilized in the final system design.

The expression, equivalent to Equation (1), that holds for the AC servo loop configuration is (see Gangi, 1969):

$$x(t) = K_K K_T \left[e_0'(t) + \int_{-\infty}^{t} e_\phi(T) f(t - T) \, dT \right] \tag{7}$$

where $e_\phi(t) = A\alpha K_p K_D I_0 D \quad \sin(4\pi/\lambda)[x_0 + x(t) - y(t)] + e_{\phi_0}(t) =$ voltage from the phase detector; $A =$ voltage amplitude to the phase detector; $D = 2\pi d_0/\lambda \ll 1$; $d_0 =$ amplitude of vibration of the displacement transducer induced by the oscillator; $e_{\phi_0}(t) =$ drift voltage from the phase detector; $e_0'(t) =$ drift voltage from the high voltage amplifier and the filter.

The drift voltage from the filter and high voltage amplifier had been neglected in the DC analysis since they were negligible compared to the drift voltages considered there. The combined drift voltage from the phase detector, the filter and the high voltage amplifier (all referenced to the high voltage amplifier input) is less than 0.1 mV. With $K_T = 3 \times 10^{-3}$ microns/volt and $K_K = 200$, the error displacement due to this drift voltage is 0.6×10^{-10} m. This corresponds to an error strain (for a 1 m arm length) of 0.6×10^{-10} which is below the sensitivity threshold of the strainmeter.

While the AC configuration eliminates the drift problems associated with the photo-detector sensitivity K_D, the preamplifier gain K_p, the preamplifier bias voltage e_B and the LASER intensity I_0, the error sources associated with thermal variations, LASER frequency variations, thermal noise and index of refraction changes remain.

C. INDEX OF REFRACTION VARIATIONS

In the preceding we have assumed that the indices of refraction of the light paths were constant. When they are not, the difference in the optical path length is given by

$$2 \int_0^{x_1} n(x_1) \, dx_1 - 2 \int_0^{x_2} n(x_2) \, dx_2 = \Delta. \tag{8}$$

The index of refraction of air varies both due to temperature and pressure differences along the two path lengths. We will assume the temperature effects are small since we will assume we are in a temperature controlled environment. Also, the errors due to temperature induced variations in the index of refraction are negligible compared to the errors due to temperature induced physical path lengths.

From Equation (8) we see that if the changes in the index of refraction are rapidly varying with distance over the path lengths, they will tend to cancel or average out if these rapid variations have zero mean. If the changes in the index of refraction are common to both paths (say due to atmospheric pressure changes) and the path lengths are equal, these differences will cancel also. Thus the pressure fluctuations which will cause difficulties are those associated with random turbulence along the two paths.

To obtain some idea of the problem involved, we note that the index of refraction of air at $0°C$ and 760 mm pressure (one atmosphere) is 1.000293. If we assume a linear variation in the index of refraction with pressure and use the fact that the index of refraction for vacuum is 1, we have a change of 293×10^{-6}/atmosphere of pressure. The pressure fluctuations associated with turbulence are of the order of microbars (10^{-6} atmospheres).

If we assume there is a difference in the average index over each path length, the apparent path length change due to this difference would be (for $X_1 = X_2$)

$$2X_1(n_1 - n_2) = \Delta. \tag{9}$$

The apparent strains due to these variations would be just $2(n_1 - n_2)$. To have these strain variations less than the strain threshold of 10^{-10}, we require $2(n_1 - n_2) < 10^{-10}$. This means the pressure fluctuations must be kept below

$$2 \times 293 \times 10^{-12} \Delta P < 10^{-10} \, (P \text{ in microbars}) \tag{10}$$

or $\Delta P < 0.16$ microbars.

While it may be possible to keep the turbulence pressure variations below this value by using various turbulence inhibitors, the approach that has been taken with the instrument is to provide a partial vacuum for the light beam paths. This would also eliminate any errors due to temperature induced changes in the index of refraction of the light paths.

D. THERMAL NOISE ERRORS

The thermal noise error sources in the servo-loop are negligible compared to the DC drift errors. The most important thermal noise source is at the photo-detector since these noise signals are amplified in the AC amplifier, the phase detector, the filter amplifier and the high voltage amplifier. The rms thermal noise from the photo-detector is given by

$$v = \sqrt{4kTBR}$$

where k = Boltzmann's constant = 1.38×10^{-23} watt-sec/K; T = photo-detector noise temperature in K; B = the noise bandwidth = 1 cps; R = the source resistance of the detector = 10^4 ohms.

If we assume a conservatively high noise temperature of 3000 K, the rms thermal noise from the photo-detector is about 0.04 microvolts. Since the gain of the preamplifier, the phase detector and filter amplifier is approximately 100, the rms noise voltage at the input to the high voltage amplifier will be approximately 4 microvolts. This noise voltage is smaller by a factor of 25 than the DC drift voltage from the phase detector, filter amplifier and high voltage amplifier. The 4 microvolt noise signal will give extraneous strains of the order of 2×10^{-12} which are well below the strain sensitivity threshold.

The thermal noise voltage at the input to the preamplifier is given by the manufacturer as 0.01 microvolts in a 1 cps bandwidth for frequencies above 300 cps. For frequencies above 300 cps, the preamplifier has a flat noise spectrum; for frequencies below 300 cps, the preamplifier has a $1/f$ noise spectrum. This fact illustrates another advantage to using the AC configuration of the servo system. The 0.01 microvolt thermal noise at the preamplifier input is smaller than the thermal noise from the photo-detector which is also at the preamplifier input. Consequently, the thermal noise from the preamplifier is also negligible.

The most serious thermal noise error sources are those in the phase detector output, filter amplifier and high voltage amplifier. This is due to the fact that the servo-loop signals are at low frequency (from DC to 1 cps) in these parts of the servo-loop and the $1/f$ noise spectrum of these devices dominates. However, we have analyzed the effects of these error sources in the previous section on DC drift errors.

E. TEMPERATURE VARIATIONS

The most serious error source for strainmeters is that due to temperature changes in the strainmeter's environment. The instrument design given here has an advantage over most strainmeters in that it is insensitive to temperature changes to first order. Even so, temperature effects represent the largest error source in this instrument.

The change in arm length difference due to a temperature change of the arm lengths is given by

$$\Delta = a_1 X_1 (T_1 - T_0) - a_2 X_2 (T_2 - T_0) \tag{12}$$

where $a_1, a_2 =$ coefficients of linear thermal expansion of the arms X_1 and X_2 respectively; $T_0 =$ equilibrium temperature when both arm lengths are equal; $T_1, T_2 = average$ temperature of the arms X_1 and X_2 respectively.

From the above expression we can see that there is no error if the two arm lengths are exactly equal, are always at exactly the same average temperature, and have exactly the same thermal coefficient of expansion. However, it is quite unlikely that both arms would have exactly the same average temperature or have expansion coefficients equal to the degree necessary to have the thermal effect negligible. For rock materials, concrete or metals, the coefficients of linear thermal expansion are of the order of $10^{-5}/°C$. If we assume the two arm lengths have matched coefficients to within 1 % (which is probably optimistic) and assume equal arm lengths, the extraneous strain introduced by the temperature variations is

$$\Delta/X = a_1 [T_1 - T_0 - (a_2/a_1)(T_2 - T_0)]$$
$$= a_1 [T_1 - T_2 + (T_2 - T_0)(a_2 - a_1)/a_1]. \tag{13}$$

Equation (13) is equally valid for a Benioff quartz strainmeter or a Fabry-Perot interferometer strainmeter (see Vali et al., 1965). For the quartz strainmeter $X_1 = X_2$, a_2 is the thermal expansion coefficient for quartz, T_2 is the average temperature of the quartz, a_1 is the thermal expansion coefficient of the ground and T_1 is the average temperature of the ground below the quartz rod. For the Fabry-Perot interferometer strainmeter, a_2 is the thermal coefficient of vacuum (and therefore zero), and a_1 and T_1 are the thermal expansion coefficient and average temperature of the ground below the LASER beam.

From Equation (13) we see that even if the average temperatures of the two arms are equal (i.e., $T_1 = T_2$) and the thermal coefficients are matched to within 1% (i.e., $|1 - a_2/a_1| < 0.01$), the temperature change $(T_2 - T_0)$ must be less than 10^{-3}°C to obtain a threshold strain sensitivity of 10^{-10} for $a_1 = 10^{-5}/$°C. It also shows that the average temperatures of the two arms $(T_1 - T_2)$ must not differ by more than 10^{-5}°C to obtain extraneous strains lower than the threshold strain sensitivity. We see these are quite severe requirements on the temperature stability (represented by $T_2 - T_0$) and on the temperature gradients (represented by $T_1 - T_2$) in the instrument environment. This explains why such instruments are installed in deep mines, tunnels and caves where temperature variations and gradients of this order or smaller may be obtained.

3. Summary

An error analysis has been performed for a small (1 m²) servo-controlled LASER interferometer strainmeter with a strain threshold of 10^{-10}. The error sources considered are variations in the LASER frequency, DC drifts in the servo-loop, variations in the index of refraction along the LASER beam path due to pressure fluctuations, thermal noise in the servo-loop and temperature variations in the instrument environment.

It is possible to eliminate the error strains due to LASER frequency variations by making the arm lengths of the interferometer equal.

The major error strains due to DC drift voltages can be eliminated by using the AC servo-loop configuration. This configuration eliminates the DC drift voltages due to changes in the photo-detector sensitivity and 'dark' voltage, the LASER light intensity and the gain and bias voltage in the preamplifier. Drifts in the phase detector, the DC filter amplifier and the high voltage amplifier will remain, but the signal level is sufficiently high before these component parts that their drifts cause less severe problems.

The effects of thermal noise in the servo-loop are shown to be negligible for the desired strain threshold. The effects of pressure induced changes in the index of refraction are made small by evacuating the LASER beam light paths.

The most severe problem is that associated with the temperature stability and temperature gradients in the instrument environment. While it may be possible to match the thermal expansion coefficients of the two arms of the strainmeter (by using differing amounts of materials with different thermal expansion coefficients), this will not eliminate the problem associated with the temperature gradients. The most reasonable way to eliminate the effects of the thermal fluctuations is to place the instrument in a temperature stable environment. With small temperature variations, the temperature gradients will be small also. It is seen that the temperature fluctuation problem is *not* uniquely associated with this particular instrument, but that this would be (and is) the most severe problem for any sensitive strainmeter irrespective of its length or configuration.

It is concluded that a 1 m strainmeter can be constructed with a threshold strain sensitivity of 10^{-10} provided a sufficiently stable temperature environment is provided. Such stable temperature environments are generally found only in deep mines, caves and tunnels in the Earth.

References

Benioff, H.: 1935, 'A linear Strain Seismograph', *Bull. Seism. Soc. Amer.* **25**, 283–309.

Gangi, A. F.: 1969, 'Analysis of a Small LASER Strainmeter' in *Proceedings of Symposium on LASER Applications in the Geosciences*, Meeting held at Douglas Advanced Research Laboratories, Hungtington Beach, Calif., June 30 to July 2, 1969 (to be published).

Salisbury, M. and Gangi, A. F.: 1967, 'A Servo-controlled Michelson Interferometer LASER Strainmeter', (Abstract), *Trans. Amer. Geophys. Un.* **48**, 203–204.

Siegman, A. E., Daino, B., and Manes, K. R.: 1967, 'Preliminary Measurements of LASER Short-Term Frequency Fluctuations', *IEEE J. Quantum Electron.* **QE-3**, 180–189.

Vali, V., Krogstad, R. S., and Moss, R. W.: 1965, 'LASER Interferometer for Earth Strain Measurement', *Rev. Sci. Instr.* **36**, 1352–1355.

Discussion

Runcorn: Have sufficient measurements of strain been made in what are otherwise seismically quiet areas to know whether the pattern of strain is constant over these very large areas? The ideas now being developed about plate tectonics seem to indicate that the stresses, and therefore the strains, should be fairly uniform over large areas of the Earth's crust.

Gangi: Very few measurements have been made because there are not enough strain instruments around. One thing I should point out is that with a highly sensitive small instrument you have a greater chance of being able to find one solid piece of rock, about 1 m square, that you could put the instrument on and be able to say that the rock was uniform and that you would therefore be measuring the strain in that rock. With the larger instruments there was always the problem that you might site it across an unknown fault. Of course, even in the small sample of rock there might be cracks so that one would require a large number of instruments in the area to find the true strain of that area. The only way we could hope to do this is to have inexpensive instruments so that we could afford to put out a large number of them.

Romig: We have about 10 strainmeters within 45 km of an active area near Denver. The active area is about 5 km wide and buried at a depth of about 5 km. Around this area the secular strain rate varies by two orders of magnitude. Though this is admittedly based on just one observation it does vary radically in a very short distance in this area.

Runcorn: The point I was trying to make is that up to now our attention has been directed towards seismically active areas but that for fundamental geophysics we should perhaps be doing much more in the apparently inactive areas.

Bender: We have been looking at Earth tides with our 30 m strain laser but even though we expected to avoid troubles with the secular strain we still see something like 40 or 50% larger amplitudes for the Earth tides than do Romig and Major at their site 30 or 40 km away from ours. It therefore really seems that there are some very serious site problems.

Runcorn: It is almost certain that the area is broken up into a large number of rather smallish blocks. What one really wants to look at is an area where there has not been recent mountain building and large scale tilting, where you might expect the relationship between the underlying part of the crustal plate and what you see at the surface to be reasonably coherent.

Bender: A shield area would certainly be desirable.

SPURIOUS LOCAL EFFECTS ASSOCIATED WITH
TELESEISMIC TILTS AND STRAINS

F. D. STACEY and J. M. W. RYNN*

Physics Department, University of Queensland, Brisbane, Australia

Abstract. Records of strain and tilt-meters operated in S. E. Queensland, a seismically quiet area, have been examined for evidence of residual strains associated with distant earthquakes. Although these effects have been observed on a number of occasions, more often on tilt than on strain meters, there appears to be no instance in which they can be positively identified as genuine. On one striking occasion separate permanent offsets of a tilt-meter trace, all in the same direction, were observed to accompany the arrivals of P, S and surface waves. Subsequent investigation showed that the directions of observed offsets normally occurred in the direction of any secular trend on a record. It appears that at least the majority of residual strains are local effects, stimulated by seismic waves.

1. Introduction

Press (1965) drew attention to the importance of observing the extended displacement fields of large earthquakes. Static (residual) strains at distances large compared with the dimensions of a slipping fault face give a direct measure of the moment of fault movement ($\int$(displacement) d(area)), which is particularly relevant to the topic of the present conference.

Records from two mercury-level tilt-meters and an interferometric strain-meter, operated at different sites near Brisbane, Australia, have been examined for evidence of residual teleseismic strains. The instruments are developmental and unbroken records for long periods have not been obtained, but 139 earthquakes have been identified in the records and 33 have been examined in detail. Five tilt offsets have been found to accompany arrivals of seismic waves from earthquakes of magnitudes 5 to 6 in the range $\Delta = 17°$ to $36°$. No residual tilts have been observed to accompany more distant shocks, even including one of magnitude 7.8 at $\Delta = 68°$. This encourages the supposition that the observed effects are genuine. However, the only observed strain effect accompanied the magnitude 7.8 earthquake. Further, the offsets appear to be correlated with local secular changes.

2. Instruments and Siting

Two mercury level tilt-meters, prototypes of the design described by Stacey *et al.* (1969), have been operated N-S and E-W on a pier in the University of Queensland's seismic vault at Mt. Nebo, 25 miles NNW of Brisbane. The rocks of the area are lower Paleozoic geosynclinal sediments (the Nereanleigh-Fernvale group of the Brisbane Metamorphics), being predominantly cherts and jaspers. This provides a convenient site in what was believed to be a seismically stable area.

* Now at Lamont-Doherty Geological Observatory of Columbia University, New York, U.S.A.

L. Mansinha et al. (eds.), Earthquake Displacement Fields and the Rotation of the Earth, 230–233.
All Rights Reserved. Copyright © 1970 by D. Reidel Publishing Company, Dordrecht-Holland

Shear strain was measured in a chamber of the University mine at Indooroopilly in outer Brisbane by means of a Michelson interferometer with equal arms of length 2 m (Shamsi and Stacey, 1967). The chamber is in a rhyolitic dyke intruding Brisbane schist. The floor was dug out to a depth of about 18 inches to provide sound bases for the 3 concrete piers on which the interferometer components were mounted. Fringe positions were recorded continuously on a moving film.

3. Summary of Observed Tilt and Strain Increments

Earthquake:			Tilt Increment		
Region	Distance	Magnitude	Direction	Magnitude	Secular trend:
Solomon Is.	18.6°	5.4	South	2.5×10^{-8}	South
Bismarck Sea	24.1°	4.7	South		South
E. New Guinea	17.2°	5.3	South	8.4×10^{-8}	South
Fiji	26.0°	5.6	North	4×10^{-8}	Nil
Flores Is.	35.8°	5.9	South	8.5×10^{-8}	South
Japan	68.4°	7.8	Shear strain increment approx. 1.3×10^{-8}		

The N-S tilt-meter record of the third earthquake, in which separate offsets appear to accompany different wave arrivals is reproduced in Figure 1.

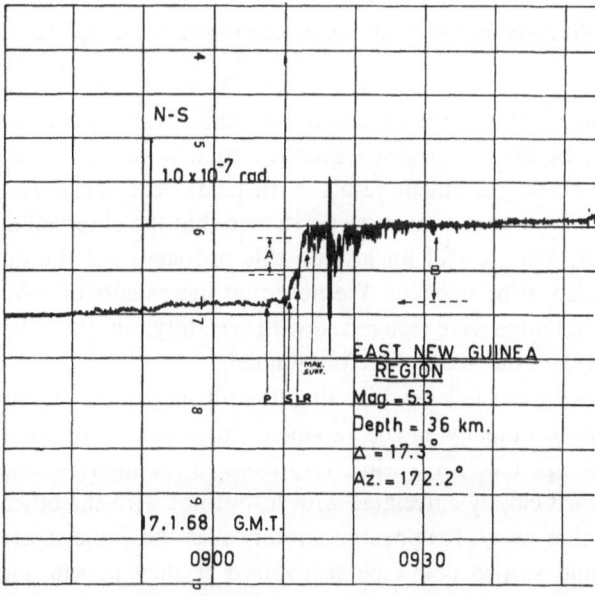

Fig. 1. Separate offsets accompanying different wave arrivals on N-S tilt-meter record for East New Guinea earthquake.

4. Discussion

The number of observations is too small for statistical significance to be ascribed to them. However, except for the Japanese shock, the observed effects appear distinctly too large for moderate earthquakes distant 20° or so and the correlation with the secular trend is suggestive of a local interpretation. It appears that, at least in some cases, arrival of earthquake waves accelerates the trend in the manner idealized in Figure 2, although the secular change itself is not sufficiently regular to be sure of this.

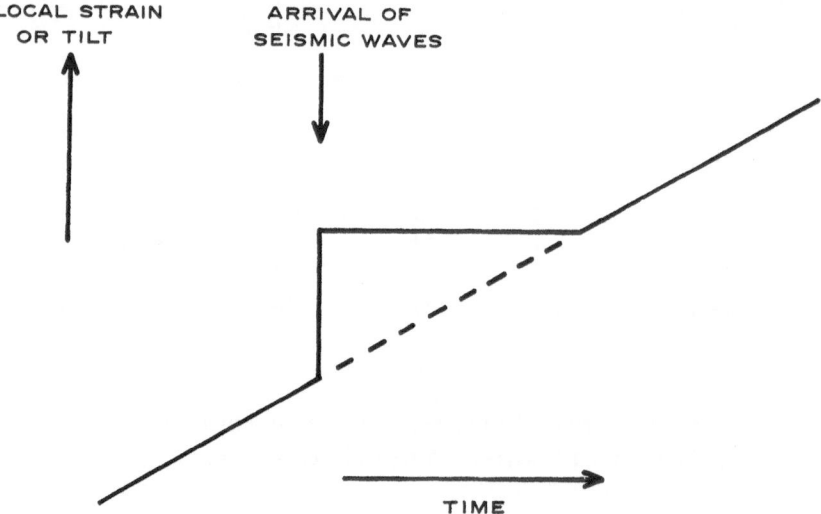

Fig. 2.   Schematic indicating how earthquake wave arrivals accelerate tilt trends.

The occurrence of tilt increments does not appear obviously correlated with seismic wave amplitude; by far the largest waves recorded were from the Japanese earthquake, from which no increment resulted. Instead, near earthquakes appear more effective, so that we were for some time persuaded that the observed tilts were genuine teleseismic effects. We are still unclear on this problem, but the direction of wave arrival appears also to be relevant. We note that increments of E-W tilt were fewer and smaller, so that none were recognized with certainty, but the E-W record suffered more breaks and the evidence is therefore limited.

The interferometer was less sensitive than the tilt-meters by a factor 10 and the reported increment accompanying the Japanese shock was at the limit of resolution. A third tilt-meter operated for a while very close to the interferometer, producing a secular trend record closely correlated with it, but not with the other two tilt-meters which were 25 miles away. It appears therefore that these effects are local and that genuine teleseismic strains could be recognized if they appear identically on instruments a few tens of miles apart, but not otherwise.

If really remote observations of strain cannot be made reliably, an extrapolation of near field strains, using elasticity theory, may be useful. Assuming teleseismic strains to fall off with distance r as r^{-3} in an elastic half space, then it appears that significant contributions to changes in moments of inertia occur at indefinitely remote distances and the limitation is imposed by the boundary conditions. Integrating out to a distance R, the total change in moment of inertia due to an average displacement $\bar{x}$ on a fault plane $d_1 \times d_2$ is roughly

$$\Delta I \approx \varrho (d_1 d_2)^{3/2} R\bar{x} \tag{1}$$

for a medium of density ϱ. If d_1 is the smaller dimension and the strain release is ε, then

$$\bar{x} = \tfrac{1}{4}\varepsilon d_1 . \tag{2}$$

Taking $R = 3000$ km, $\varrho = 3$ gm/cm^3, $\varepsilon = 10^{-4}$ (corresponding to a stress release exceeding 100 kg/cm^2), $d_1 = 200$ km and $d_2 = 800$ km for the Alaska earthquake we obtain $\Delta I \approx 3 \times 10^{34}$ gm cm^2 and the maximum corresponding pole shift is

$$\Delta\alpha \approx \Delta I/(C - A) \approx 10^{-8} \text{ radian} = 0.002 \text{ arc sec}, \tag{3}$$

which is not significant.

The summation over many earthquakes would not increase $\Delta\alpha$ very much, even if they were synchronized, because earthquake energy release is dominated by the very few large shocks and ΔI depends even more strongly upon magnitude. The possibility of appreciable wobble excitation by earthquakes thus requires that Equation (1) underestimate ΔI by a factor 20 to 50. That this is so is indicated by the strain steps reported by Major and Wideman to this conference, but their strain steps look suspiciously like transient 'strain steps' which we observed when our amplifiers were imperfectly linear and thus provided a rectified output during surface wave arrivals, with a decay corresponding to the time constant of the electrical system. Their Benioff tuned circuit type of capacitative transducer is particularly beset by this nonlinearity.

Acknowledgements

This work is supported by the Australian Research Grants Committee. Collaboration of Dr. J. P. Webb and Mr. P. Gaffy in the use of the seismic vault and in correlating our data with conventional recordings is gratefully acknowledged. Our use of the mine facilities was by cooperation of Mr. W. E. Vance.

References

Press, F.: 1965, 'Displacements, Strains and Tilts at Teleseismic Distances', *J. Geophys. Res.* 70, 2395.
Shamsi, S. K. and Stacey, F. D.: 1967, 'Michelson Interferometer as an Earth Strain Sensor', *Earth Plan. Sci. Lett.* 3, 466.
Stacey, F. D., Rynn, J. M. W., Little, E. C., and Croskell, C.: 1969, 'Displacement and Tilt Transducers of 140 db Range', *J. Sci. Instrum.* 2, 945.

EARTHQUAKE PREDICTIONS FROM FAULT MOVEMENT AND STRAIN PRECURSORS IN CALIFORNIA

RENNER B. HOFMANN

Supervisor Earthquake Engineering, Division of Resources Development, Dept. of Water Resources, The Resources Agency, State of California, U.S.A.

Abstract. Fault movement on the San Andreas system has been observed to vary in time and place. Variations in movement with time at a particular place along the fault have been shown to often precede earthquakes of magnitude 4.5–5.5. Where the fault moves regularly, the variations consist of the fault sticking until several centimeters of strain are stored. The amount of strain stored and released corresponds reasonably well with other observations of magnitude vs. strain. Where the fault has not moved regularly during the past ten years, evidence from several types of surveys indicate general compression of the area. Compression events of about six months duration and of about 10–15 cm over about 20 km have been observed to precede earthquakes in these areas.

1. Introduction

Active faults in California have been monitored by Geodimeter distance measurements for about 10 years. Both strain and creep have been observed along the San Andreas fault system at different places and times. Fault movements which are atypical of those usually observed at a particular place along the fault have been observed to precede earthquakes of magnitude 4.5 or greater. This paper is a summary of these observations. Additional detail for all but the most recent work is in the California Department of Water Resources Bulletin No. 116-6, 'Geodimeter Fault Measurements in California', May 1968.

2. Measurements Method

A Model 2A Geodimeter (Figure 1) using a mercury arc sends a beam of light, intensity modulated at 10 MHz, from a station mark on one side of a fault – obliquely across – to a retrodirective prism assembly set up over another station mark on the opposite side. The returning beam is intercepted by a photomultiplier in the geodimeter and the phase relationship of the outgoing and received modulated beams measured electronically. In effect, the transit time of the light beam is computed.

Although the mercury arc produces a wide spectrum of colors, there are predominant bands. The band-pass characteristics of the photomultiplier tube further limit the effective band width of usable light frequencies resulting in an effective wavelength of 0.546 μ. The speed of this wavelength of light can be determined if the refractive index of air is known.

Temperature, pressure and humidity, in that order, affect the atmospheric refractive index for visible light most strongly. Pressure and humidity are adequately sampled at ground level during the measurements. Because temperatures are often influenced

L. Mansinha et al. (eds.), Earthquake Displacement Fields and the Rotation of the Earth, 234–245.

by near-surface inversions at night, they are usually obtained from calibrated thermistors at the top of 50-foot poles at each of the station marks. The marks are usually on mountain peaks on either side of the fault valley; consequently, the temperature is often sampled midline with a balloon and radiosonde. Temperatures are measured to 0.1 °C.

Fig. 1.   Model 2A Geodimeter.

Aircraft flown along the measured distance have been used as an alternative in sampling temperatures. Most measured distances are between 8 and 20 km. For many, if not most, of the distances measured, the potential error is estimated at about ½ cm. Particularly poor weather conditions or steeply-sloping lines which commonly penetrate several temperature inversions reduce accuracy to 2 cm or more. This level of accuracy is in part obtained by multiple measurements (often as many as 24 over a period of two nights' work) which are later combined statistically.

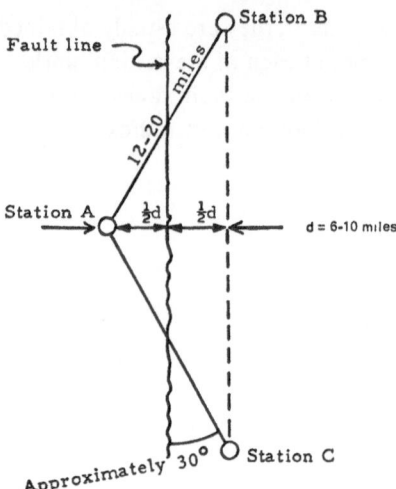

Fig. 2.   Criteria for geodimeter lines.

3. Area Measured

The Fault Monitoring Program was initiated to provide design data for the California Water Project which must cross or closely parallel the San Andreas fault system from the San Francisco Bay area to about 20 miles southeast of San Bernardino (Figure 3). This is the general area which has been monitored since 1959. The San Andreas, Hayward, Caliveras, White Wolf, Garlock, San Jacinto, and several smaller faults are monitored over this 400-mile-long area. All faults named have shown some indications of activity at some time or place along their lengths during the period of observation.

4. Average Regional Movements

In the San Francisco Bay area, movement totals over 4 cm/yr on the Hayward, San Andreas and Caliveras faults. The greatest percentage usually occurs on the Hayward (averages about 2 cm/yr) but the amount carried by each fault at any particular time is highly variable. Farther south where these faults have come together to form a single San Andreas zone south of the town of Hollister, the same 4+ cm/yr rate is observed. This is the area where the Almaden Winery astride the most obvious trace in the zone shows visible damage from creep. A previous structure at this site was torn down because of similar damage.

Farther south along the trend of the San Andreas, the annual creep diminishes until it is unmeasurable in the Carrizo Plains, a few miles south of the town of Cholame. From 1959 through 1965 this reduction in fault movement was regular and progressive as a function of distance along the fault south from Almaden Winery, with one exception. A small segment of the fault between the towns of Parkfield and Cholame had consistently indicated a lower movement rate than adjacent segments (Figure 4).

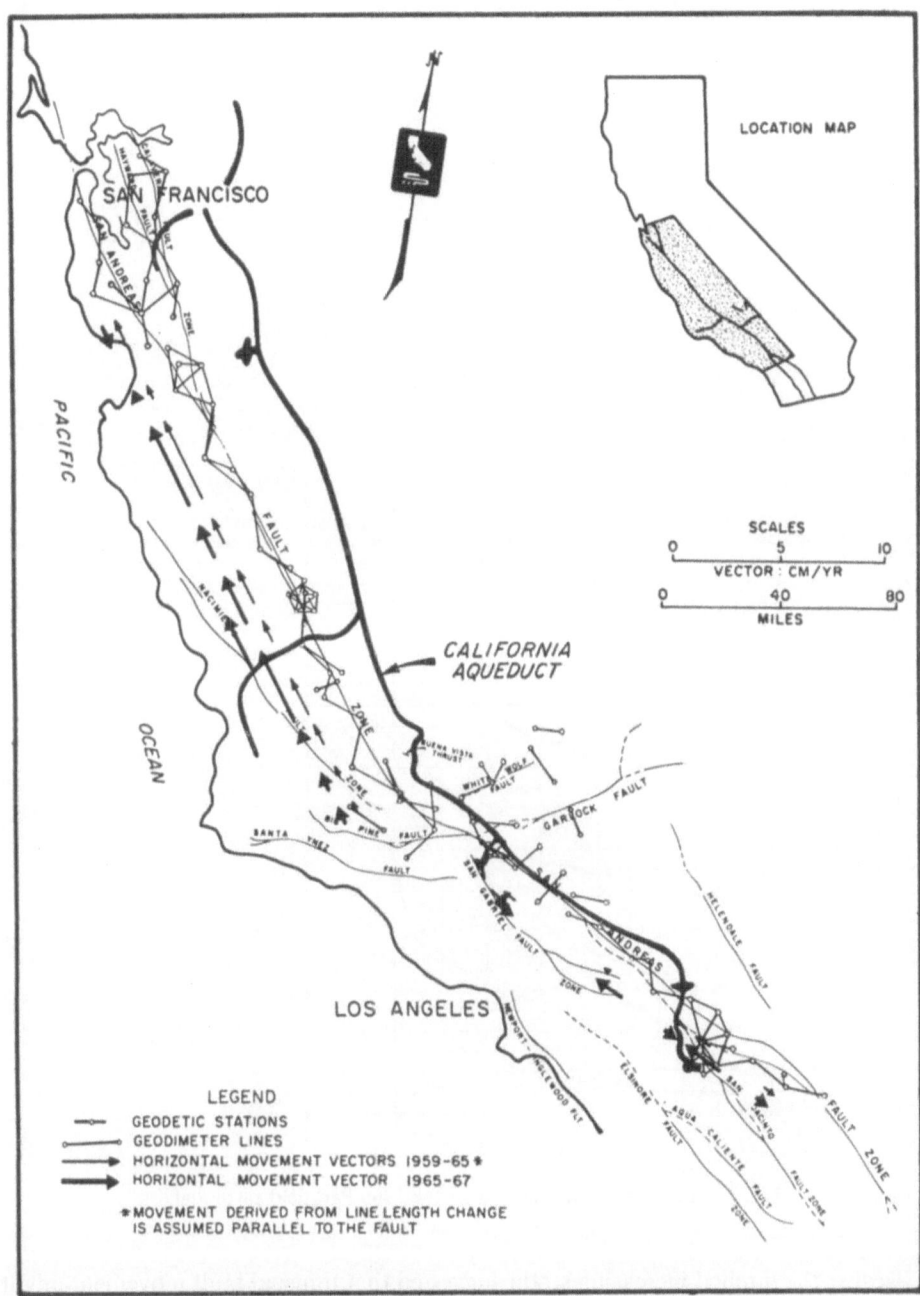

Fig. 3.   Average annual fault movement, 1959–65 and 1965–67.

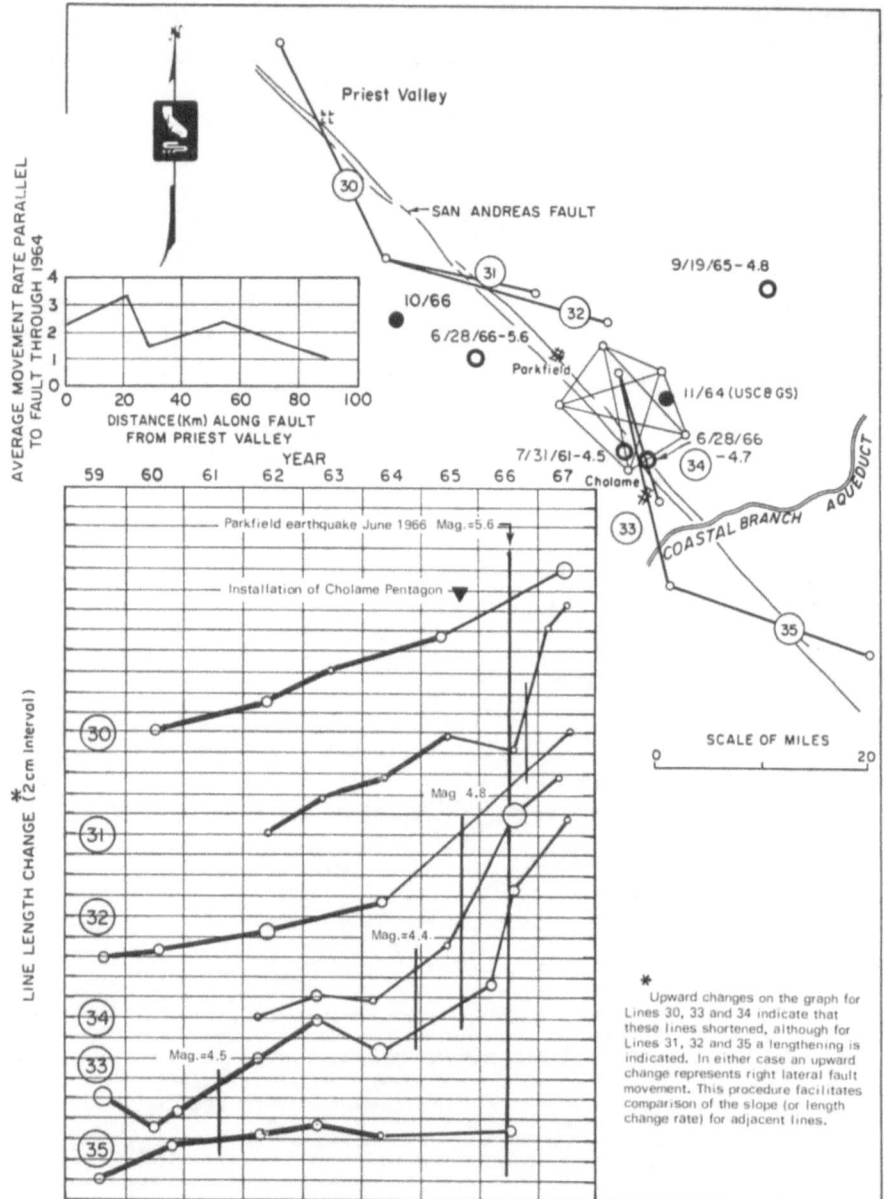

Fig. 4.   Movement rates preceding the 1966 Parkfield earthquake.

Because the implied increasing strain suggested that unusual fault movement might follow, a pentagon-shaped geodetic network was installed there in October 1965. In June 1966 a 21 cm fault displacement accompanied by a magnitude 5.6 earthquake increased the average fault movement in the lagging segment to slightly more than the adjacent ones (Figure 5).

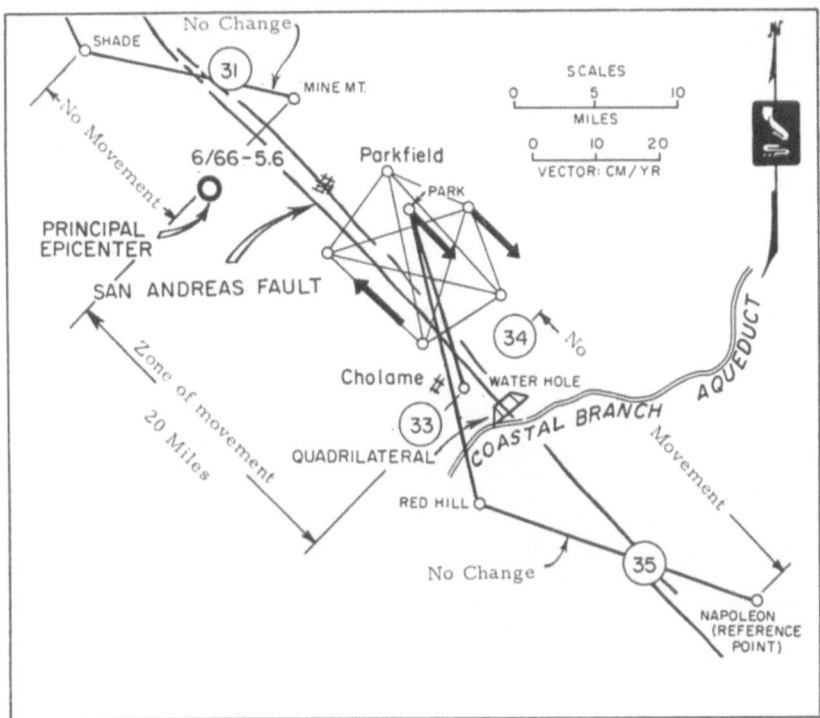

Fig. 5.   Relative motion, June 1966 Parkfield earthquake.

The zone initially affected was about 20 km long. Since the time immediately following the shock, subsequent small movements have extended the affected zone to about 30 km.

From the Carrizo Plains farther south and east along the San Andreas to the vicinity of the Tehachapi Mountains, movements are small, or nonexistent.

In the Tehachapis, fault slip is not obvious; but a plot of geodimeter line extensions (the change in length between station marks divided by the length) vs. azimuth, suggests N-S compression and E-W extension.

Still farther south and east along the fault past San Bernardino to Palm Springs a similar plot suggests N-S compression unaccompanied by E-W extension (Figure 6). Possible movement on the many thrusts in the area may account for this.

Still farther south in the Imperial Valley, out of the area of our investigations, others (e.g., Whitten, 1956; Brune and Allen, 1966) indicate slippage at a relatively high rate is occurring.

Small, consistent movements (about 1.2 cm/yr) have been observed on the White Wolf fault. Recently, some movement has been observed on the Garlock fault which was confirmed by measurements (made by the USC&GS as a part of our cooperative program) on a small fault quadrilateral astride the fault where it is crossed by a California Water Project pipeline.

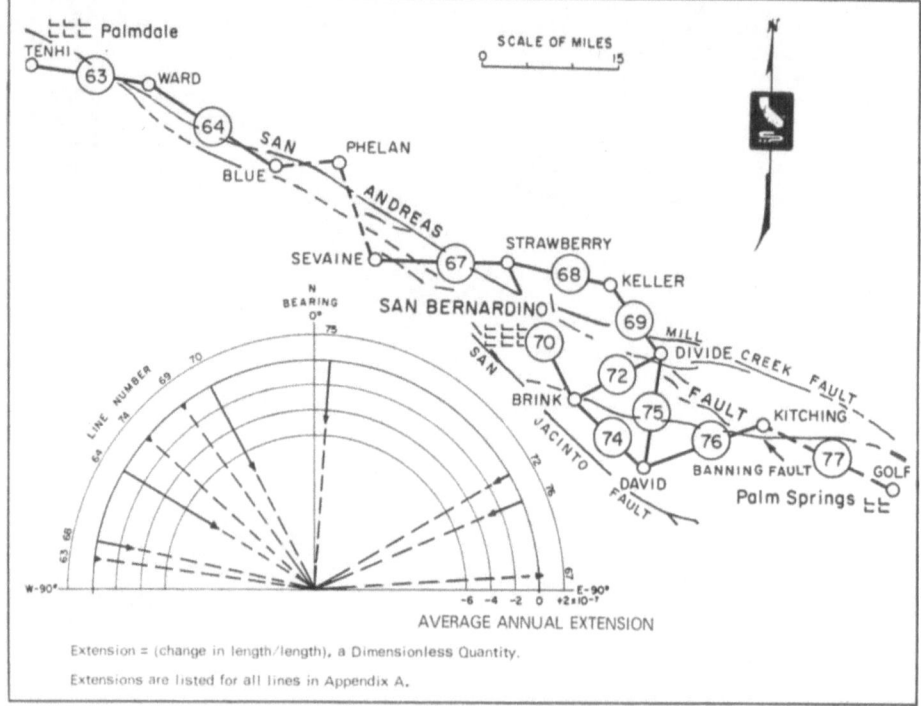

Fig. 6.   Extension rate vs. line direction, Palmdale to Palm Springs.

5. Anomalous Movements Preceding Earthquakes

Deviations from the usual trend of movement in a given area were noted to be more frequent where earthquakes are more prevalent than elsewhere. Further investigation showed that, where measurements were considerably more frequent than earthquakes, these movements often preceded earthquakes.

In the northern area of the San Andreas system described, the anomalies are apparently caused by fault-sticking. Sometimes movement is transferred to an adjacent splinter fault (Figures 7 and 8). After the strain accumulates for a year to 18 months, the main fault suddenly moves – becomes unstuck – resulting in a small-to-moderate earthquake. In 18 months' time, from 3–6 cm of strain accumulates which, when released, results in a magnitude 4.5–5+ shock.

In the Tehachapi Mountains and farther south and east near San Bernardino, sudden shortening of distances between several pairs of station marks may herald the coming of similar earthquakes. Sometimes the lines extend to their normal length before the shock occurs (Figure 9).

A computer program was developed to identify these anomalies in order to reduce the potential bias of human judgment.

In principle, the program establishes a best-fit line through points whose position is

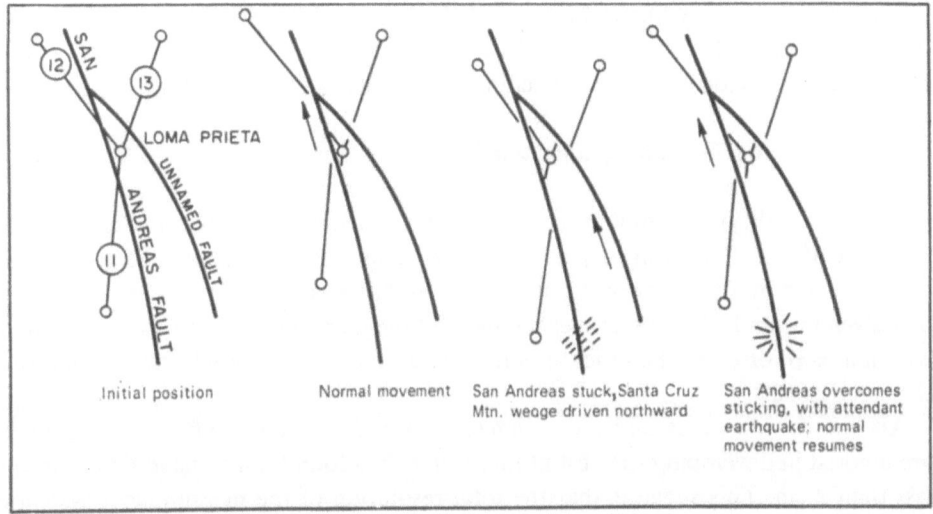

Fig. 7.   Possible sequence of fault movements. Amount of movement greatly exaggerated.

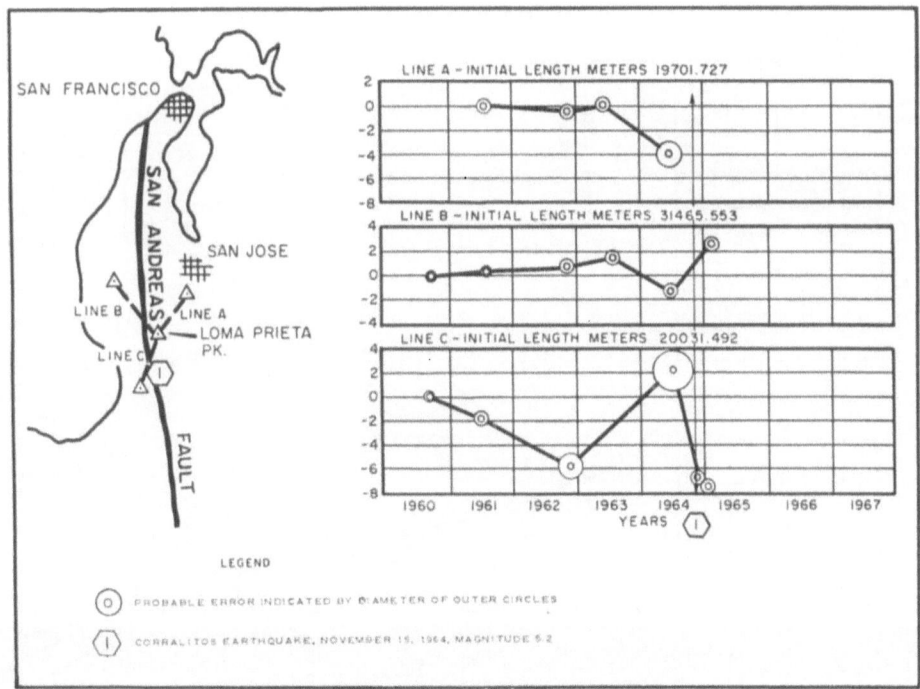

Fig. 8.   Crustal movement and Corralitos earthquake.

determined by the time of measurement and change in distance between each pair of stations. This line is projected to the time of a new measurement. If the new measurement fails to fall on the line by an amount X, the program identifies the new measurement as an anomaly. The amount X was originally determined by a recursive fit to factors characteristic of obvious precursive anomalies and earthquakes. This resulted in:

$$X = 0.079 \times (\text{line quality}) \times \sigma^{1/2}$$

where (line quality) is a number from 1–6 arbitrarily assigned by the staff to designate the difficulty of measuring a particular line. This was later shown to be roughly approximated by $PE - 2$ where PE is the 50% probable error of pairs of measurements of a given line and where error measurement of the pair was taken on a different night within a week or so of the other. σ is the standard deviation of all prior points from the best fit line.

All known anomalies and measurements taken immediately after an earthquake are ignored in developing the best fit line and σ. X is found not to have significance if less than 2 cm. This suggests that the total resolution of the measurement system is not better than 2 cm or that there exists a background strain noise of 2 cm for lines in

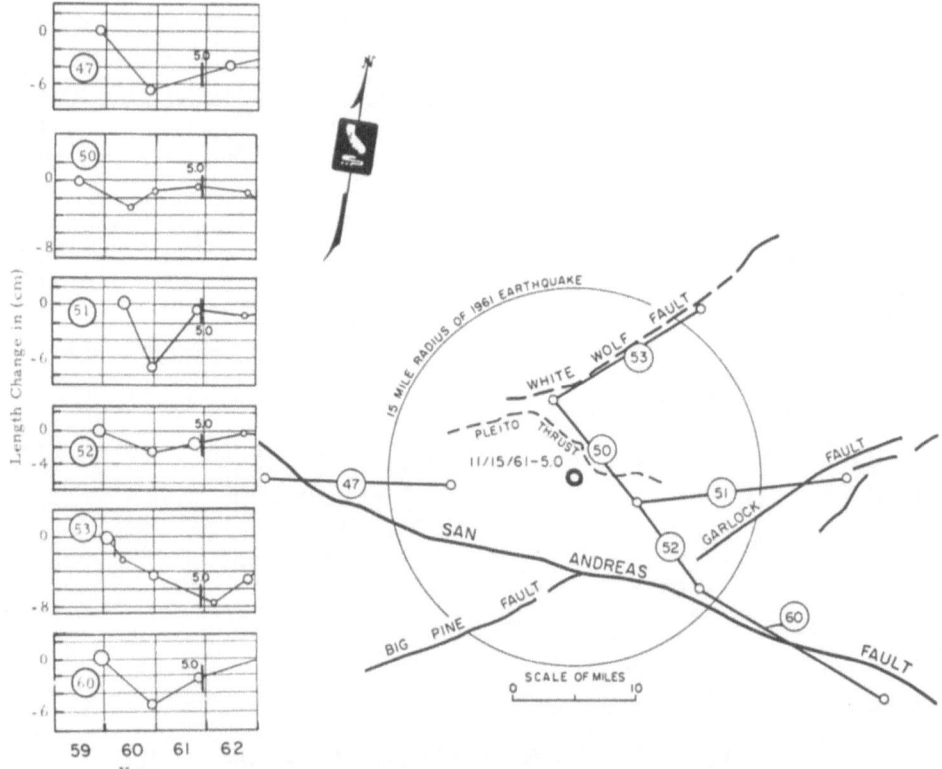

Fig. 9.   Fault movement and the Wheeler ridge earthquake of 1961.

the 8–20 km range. Both may be possible explanations in various segments of the fault system. Calibration data and excellent closures for closed geodetic networks indicate a resolution of near $\frac{1}{2}$ cm in some areas. This suggests a possible long-period noise perhaps on the order of months which would not be significant during the short time usually required to measure all the lines in a closed geodetic network.

The program has been refined to not identify movement as an anomaly if the deviation of the new measurement from the line is less than 2 cm or if:

(a) Line quality is five or six (this eliminates six lines as being useful for prediction).

(b) Fewer than four measurements have been made on the line.

(c) The new measurement follows an earthquake of M ⩾ 4.5 by nine months or less.

(d) Less than two measurements have been made following the most recent earthquake near that line.

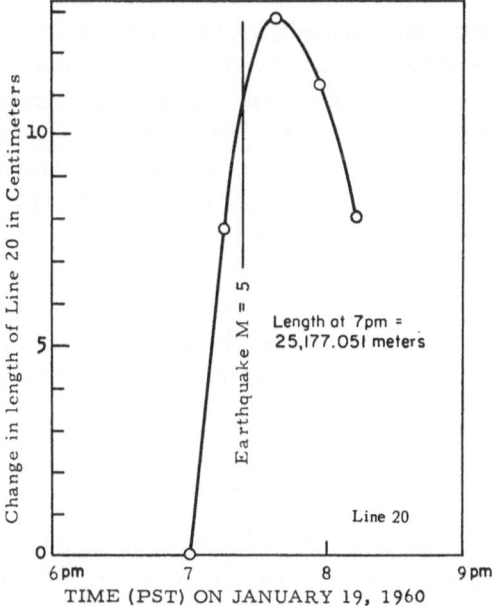

Fig. 10.   Short-range anomaly.

Anomalous movements sometimes occur which are not followed by earthquakes. Eight fault movement anomalies have been indicated by two or more geodimeter lines.

Three of these were not followed by significant nearby earthquake activity – one of these was caused by shifting of movement between the San Andreas, Hayward, and Caliveras faults in the Bay area. An earthquake did occur, a consequence of the changing strain pattern; but it was some distance from the area of the indicated anomaly. This points out the need for a more thorough understanding of the strain patterns and the application of this knowledge to a predictive computer routine. There were three earthquakes following anomalous movements indicated by only one geodimeter line but the computer flagged 14 similar anomalies which were not followed by earthquakes. For all data, 8 earthquakes were predicted, and 17 predictions were not followed by an earthquake. However, if only anomalies observed on two or more geodimeter lines are considered, 5 were accurately predicted and 3 predictions were not followed by an earthquake.

Several anomalies were identified just before termination of the program in 1968. Time is soon to run out on some of these (most shocks occurred within 12–18 months from the first observation of an anomalous movement) so these statistics will soon change.

Anomalous movements have been observed prior to every shock over magnitude 4.5 (where measurements were adequate) since the program began in 1959. Adequate measurements have not been always possible. For example, in the Hollister area, earthquakes over magnitude 4.5 are several times more frequent than the annual distance measurement in the area. Consequently, there seems to be significant promise for the development of a predictive program, but some false alarms will have to be tolerated. This is generally true of two-dimensional warning systems (space and time) for natural phenomena. Many more tsunamis, hurricanes and tornadoes are predicted than actually occur, but these warnings have saved lives.

Increased accuracy, a more dense network of lines and more frequent monitoring of each area along the faults may well increase the reliability of the system. A series of measurements taken early in the program were fortuitously made during the time of an earthquake (Figure 10). A large anomaly was in progress within the hour prior to the shock. If such short-term anomalies generally occur, a very useful improvement in warning capability might be developed by continuous or near-continuous monitoring at sites where regular monitoring at intervals of several months indicated unusual movements were taking place.

Acknowledgements

The California Department of Conservation working in cooperation with the U.S. Geological Survey provided funds to carry on the program for a few months in 1969 after it was dropped by the Department of Water Resources in June 1968. An evaluation of a field demonstration of new laser-ranging devices during this time has indicated that daytime work, comparable to and perhaps exceeding the accuracy of the Model 2A Geodimeter, is possible with such devices as the Model 8 Geodimeter and the Geodolite. Dual-color devices such as the one being developed by North American-Rockwell, Inc., show great promise for 1 mm accuracies without the necessity for atmospheric data sampling. The Department of Conservation is actively seeking additional funds to carry on the program this fiscal year. The late Dr. Hugo Benioff originally recommended in 1958 that the work be carried out.

Mr. D. B. Crice and Mr. E. E. Hagen have contributed much original work in data analysis and refinements in equipment. The program bears the mark of many substantial contributors. The reader is referred to DWR Bulletin No. 116-6 for a more thorough acknowledgement.

References

The following is not intended to be a complete list of pertinent material. A more thorough list is in Hoffman, R. B.: 1968, California Department of Water Resources, Bulletin No. 116-6, 'Geodimeter Fault Measurements in California'.

Brune, J. N. and Allen, C. R.: 1966, 'A Micro-Earthquake Survey of the San Andreas Fault System in Southern California', *Bull. Seism. Soc. Amer.* **57**, 277-296.

Hofmann, R. B.: 1966, 'Changes in Rate of Fault Movement Preceding California Earthquakes', in *Proceedings of the Second United States-Japan Conference on Research related to Earthquake Prediction Problems*, published by Lamont Geological Observatory, Palisades, New York.

Hofmann, R. B.: 1968, 'Recent Changes in California Fault Movement', in *Proceedings of Conference on Geologic Problem of San Andreas Fault System*, Stanford University.

Hofmann, R. B.: 1969, 'Earthquake Prediction from Fault Movement Monitoring in California' in Summaries of Joint U.S.-Japan Conference – Premonitory Phenomena Associated with Several Recent Earthquakes and Related Problems, *Trans. Amer. Geophys. Union* **50**, No. 5.

Whitten, C. A.: 1956, 'Crustal Movements in Calfornia and Nevada', *Trans. Amer. Geophys. Union*, **37**, 393-398.

Discussion

Mansinha: Did you try to monitor any vertical movements in your bench marks?

Hofmann: They are 8 to 20 km apart, this is too far for conventional leveling. Some leveling is carried out from time to time in cooperation with the Coast and Geodetic Survey on a number of networks in the central valley. Some level changes have been observed which I expect you will hear about from Whitten.

Mansinha: Did you observe any significant movements normal to the faults and were there any diagnostic changes in them before an earthquake?

Hofmann: In almost every case there is some strain. In the northern part of the fault system it is primarily slip. In the southern part of the state we observe compressional strain.

Reiter: At what distances from the fault do you measure the changes in displacement?

Hofmann: The entire fault zone is something like 2 km wide in some areas but our stations are approximately 2½ km away from the fault.

Garg: Since we know that the width of a fault zone decreases as we go deeper into the Earth's crust, is there any problem in building underground installations when you want to cross a fault zone?

Hofmann: The difficulty is that once the structure is severely damaged by fault movement you have to get men and equipment into a tunnel to clear it out. This is a very difficult proposition as we have discovered in the past. It is much easier if you have the structure on the surface even though you have a much wider zone to contend with because the transport of men and materials to the site is a much simpler problem.

MEASUREMENT OF MOVEMENT ON THE SAN ANDREAS FAULT

ROBERT D. NASON and DON TOCHER

U.S. Environmental Science Services Administration, Earthquake Mechanism Laboratory,
San Francisco, Calif., U.S.A.

Abstract. The San Andreas fault is a major earthquake source in California. Its geologic history indicates continuing displacement at about 1.2 cm/yr over the past 25 million years, which is less than the estimates in current theories of global plate tectonics. Current movement on the fault is by fault displacements in earthquakes and by gradual fault creep slippage not directly related to earthquakes. Fault slippage in the historic period 1848 to 1968 has been equivalent to about 2.5 cm/yr average fault displacement.

1. Introduction

The San Andreas fault is the major active earthquake fault of coastal California. Historically, it has been responsible for several major earthquakes and many lesser earthquakes in the past 150 years. The most famous of these was the 1906 'San Francisco' earthquake, although an earthquake on the San Andreas fault in Southern California in 1857 was apparently as large. Important earthquakes on the San Andreas and associated faults include the 1836 Oakland East Bay, 1838 San Francisco Peninsula, 1857 Southern California, 1865 San Francisco Peninsula, 1868 Hayward, 1890 San Juan Bautista, 1906 Northern California, 1922 Parkfield, 1934 Parkfield, 1940 Imperial Valley, and 1966 Parkfield earthquakes, all of which were apparently accompanied by surface ground displacements on the faults. These earthquakes are well discussed in the general seismology textbook of Richter (1958).

2. The San Andreas Fault

The San Andreas fault, like other faults, is a discontinuity or 'fault' in the earth's crustal rocks across which rocks are not continuous and along which slippage has taken place. On many faults the relative movement is in a vertical sense, with one side becoming higher than the other side. On the San Andreas fault the slippage is mostly in a sideways, horizontal direction, a type of horizontal shear. This horizontal slippage is shown by the offset of fences, of streams, and of various sedimentary rocks.

That the horizontal slippage has been going on for millions of years is shown by the progressively greater offset of older sedimentary rocks across the fault. For instance, sedimentary rocks of late Miocene age (about 15 million years ago) west of Bakersfield, Calif., appear to have been offset by 120 km at the San Andreas fault (Hill and Dibblee, 1953). The horizontal slippage has amounted to perhaps several hundred kilometers over several tens of millions of years, at an average rate of about 1.2 cm/yr over the past 25 million years (Figure 2; Grantz and Dickenson, 1968).

The crustal rocks are very different on the two sides of the San Andreas fault, probably as a result of the long offset over millions of years (Figure 1). The most

L. Mansinha et al. (eds.), Earthquake Displacement Fields and the Rotation of the Earth, 246–254.
All Rights Reserved. Copyright © 1970 *by D. Reidel Publishing Company, Dordrecht-Holland*

outstanding difference is in Central California, where for 400 km from Bodega Bay to Cholome a granitic crust is exposed west of the fault, but the crust just east of the fault is non-granitic, consisting of eugeosynclinal oceanic deposits (Franciscan assemblage) on a probably oceanic crust. This granitic vs. non-granitic crustal rock distribution may indicate more than 400 km of right-lateral strike-slip displacement on the San Andreas fault since mid-Cretaceous (Hill and Dibblee, 1953).

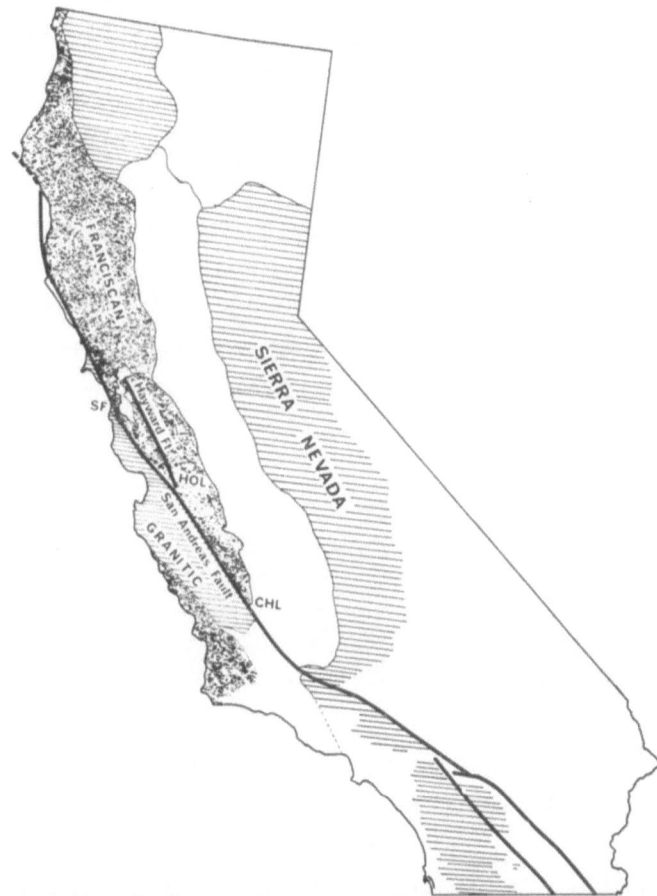

Fig. 1.   Generalized map of California geology, showing San Andreas and related faults and area of granitic (lined) and Franciscan (speckled) crustal rock. HOL is Holister, Calif., and CHL is Cholame, Calif.

The San Andreas fault is 950 km long as a continuous feature from Cape Mendocino south to Banning, and related faults continue the San Andreas fault system another 1600 km southward to the mouth of the Gulf of California. At the Gulf of California the fault system intersects the East Pacific Rise, which is the major tectonic feature of the eastern Pacific Ocean basin and part of the global mid-oceanic ridge/rise system. At its north end, the fault intersects the Mendocino fracture zone and the Gorda rise,

which is similar to the East Pacific Rise. The fault has been suggested to be a transform fault between the East Pacific Rise and the Gorda Rise in many current theories
of sea floor spreading and global plate tectonics (Wilson, 1965; Isacks *et al.*, 1968).
However, the rate of displacement along the fault of about 1.2 cm/yr (Figure 2) is less
than the almost 6 cm/yr movement shown by magnetic anomalies at the Juan de Fuca
Ridge crest north of Cape Mendocino (Vine, 1966). Thus much of the interpreted
right-lateral shear from sea floor spreading must be occurring on other features, such
as the Basin and Range province of Nevada and Utah east of the San Andreas fault.

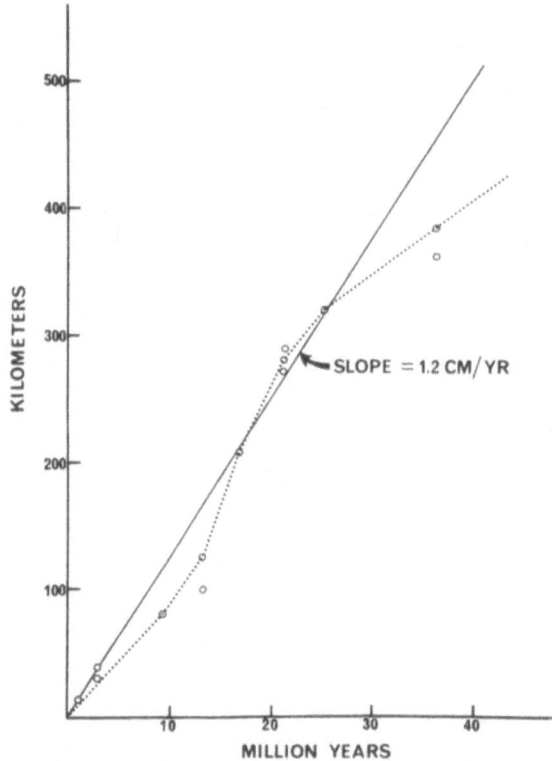

Fig. 2. Apparent offsets of geologic markers of different ages at the San Andreas fault. Older sediments appear to have greater offsets, indicating an average rate of displacement of 1.2 cm/yr for the
past 25 million years. (After Grantz and Dickinson, 1968.)

3. Current Fault Movement

Current movement on the San Andreas fault is occurring in two distinct ways – by
seismic fault slippage during earthquakes and by fault creep slippage without earthquakes.

The seismic fault slippage in earthquakes on the San Andreas and other faults has
been well known for many years. Examples are the 1857 and 1906 earthquakes which

had seismic fault slips of as much as 6 m, and the many earthquakes with lesser fault slips. The clear connection between earthquakes and fault slippage in California was a basis for formulating the important elastic-rebound theory of earthquakes (Reid, 1910).

The majority of seismic fault slippage on the San Andreas fault has occurred in the two major earthquakes of 1857 and 1906. Additional fault slippage has occurred in the many lesser earthquakes and in the non-seismic fault slippage. The historic fault slippage (of the past 120 years) can be calculated by summing the effects of the earthquakes and distributing the earthquake fault slippage along the length of the fault, as shown below.

The fault slippage occurring in an earthquake can be approximated by one-half the maximum observed displacement along the observed length of fault breakage, although the average may exceed half the maximum in many instances. Thus the average fault slippage in the 1906 earthquake can be considered as about 3 m (one-half the reported 6 m maximum) over a fault length of 420 km from San Juan Bautista north to Cape Mendocino. This is equivalent to a fault slippage of 130 cm along the total fault length of 950 km. The 1857 earthquake is much less well known, but was similar to the 1906 earthquake in many ways. An estimated average displacement of about 3 m and a fault length of about 400 km for the 1857 earthquake is equivalent to a fault slippage of about 125 cm along the 950 km total fault length.

The contribution of smaller earthquakes such as those in 1890, 1922, 1934 and 1966, can also be estimated using an average slippage of about 15 cm for a fault length of perhaps 30 km in each instance. Such an earthquake is equivalent to about 0.45 cm of slip on the total fault length. The contribution from perhaps ten such earthquakes in the past 120 years would total only about 5 cm for the 950 km fault length. Even if the calculation is doubled or tripled to include all the uncertainties, it is evident that the contribution of smaller earthquakes to total fault slippage is much smaller than the contribution from the large earthquakes.

The summed contribution of seismic fault slippage of earthquakes of the past 120 years in California is perhaps 260 cm, mostly in the 1857 and 1906 earthquakes. 260 cm in 120 years is about 2.2 cm/yr average slippage on the San Andreas fault in the historic time period 1848–1968. Adding in the effects of fault creep slippage as described below may raise the average fault slippage to about 2.5 cm/yr. This slippage is not evenly distributed along the fault in the historic period, but may even out over longer time periods. The 2.5 cm/yr average is about twice the 1.2 cm/yr average of the past 25 million years, showing that the historic period has been much more active than the long-term average movement of the San Andreas fault.

4. Fault Creep Slippage

A second form of fault slippage, non-seismic fault creep slippage, is a less-known phenomenon discovered on the San Andreas fault only recently (Steinbrugge and Zacher, 1960; Tocher, 1960). It was discovered at a winery south of Hollister, Calif., where the main winery building was being internally torn by fault displacement of about 1.2 cm/yr. The 8-year old building was already offset by 10 cm when discovered in 1956, and the offset has grown to 25 cm in 1969. Most of this movement occurred independently of local earthquakes.

Since 1956, fault creep slippage has been found more widely on the San Andreas and related faults in central California (Nason, 1969). It offsets curbs and sidewalks in Hollister, Calif., underground tunnels in Berkeley, Calif., and fences, bridges, and other features in other areas (Figures 3 and 4).

The pattern of fault creep slippage in central California is now fairly well known. It occurs along 200 km of the San Andreas fault from Cholame north to the Hollister area (Brown and Wallace, 1968) and on the Calaveras and Hayward faults from Hollister north to Richmond (Cluff and Steinbrugge, 1966; Rogers and Nason, 1967). Fault creep slippage apparently is not occurring on the San Andreas fault north of a point west of Hollister in the area of the 1906 earthquake fault break, or south of Cholame in the area of the 1857 earthquake fault break (Brown and Wallace, 1968).

Field studies have shown the fault creep slippage to be from as much as 30 cm to more than 1 m in the last 60 years. The fault creep slippage rate is as much as 2.5 cm/yr

Fig. 3.   Sidewalk in Hollister, Calif., offset by 25 cm of gradual fault slippage. Note date of 1928 in corner cement.

Fig. 4.   Curb in Hollister offset by gradual right-lateral fault slippage.

along the central part of the San Andreas fault, dying off to zero at the ends of the creep slippage zones. Using an average of 1.5 cm/yr creep slippage on a 200 km fault length is equivalent to 0.3 cm/yr along the total fault length. The 0.3 cm/yr from fault creep slippage is much smaller than the approximately 2.2 cm/yr from large earthquakes, but is nonetheless significant. It should be noted also that this average creep rate may vary greatly with time.

5. Survey Measurements

The U.S. Coast and Geodetic Survey maintains networks of triangulation in active areas which are regularly resurveyed to determine changes. These networks show sizable regional distortion near to and across the San Andreas and other faults over definite time periods. The regional distortion includes both elastic and permanent deformations, with the elastic portion of the deformation supposedly being recoverable in future fault movements. One network in the San Francisco Bay Area shows a difference in movement across the Hayward fault of about 20 cm in the 12 years 1951–63 (Pope *et al.*, 1966). This is much greater than the probable fault creep slippage in the same time period at this locality (about 8 cm in 12 years) and may indicate elastic strain buildup in the region. Another USC & GS network south of Hollister was resurveyed after 20 years in 1963 (Figure 5; Meade, 1966). The difference in movement across the San Andreas fault is about 60 cm in 20 years, averaging about 3 cm/yr. This is very close to the amount of known fault creep slippage in the area (about 2.5 cm/yr on the San Andreas fault) and indicates little or no elastic strain buildup in the area. Similar regional measurements by geodimeter are described by Hoffman (1968).

The fault movement can also be monitored by establishing a line of survey marks across the fault and measuring the offset of the line. Figure 6 shows the results from

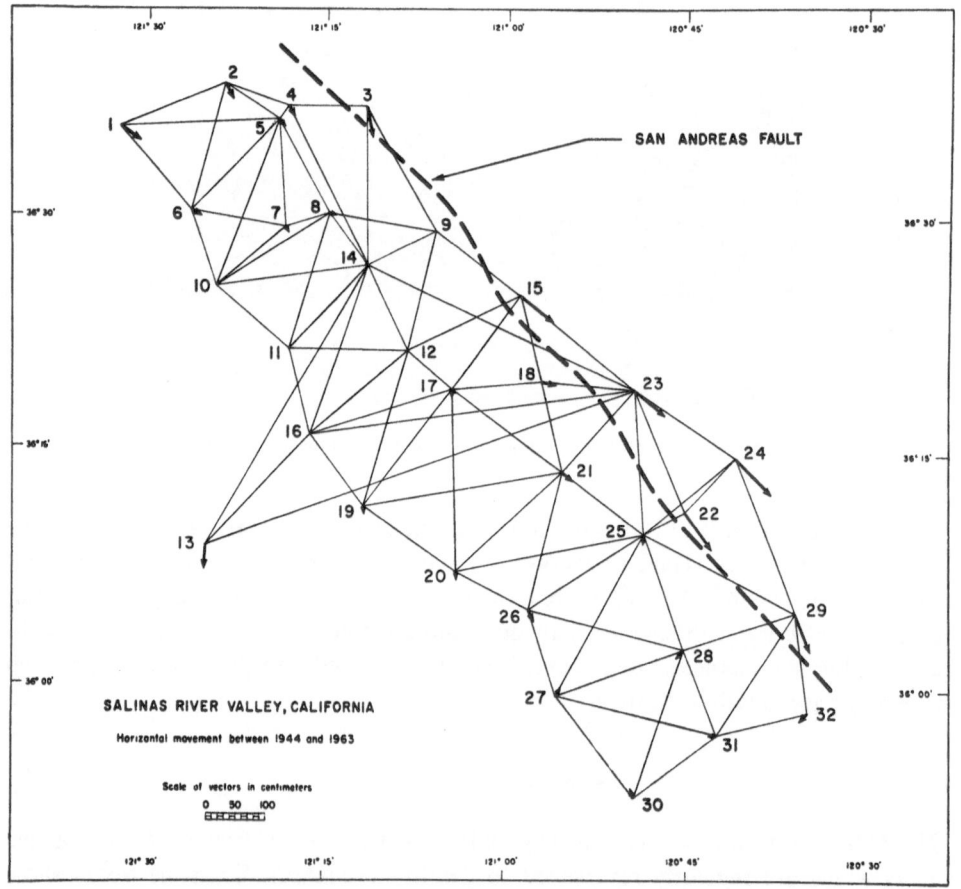

Fig. 5.   U.S. Coast and Geodetic Survey network south of Hollister, Calif. Original survey in 1944 and resurvey in 1963. Vectors show change in survey monument position due to crustal strain and distortion.

one such line in Hollister, Calif. The line was installed straight in September, 1967 and was offset 6 mm in February, 1968 and 10 mm in June, 1968. The survey line also shows the width of the slippage zone to be 5 m here, similar to the 2–5 m wide slippage zones found at many other localities. The survey lines have shown that the rate of slippage varies from month to month.

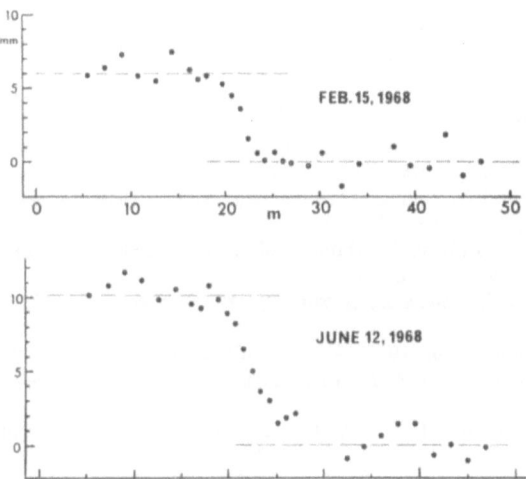

Fig. 6.   Offset of straight line of closely spaced survey marks across the Calaveras fault in Hollister, Calif. Line installed straight in September 1967.

6. Conclusions

The San Andreas fault is a major geologic feature along which horizontal displacement of about 1.2 cm/yr has apparently been occurring for the last 25 million years. This rate of displacement is significantly less than that interpreted from magnetic anomalies and sea floor spreading, and indicates complexity of movement in western North America.

Current movement on the San Andreas fault is occurring in two distinct ways – by earthquake fault slippage and by non-seismic fault creep slippage. The earthquake fault slippage in the historic period from 1848 to 1968 has totaled about 260 cm distributed along the total fault length, for an average of about 2.2 cm/yr. This is greater than the long-term average of 1.2 cm/yr over the past 25 million years.

Geodetic surveys show the sizeable regional distortions near the San Andreas fault and indicate an elastic strain buildup in many areas.

References

Brown, R. D. Jr. and Wallace, R. E.: 1968, 'Current and Historic Fault Movement along the San Andreas Fault between Paicines and Camp Dix, California', in *Stanford University Publications in Geological Sciences*, vol. XI, pp. 22–41.

Cluff, L. S. and Steinbrugge, K. V.: 1966, 'Hayward Fault Slippage in the Irvington-Niles Districts of Fremont, California', *Bull. Seism. Soc. Am.* **56**, 257–279.

Grantz, A. and Dickinson, W. R.: 1968, Indicated Cumulative Offsets along the San Andreas Fault in the California Coast Ranges', in *Stanford University Publications in Geological Sciences*, vol. XI, pp. 117–120.

Hill, M. L. and Dibblee, T. W., Jr.: 1953, 'San Andreas, Garlock, and Big Pine Faults, California', *Geol. Soc. Am. Bull.* **64**, 443–458.

Hoffman, R. B.: 1968, 'Geodimeter Fault Movement Investigations in California', *Calif. Dept. of Water Resources Bull. 116–6*.

Isacks, B., Oliver, J., and Sykes, L. R.: 1968, 'Seismology and the New Global Tectonics', *J. Geophys. Res.* **73**, 5855–5899.

Meade, Buford, K.: 1966, 'Horizontal Crustal Movements in the United States', *Acad. Sci. Fennicae Annales*, Series A, III, *Geologica–Geographica*, No. 90, pp. 247–266.

Nason, R. D.: 1969, 'Preliminary Instrumental Measurements of Fault Creep Slippage on the San Andreas Fault, California', *Earthquake Notes* **40**, 6–10.

Pope, A. J., Stearn, L., and Whitten, C. A.: 1966, 'Surveys for Crustal Movement along the Hayward Fault', *Bull. Seism. Soc. Am.* **56**, 317–323.

Reid, H. F.: 1910, 'The California Earthquake of April 18, 1906 – Mechanics of the Earthquake', *Carnegie Inst. Washington Publ. 87*, V.3.

Richter, Charles F.: 1958, *Elementary Seismology*, W. H. Freeman & Co., San Francisco, Calif., 738 pp.

Rogers, T. H. and Nason, R. D.: 1967, 'Active Faulting in the Hollister Area, in *Guidebook to the Gabilan Range and Adjacent San Andreas Fault*, Am. Assoc. Petroleum Geologists, Pacific Section, pp. 102–104.

Steinbrugge, K. V. and Zacher, E. G.: 1960, 'Creep on the San Andreas Fault – Fault Creep and Property Damage', *Bull. Seism. Soc. Am.* **50**, 389–396.

Tocher, D.: 1960, 'Creep on the San Andreas Fault – Creep Rate and Related Measurements at Vineyard, California', *Bull. Seism. Soc. Am.* **50**, 396–404.

Vine, F. J.: 1966, 'Spreading of the Ocean Floor: New Evidence', *Science* **154**, 1405–1415.

Wilson, J. T.: 1965, 'Transform Faults, Oceanic Ridges, and Magnetic Anomalies Southwest of Vancouver Island', *Science* **150**, 482–485.

Discussion

O'Hora: Is there any way in which possible motion of the Ukiah observing station could be deduced? This has a very large bearing on the secular motion of the pole being derived from the ILS stations; instead of the pole moving towards North America it may well be that an observing station in North America is moving towards the pole.

Nason: In the theory of global plate tectonics if the strain was limited to the area immediately close to the San Andreas fault, Ukiah, which is east of it, would probably be moving southward relative to Japan. However, if the strain goes all the way across the western U.S.A. to Nevada and Utah, and I believe it must, then Ukiah is already in a zone of the earth where it will be moving northward relative to the eastern U.S.A. not as fast as San Francisco which is west of the fault, but still moving northward each year. That is the way I prefer to interpret the data.

Swetnick: If the data indicate differential slip motion between two fault lines, could not these same data also be interpreted as a rotation of the region between the faults or a combination of translation and rotation. Do you take this into account?

Nason: This is very difficult to resolve because the measurements are not precise enough. Shear will show as a change of shape of the geodetic triangle and this can be identified; however, you cannot identify rotation by itself.

CRUSTAL MOVEMENT FROM GEODETIC MEASUREMENTS

CHARLES A. WHITTEN

Chief Geodesist, Office of Geodesy and Photogrammetry, U.S. Department of Commerce,
Environmental Science Services Administration, Coast and Geodetic Survey, Rockville, Md., U.S.A.

Abstract. Large horizontal and vertical surface displacements which are associated with major earthquakes may be measured by comparing pre- and post-earthquake survey data. A brief review of the
investigations made after several such earthquakes is given. Special types of surveys are used to monitor the continuing slippage or slow creep along some of the major faults as well as the regional deformation in the same general area. The annual rates of creep, horizontal or vertical, vary from one or
two millimeters to 10 or 12 mm. The annual rates of accumulation of strain in these regions are
frequently two or three parts per million. For global studies, rates of rotation of continental blocks or
'plates' can be computed from long series of latitude observations. For North America, such a rate is
$-4\frac{1}{2}°$ in 10 million years.
   Geodetic measurements are also used to monitor surface changes resulting from the activity of man
in economic exploitation. Instances of subsidence, with the associated horizontal movement, in areas
of high production of petroleum or the withdrawal of water, gas, sulphur or other minerals, as well as
cases of extreme compaction of soils from irrigation, are cited.
   Both types of investigations contribute to a national research program relating to earthquake
hazards or crustal collapse.

Geodesy is a science which is engaged in the never ending task of determining a more
exact size and shape of the Earth, and which is concerned with the establishment of
precise networks of horizontal and vertical control points. This second phase of
interest is frequently referred to as an engineering operation and considered by many
to lack any significant scientific challenge. However, because of the time-varying
aspects of geodesy, there are small changes in the horizontal or vertical coordinates
of these precisely located points. These variations of geographic position or elevation
provide geometrical data for studies of the deformation of the Earth's crust.

   Some of these variations are periodic, such as Earth tides or the combined annual
and Chandler wobble of the pole. Other variations, non-periodic in nature, are the
slow continuous drift, rotation, creep, or similar effects of the Earth's crust. Occasionally, there are sharp discontinuities in these slow movements. These are the
displacements which occur at times of earthquakes. In this brief summary of the
applications of geodetic measurements to the study of crustal deformation, I will
select various types of measurements which have been made during the past 100
years to illustrate the type of deformation that thas occurred and to indicate the
magnitude or rates of change.

   I assume that all of you are fully appreciative of the precision which is inherent in
making all types of geodetic measurements. At times though, the 'noise' of the measurement is approximately equivalent to the 'signal' or amount of movement. I will
identify those cases in which the signal to noise ratio is unfavorable.

   Within recent years, geophysicists have focused much of their attention on global
tectonics – continental drift, plate rotation, and expanding or non-expanding Earth.
When considering the work of Hess, Dietz, Morgan, LePichon, Sykes, Oliver, Isacks,

L. Mansinha et al. (eds.), Earthquake Displacement Fields and the Rotation of the Earth, 255–268.
All Rights Reserved. Copyright © 1970 by D. Reidel Publishing Company, Dordrecht-Holland

Talwani, and many others, I sought for some way in which geodesists could provide adequate data to support their research. Techniques for measuring intercontinental distances to a resolution of a small fraction of a meter are being developed. Radio interferometry and satellite laser measurements indicate that in a few years definitive results should be available. Measurements of this type should confirm the rates of continental drift determined from other geophysical data. There is another type of measurement which merits investigation. For almost 70 years, geodesists have been measuring latitude in a very systematic way for the purpose of determining polar motion. The results show evidence of secular movement. It is agreed that we cannot separate the effects of crustal movement from those of secular motion. There is no question on this point, but the rates of polar secular motion which have been published assume that the crust is fixed.

If we assume that 'plate rotation' offers an explanation to the differential change in latitude at the five International Latitude Service Observatories, the results are interesting; in fact, in close agreement with the rates of rotation suggested by geologists and geophysicists. I have used data published by Markowitz (1968) in which the effects of the annual and Chandler wobble have been filtered out. By using the differential change between two observatories on the same continental block and by using the distance between observatories as a base line, I have calculated the small angle of rotation with respect to the axis of the Earth. For purposes of comparison, I have expressed these rates of rotation in degrees per 10 million years. For North America the rate is 5° counterclockwise, for Europe 6½° clockwise, and for Asia 4° clockwise (Figure 1). These rates of rotation agree quite closely with those which have been suggested by many of the geophysicists. As I stated earlier, there is no way, at

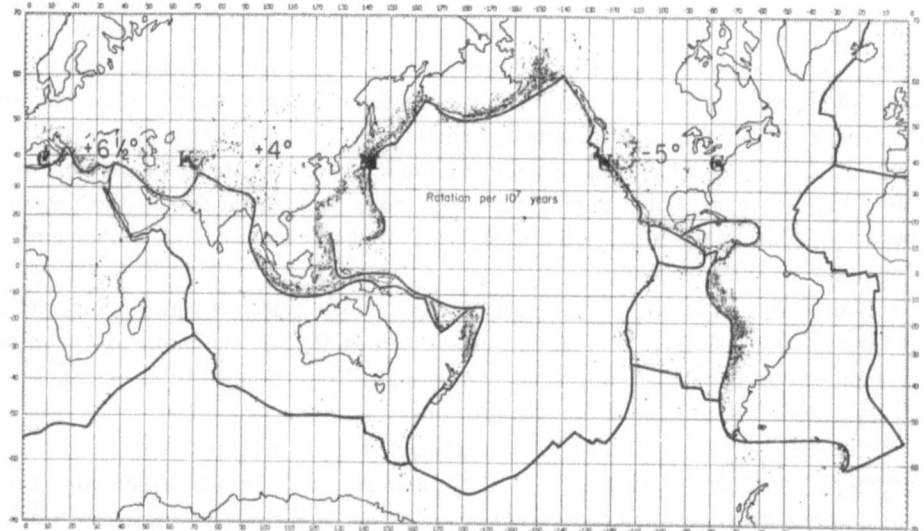

Fig. 1.   Rates of rotation for N. Amerika, Europe and Asia.

this time, that we can separate the secular motion of the pole from plate rotation or other crustal movement effects, but we cannot ignore this interesting correlation.

Another type of data which can be applied to continental deformation studies is the long series of tidal measurements. The changes in sea level, or uplift or subsidence of the crust along the East and West Coasts of the United States indicate the magnitude of these vertical movements (Figures 2, 3, and 4). This is another case in which we cannot clearly identify the variable. Sea level may be rising, the changes may be due to crustal tilting, or combinations of the two. Comparable vertical changes are known to exist in the interior of continents, but, unfortunately, we do not have extensive relevelings on a broad continental scale across the United States. Crustal deformations of this type have been measured in other countries. Several European countries are now working on their third or fourth relevelings. With the cost of leveling at approximately $100 per km, you can appreciate the economic problem associated with such investigations in a country the size of the United States. It is a much easier task to monitor this vertical movement where we can use the level of the sea as a

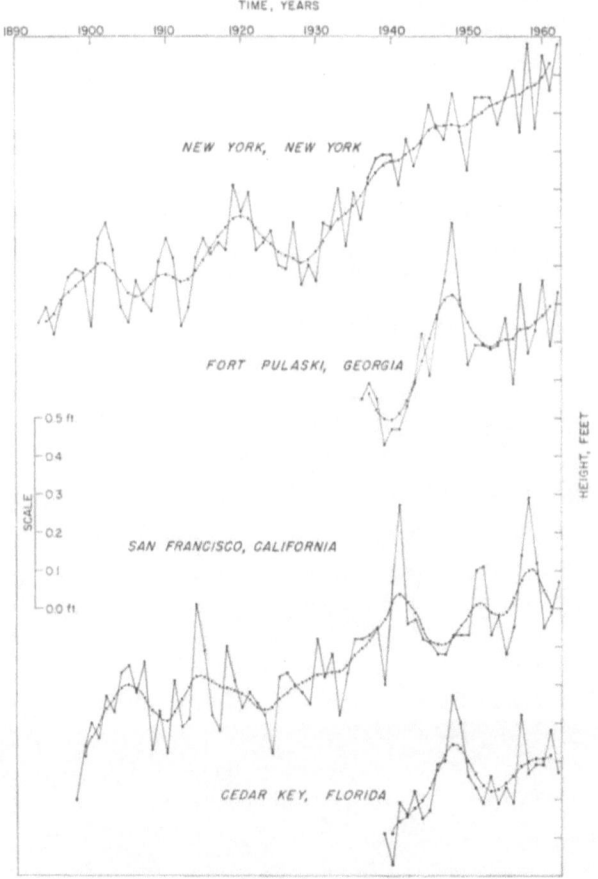

Fig. 2.   Results of tidal measurements.

reference surface. It may be of interest that one of the points of greatest continuing uplift is in North America. In the general region of Glacier Bay in Southeast Alaska, the crust is rising at nearly 4 cm per year (Figure 5). This is more than double the rate of post-glacial uplift in the Hudson Bay region or Fennoscandia. There are instances of much greater vertical movement, but these are in areas of subsidence where the movement is the result of action taken by man. I will discuss these in a later part of this review.

The primary geodetic networks of triangulation and leveling serve as reference frameworks for regional studies of crustal deformation. The most direct use of these networks has been to provide the base from which the crustal movements and displacements which occur at the time of an earthquake can be measured. After every earthquake of magnitude 6 or greater, resurveys are made over whatever marks had been in existence prior to the earthquake. In the larger earthquakes, the changes in

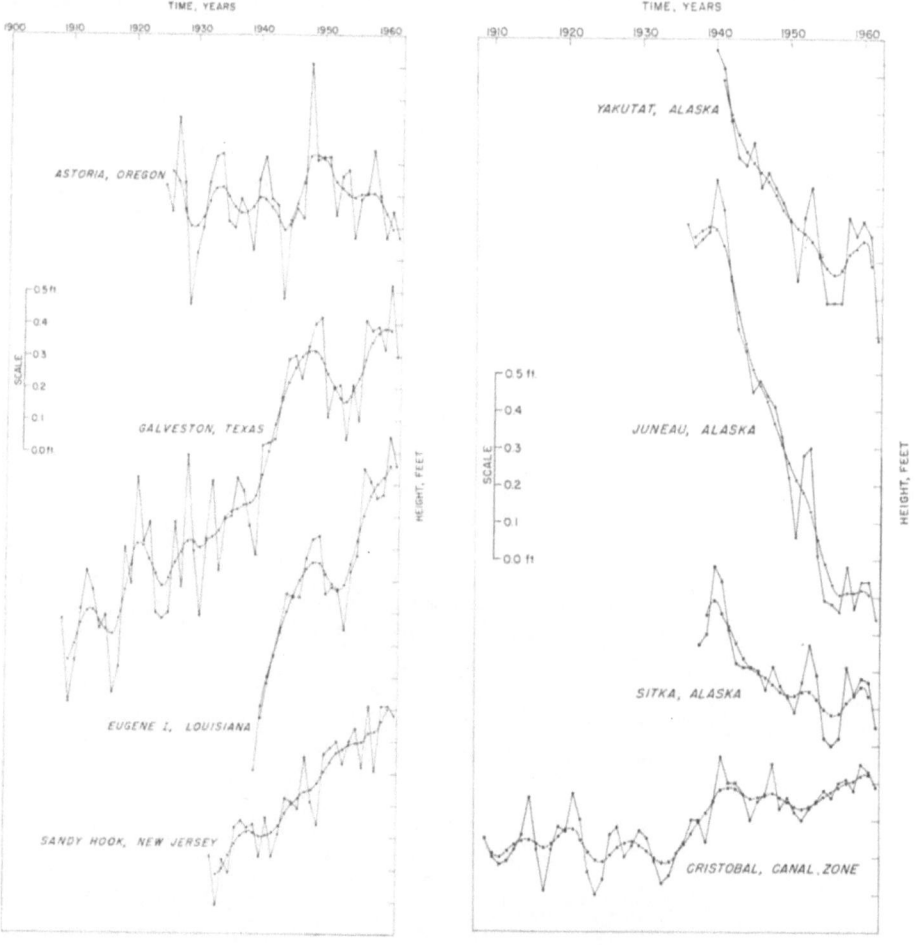

Figs. 3 and 4.   Results for tidal measurements.

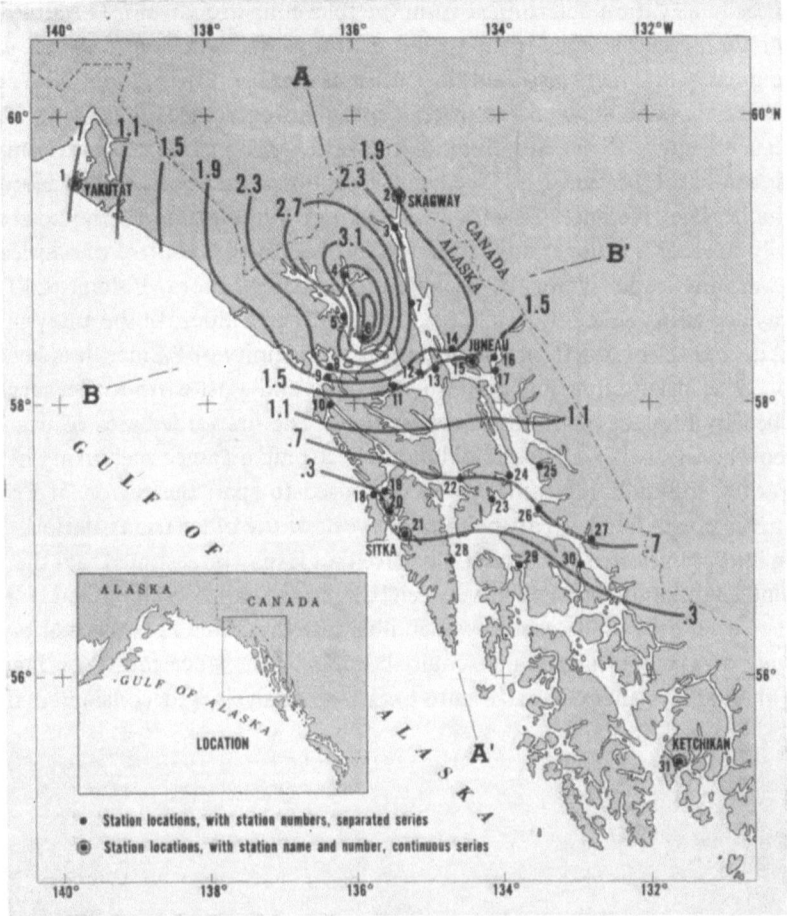

Fig. 5. Crustal uplift in S.E. Alaska.

position and elevation are quite dramatic. Geodetic surveys are hardly needed to determine the displacements along the faults, but are needed to determine the breadth of the fault zone and adjacent areas which were disturbed at the time of the earthquake. You are familiar with the literature in which details of these surveys have been published. These include the San Francisco earthquake of 1906, the Imperial Valley earthquake of 1940, the Kern County earthquake of 1952, the Dixie Valley earthquake of 1954, and the Alaskan earthquake of 1964.

The results of the resurveys made after the Alaskan earthquake were published within the last few months. Because of the size of the movements, there have been many inquiries. Some people have expressed doubt, others have sought interpretation. The geographical area involved is the largest in extent for which studies of this type have ever been made. I wish to add my personal interpretation of the published data.

The geodetic releveling was limited to regions where bench marks had been established in earlier years. These lines of leveling are along highways and railroads. The differences in elevation determined from the releveling are far more accurate than we need for the overall study. In the region of Prince William Sound, where the uplift was the greatest, we had to rely on the surface of the sea. Hydrographic surveys made after the earthquake showed the water depths had decreased by 15 m or that there had been an uplift of the sea floor of the order of 15 m. Any uncertainty in this amount can hardly exceed 1 m. The changes of horizontal position are more difficult to evaluate. The reference network which existed prior to the earthquake consisted primarily of second-order triangulation established for the control of nautical charts. There were first-order chains of triangulation from Anchorage along the Glenallen Highway and south along the Richardson Highway to Valdez. In the analysis made by Parkin, a single point north of Anchorage, in the vicinity of Palmer, was held fixed in position. The orientation and scale of the pre- and post-earthquake surveys were controlled by Laplace azimuths and base lines. The first-order arcs of triangulation were reobserved insofar as possible, but an electronic distance measuring network of trilateration, triangulation, and traverse was used to span the region of Prince William Sound rather than repeat the total network of the older triangulation.

In an interpretation of the results, I believe that it may be assumed that there was no appreciable horizontal movement between Palmer, Glenallen, Valdez, and Homer, with respect to each other. The vectors which illustrate the differences in position at these points are of a magnitude which could be due to the uncertainty or inaccuracies within the surveys themselves (Figure 6). If the analysis had considered that these

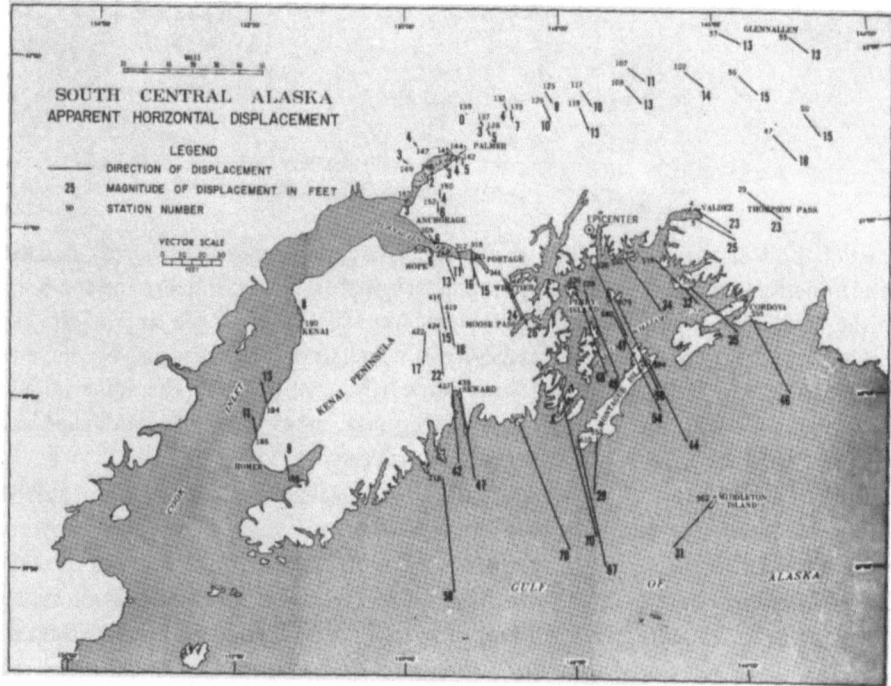

Fig. 6.   Apparent horizontal displacement following 1964 Alaskan earthquake.

four points were essentially fixed with relation to each other, the vectors then would have shown an extension of approximately 1 m across the Matanuska Valley with a general southward movement increasing rather gradually to a maximum amount of the order of 15 m for the region near Montague Island in Prince William Sound. The mountains on Kenai Peninsula also moved southward, perhaps as much as 8 to 10 m. Resurveys in 1967 across Cook Inlet and Shelikof Strait to Kodiak, Afognak, and Ushagat Islands also confirmed this general movement of the axis of this mountain chain. At the present time, we in the Coast and Geodetic Survey are confronted with the problem of which geographic positions to furnish engineers, cartographers, and others, for local control. The existence of two sets of coordinates, with some degree of uncertainty concerning their relationship, must be resolved. We are recomputing all of the post-earthquake survey measurements, holding the coordinates at these four points, Homer, Palmer, Glenallen, and Valdez. When this adjustment work has been completed, a supplemental report will be issued which can then be studied in conjunction with data already published.

During the past 40 years, the Coast and Geodetic Survey has developed an intensive program for repeating geodetic surveys to monitor slow crustal movements and deformation. In the first years of this program, a comprehensive network was established in California primarily to be able to measure displacements following anticipated earthquakes. After several of the sections of this large network had been resurveyed without an earthquake having occurred, the results showed the slow systematic movement of one side of the fault system with respect to the other side with the associated accumulation of strain across the fault zone and adjacent area. For the San Andreas system, this right lateral movement across the broad region is of the general order of three to six centimeters per year. The only exception where this type of movement has not been found is in the south-central part of the state near the junction of the Garlock and San Andreas faults.

In this particular area, a repeat survey of a very precise triangulation network, saturated with Geodimeter base lines, has shown a systematic compression in the north-south direction and some indication of extension in the east-west direction for the pie-shaped region lying between the two faults (Figure 7). The time interval between the two surveys was only six or seven years and the results are merely indicative of this type of deformation or strain accumulation. The changes in length are in millimeters. This is one of those occasions where the signal to noise ratio approaches one. However, the same region was traversed by one of the older, long chains of triangulation. This particular section has been surveyed four times – twice before the Kern County earthquake and twice following it. The strain calculations from the individual triangles of this older arc confirm the values computed from the larger, more precise network.

A recent rereading and review of the report by Hayford and Baldwin (1907) on the 1906 earthquake has suggested that further remeasurements should be made along the San Andreas fault north of San Francisco. When the 1906 study was made, it was noted that surveys made between 1850 and 1905 were not internally consistent. This is

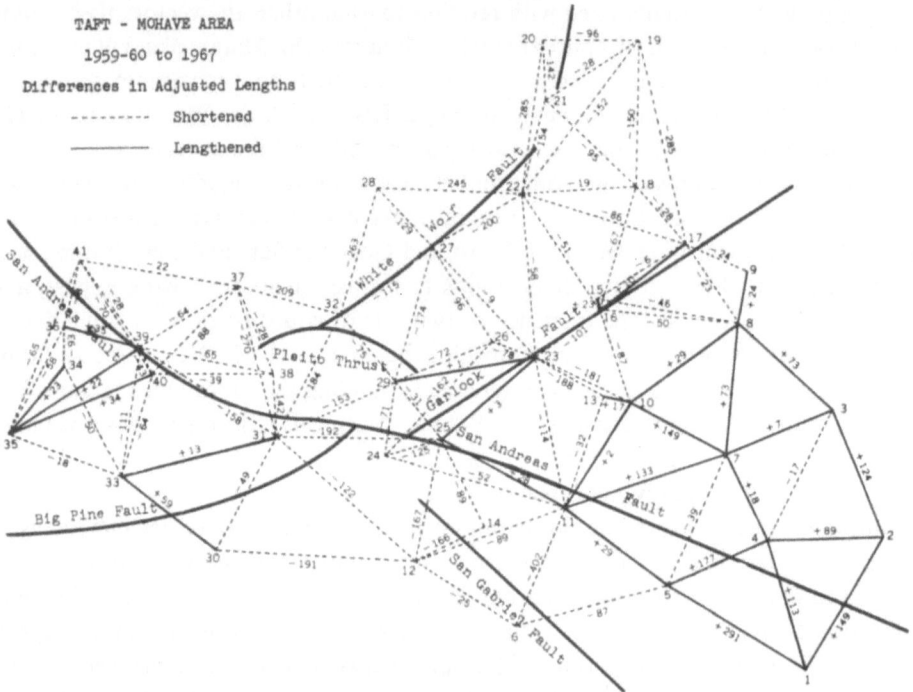

Fig. 7.   Compression and extension near the intersection of the Garlock and San Andreas faults.

comparable to some of the difficulties encountered with surveys made at different times since that earthquake. The pre-1906 data were divided into two sets, and an assumption was made that there was displacement along the San Andreas fault at the time of the 1868 earthquake. This could have been the case, but I believe that it will be possible to make a new study using time as a dimension, combining all of the survey data from 1850 to 1969 with only one discontinuity – that of the 1906 earthquake. Hayford and Baldwin did not attempt to fit a theory of slow creep to the earlier data.

    Another type of measurement which defines the rate of slow movement is that of repeated astronomical azimuths. Lines of triangulation parallel to the fault do not rotate, but lines which cross the fault and are normal to it do show a slow change of azimuth or rotation with respect to the axis of the Earth. A classic line is Mt. Toro to Santa Ana, a primary line on the coastal arc of triangulation. It is east of Monterey Bay. The first azimuth determination was in 1880. The observations have been repeated at approximate 10-year intervals with an average change of 1 sec per 10 years. The length of the line is approximately 50 km. Thus, the azimuth change indicates a slow movement of 25 cm per 10 years.

    About 12 years ago, Tocher and Steinbrugge found evidence of slippage along the San Andreas fault at the Cienega Winery south of Hollister. This slippage, even though it occurs in episodes of fractions of mm, accumulates quite uniformly with

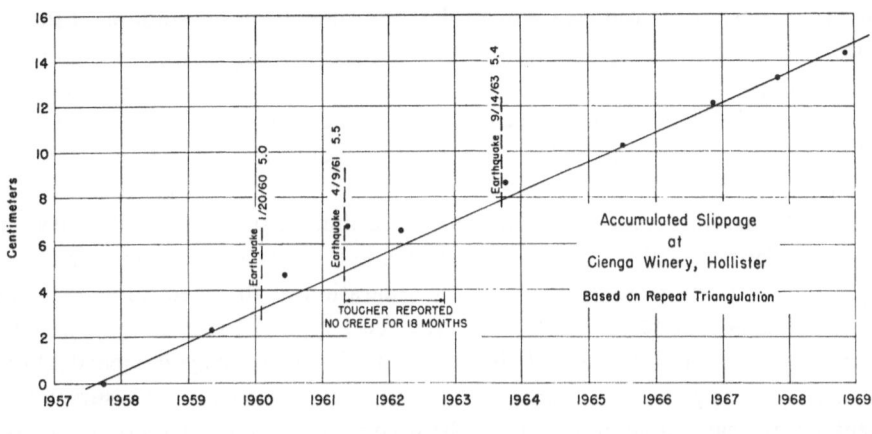

Fig. 8.

respect to time. Special geodetic surveys at the Winery have been repeated almost annually. Tocher has maintained a continuous recording instrument within the Winery. He reports that at the time of earthquakes there is an increased number of episodes with greater slippage. The graph in Figure 8 shows the accumulated slippage as determined by triangulation. The displacement computed from the repeat geodetic surveys confirms this increased slippage associated with earthquakes and shows a remarkable uniformity during the long periods of time when there are no earthquakes of appreciable magnitude.

Because of the ease with which measurements of the type used for the Winery study could be made, more than 20 similar networks have been established along various sections of the faults in California where slippage or creep has been suspected. The data from these networks have established rather clearly the rates of horizontal and vertical movement.

At one of the sites across the San Jacinto fault, southwest of San Bernardino, the data indicate an annual right lateral movement of 2 to 3 mm and an annual rate of vertical slippage of 6 mm, with the east side subsiding or the west side uplifting. At a site across the Garlock fault, a few kilometers east of the junction of the Garlock and San Andreas faults, the annual rate of horizontal movement is 6 mm left lateral and the south side subsiding at an annual rate of 10 mm. Engineers engaged in constructing the aqueduct crossing at this point have encountered some difficulties with respect to this slippage and have been in frequent communication with the Coast and Geodetic Survey to confirm the rates.

Further north, along the San Andreas fault, a special figure was established in 1964 at a proposed aqueduct crossing. This is near Cholame, south of Parkfield. In November 1966, after the Parkfield earthquake, a resurvey disclosed a right lateral displacement of 3 cm. A resurvey in September of last year disclosed an additional slippage of 1 cm or an annual rate of 5 mm for the last two years.

Near the City of Hayward, the data indicate a rate of 5 or 6 mm right lateral per year. An interesting location for a quadrilateral is the one at the stadium at Berkeley. Professor Bruce Bolt has been monitoring the slippage along the Hayward fault, which passes through the stadium, with instrumentation in a tunnel under the football field. The Coast and Geodetic Survey has placed four markers on the upper rim of the stadium. The resurveys indicate a deformation equivalent to 3 mm right lateral movement per year.

Strain calculations, based on changes of coordinates or changes of angles, have been made for practically all of the surveys made for crustal movement studies. One of the difficulties encountered involves the separation of the effect of slippage from that of strain. About 10 years ago, we developed a type of strain analysis which was based on two assumptions but did give us a value for the slippage independent of the strain. One assumption was that the strain was homogeneous over a survey figure. Essentially the same assumption has to be used in any strain calculation. The other assumption was that the direction of crustal movement was parallel to the fault. Thus, the method would not be satisfactory for dilatation perpendicular to the fault. The deformation was based on systematic differences of directions from a triangulation station to all other visible points. Essentially these would be changes of angle with reference to the axis of the fault zone. In a report published in 1960, data from 1930 and 1951 surveys near Hollister indicated slippage of 10 mm per year, data from 1932 and 1951 surveys near Cholame 3 mm per year, and 1941 and 1954 surveys in the Imperial Valley indicated 4 mm per year. The rates of slippage are essentially the same as now being obtained from the small fault crossing surveys or more recent triangulation. I cite this similarity of rates as support for the validity of the method used.

In a program which was closely coordinated with that of the Coast and Geodetic Survey, the California Department of Water Resources initiated a geodimeter fault crossing survey. This was essentially a program for remeasuring long lines which crossed the fault in a zigzag manner or like a flat, sawtooth configuration. Occasionally, networks of shorter lines were established for special regional studies. This work was under the general direction of Renner Hofmann. Hofmann recently published the results of most of this work in Bulletin 116-6, 1968, 'Geodimeter Fault Movement Investigations in California'. These repeat geodimeter measurements show, even more dramatically, the slow continuous movement. There has been some interpretation that the rates of movement are subject to change, or strictly speaking to no change, such as a lock along the fault. My personal evaluation of this is that the geodimeter measurements are subject to some uncertainty because of the index of refraction. The signal-to-noise ratio is too close to one. Techniques and instruments being used at the present time are much improved and, no doubt, within the next year or two, there will be further remeasurements in these areas where locks have been reported. This special program is being continued through the cooperative efforts of the U.S. Geological Survey, the University of California at Berkeley (with a National Science Foundation grant), the California Division of Mines and Geology, and the Coast and Geodetic Survey.

An additional special trilateration network is being established this summer near

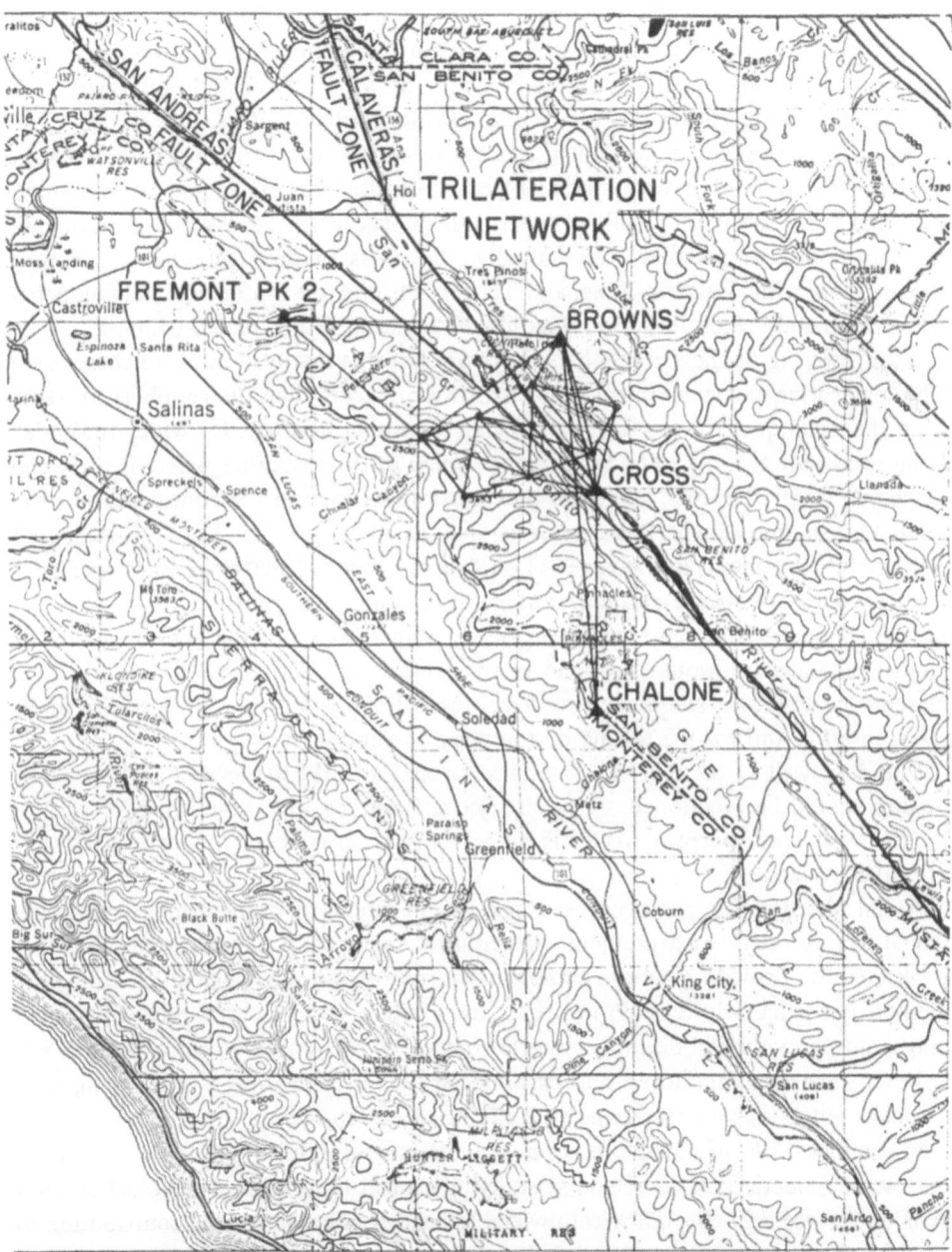

Fig. 9. Trilateration network near the Cienega Winery and Stone Canyon.

the Cienega Winery and Stone Canyon in an effort to isolate slippage from strain (Figure 9). Portions of the net cross the San Andreas and Calaveras faults which are only about 2 km apart. The side wings of the net are on supposedly homogeneous blocks in which we believe strain is accumulating. The longer lines of the net are the same as lines in the Water Resources Program, so we will have additional data to be compared with the earlier series of measurements started in 1959.

Man, through his exploitation of natural resources, has contributed to extensive crustal movement of a very special type. Frequently, the movement is subsidence resulting from the withdrawal of water, oil, or gas, with the associated horizontal movement due to collapse. Occasionally, the areas involved are quite extensive. Buena Vista Hills, Baldwin Hills, and Terminal Island are examples involving petroleum production. San Jose, San Joaquin Valley, and Houston are examples of the critical lowering of the water table. There are hazards involved in all of these cases. The engineering and economical problems are more acute than the geophysical. However, the techniques for geodetic monitoring are similar to those for which the underlying causes are tectonic. The rates of movement of these man-triggered disturbances are much larger than the 1 to 5 cm per year known to exist for continental drift or related crustal movement. Subsidence as great as 30 cm per year has been measured. Horizontal movements at the edges of such areas may be as large as 10 cm/per year. I cite these cases because of the advantage of obtaining real time data from an actual physical model where the experiment, if we wish to refer to the action as such, is controlled much as would be done in the laboratory. The signal-to-noise ratio is extremely favorable. These are excellent areas of study for the development of instrumentation and for the testing of analytical methods.

The deep well at the Rocky Mountain Arsenal and the associated Denver earthquakes should be mentioned. A precise geodetic survey was established in the area in 1968, but until a repeat survey is made, there is no geodetic evidence relative to crustal movement.

Aerial photography has been used in a few instances to supplement the data obtained from geodetic measurements. When used analytically, this technique is a very powerful tool, providing a more complete description of large scale deformation. The method is particularly feasible in urban areas where streets, curbs, sidewalks, buildings, etc., provide additional geometric control. A section of Wasatch fault in Salt Lake City has been monitored photogrammetrically. The method is also being used to detect any possible post-earthquake sliding in Anchorage.

In summary, I have endeavored to review the most significant points of interest for the use of geodetic measurements in studying crustal movement. I have deliberately avoided the use of a chronological review. Instead, I have shown the contribution of geodesy, first, as applied to the global problem, then to the continental, next to the regional, and, finally, to the limited local situation. All of the bits of information are beginning to fit together to increase our knowledge of geodesy as applied to the study of the Earth, and to improve our capability for predicting the direction and magnitude of crustal movement we may expect in the future.

References

Bowie, Wm.: 1924, 'Earth Movements in California', USC&GS Spec. Pub. No. 106.

Bowie, Wm.: 1928, 'Comparison of Old and New Triangulation in California', USC&GS Spec. Pub. No. 151.

Burford, R. O.: 1965, 'Strain Analysis across the San Andreas Fault and Coast Ranges of California', in *Proc. 2nd Int. Symp. on Recent Crustal Movements, Aulanko, Finland.*

Hayford, J. F. and Baldwin, A. L.: 1907, 'The Earth Movements in the California Earthquake of 1906', USC&GS Annual Report, Appendix 3.

Hicks, S. D.: 1968, 'Sea Level – A Changing Reference in Surveying and Mapping', *Surveying and Mapping* **28**, No. 2.

Hicks, S. D. and Shofnos, Wm.: 1965, 'The Determination of Land Emergence from Sea Level Observations in Southeast Alaska', *J. Geophys. Res.* **70**, 3315–3320.

Hofmann, R. B.: 1968, 'Geodimeter Fault Movement Investigations in California', California Dept. of Water Resources Bulletin No. 116-6.

Koch, T. W.: 1933, 'Analysis and Effects of Current Movement on an Active Thrust Fault in Buena Vista Hills Oil Field, Kern County, Calif.', *Bull. Am. Assoc. Petr. Geol.* **17**, 694–712.

Markowitz, Wm.: 1968, 'Concurrent Astronomical Observations for Studying Continental Drift, Polar Motion, and the Rotation of the Earth', in *IAU Symposium* **32**, Springer-Verlag, New York.

Meade, B. K.: 1948, 'Earthquake Investigations in the Vicinity of El Centro, Calif., Horizontal Movement', *Trans. A.G.U.* **29**, 27–31.

Meade, B. K.: 1963, Report to the Commission on Recent Crustal Movements, Int. Assoc. of Geodesy, Berkeley, Calif.

Meade, B. K.: 1965, Report to the Commission on Recent Crustal Movements, Int. Assoc. of Geod., Aulanko, Finland.

Parkin, E. J.: 1969, 'Horizontal Crustal Movements Determined from Surveys after the Alaskan Earthquake of 1964. ESSA-C&GS – The Prince William Sound Earthquake of 1964', Pub. 10-3.

Pope, A. J., Stearn, J. L., and Whitten, C. A.: 1966, 'Surveys for Crustal Movement along the Hayward Fault', *Bull. Seis. Soc. Am.* **56**, 317–323.

Rogers, T. H. and Nason, R. D.: 1967, 'Active Faulting in the Hollister Area – from *Guidebook to the Gabilan Range and Adjacent San Andreas Fault*, Amer. Assoc. Petr. Geol., Pac. Sect.

Small, J. B.: 1960, 'Subsidence in the Texas Gulf Coast Area', USC&GS Unpublished Report.

Small, J. B.: 1963, 'Interim Report on Vertical Crustal Movement in the United States', Int. Assoc. of Geod., Berkeley, Calif.

Small, J. B. and Wharton, L. C.: 1969, 'Vertical Displacements by Surveys after the Alaskan Earthquake of March 1964. ESSA-C&GS – The Prince William Sound Earthquake of 1964', Pub. 10-3.

Steinbrugge, K. V. and Zacher, E. G.: 1960, 'Fault Creep and Property Damage', *Bull. SSA* **50**, 389–396.

Tocher, D.: 1960, 'Creep Rate and Related Measurements at Vineyard, Calif.', *Bull. SSA* **50**, 396–404.

Tocher, D. and Nason, R. D.: 1967, 'Fault Creep at the Almaden-Cienega Winery, San Benito County – from *Guidebook to the Gabilan Range and Adjacent San Andreas Fault*, Amer. Assoc. Petr. Geol., Pac. Sect.

Whitten, C. A.: 1948, 'Horizontal Earth Movement, Vicinity of San Francisco, Calif.', *Trans. A.G.U.* **29**, 318–323.

Whitten, C. A.: 1949, 'Horizontal Earth Movement in California', *USC&GS J.*, No. 2.

Whitten, C. A.: 1955, Measurements of Earth Movement in California, Calif. Div. of Mines Bull. 171, Earthquakes in Kern County, Calif., during 1952.

Whitten, C. A.: 1956, 'Crustal Movement in California and Nevada', *Trans. A.G.U.* **37**, 393–398.

Whitten, C. A.: 1957, 'Geodetic Measurements in the Dixie Valley Area', *Bull. SSA* **47**, 321–325.

Whitten, C. A.: 1960, 'Horizontal Movement in the Earth's Crust', *J. Geophys. Res.* **65**, 2839–2844.

Whitten, C. A.: 1961, 'Measurement of Small Movement in the Earth's Crust', *Ann. Ac. So. Fennicae, Helsinki.*

Whitten, C. A.: 1969, 'Recent Studies by the Coast and Geodetic Survey' – from Joint U.S.-Japan Conference, E⊕S, *Trans. A.G.U.* **50**, 401–402.

Whitten, C. A. and Claire, C. N.: 1960, 'Analysis of Geodetic Measurements along the San Andreas Fault', *Bull. SSA* **50**, 404–415.

Wilt, J. W.: 1958, 'Measured Movement along the Surface Traces of an Active Thrust Fault in the Buena Vista Hills, Kern County, Calif.', *Bull. SSA* **48**, 169–176.

Discussion

Nason: Geodimeters are used for a lot of things and I wonder if you could make a quick statement as to the limits of their accuracy and what problems you have run into?

Whitten: A couple of years ago we modified the standard geodimeter by inserting a laser light source and, because of the coherent properties, our results have improved significantly. Determining the index of refraction is the critical problem and two color laser geodimeters, which are not very portable, are being developed. We have found that if we make continuous trigonometric vertical angle measurements between the terminal points, and if we know the true difference of elevation of these terminal points, then we can extract the index of refraction. Recently we measured the New Mexico line. It was 80 km in length and could be subdivided into four segments; when the four segments were measured they projected on to a total line to an accuracy of $1\frac{1}{2}$ cm. I would not say that the total line is accurate to $1\frac{1}{2}$ cm but these were the numbers we obtained. A long series of measurements were made on the long line and the range was only 6 cm.

Stacey: Will the satellite geodesy offer us any help in analysing continental drift.

Whitten: I think so. There is a world wide geodetic satellite program and the first network of 44 stations placed around the globe will be completed in 1970. The accuracy between individual points within this network will probably be within 5 m. It will not be down in the range that will help us immediately with continental drift determinations, but it will improve our overall reference framework and give some standards to which other surveys might be connected.

Hofmann: The accuracy of our data with respect to ambient noise inherent in the survey method has been questioned. I believe the data adequate for the following reasons: Our calibration of the Model 2A Geodimeter has in every instance indicated a 50% probable error of about 5 mm. There is only one chance in 20 that the real error would exceed three times the probable error. Measurements taken over short lines not across faults, indicate a similar precision when our temperature sampling methods are used. In areas where few earthquakes occur, for example south of Hollister, California, fault movement proceeds at such an even rate that each year's observation varies only a few mm from the best straight line indicating the average rate. Most of our closed geodetic networks require adjustments of only a few mm for perfect closures. Anomalies preceding earthquakes all exceed 2 cm and some are greater than 4 cm. I believe these are clearly greater than the noise level of the distance measurements. I believe that the lower fault movement rates observed by Whitten from small fault quadrilaterals stem, at least in part, from the fact that the quadrilaterals are so small they occupy only a small percent of the total width of the rift zone. This zone of crushed rock and fault gouge is usually overlain by an appreciable thickness of alluvium. Although the quadrilateral may be astride the most obvious surface lineation in the zone, I believe that the soil beneath the quadrilaterals may absorb some of the movement and that other parts of the crushed zone, outside the area occupied by the quadrilateral, may also move. Our measurements span the entire fault zone and consequently it is not unexpected that they should reflect all the movement that takes place.

GEODETIC SURVEY MEASUREMENTS TO DETERMINE MOTION
IN THE EARTH'S CRUST

HAROLD E. JONES

Department of Energy, Mines and Resources, Surveys and Mapping Branch, Geodetic Survey,
Ottawa, Canada

Abstract. For many years, the Geodetic Survey of Canada has been involved in projects to study crustal movement or continental drift, with the areas of principal interest being in the regions of the St. Lawrence Valley, Robeson Channel and the Strait of Georgia. Only one project, levelling in the Quebec – Lake St. John area, is complete enough to show any results. The other projects require future remeasurements before significant comparisons can be made. These projects, in which common instruments and techniques are used, are reported and discussed indicating the type and magnitude of movements which can be determined. New instruments and techniques which will allow more accurate determinations are also discussed.

1. Introduction

The basic purpose of the Geodetic Survey is to supply the basic survey control for mapping and for other surveys in Canada. In order to do this, it becomes involved in the broad international field of Geodesy, that is to measure the size and shape of the world, thus measuring and taking into consideration changes and shifts in the earths' crust. Discussions of the Committee on the Seismic Regionalization and Recent Crustal Movements had considerable bearing on decisions as to where studies should be made for crustal movement.

In the past several years the Geodetic Survey has been involved in some half dozen projects with a main, or subsidiary, purpose of studying crustal movement and/or Continental Drift. They have been mainly concentrated in three regions, the Upper St. Lawrence Valley, Robeson Channel and the Strait of Georgia. Only one project (levelling in the Quebec – Lake St. John area) is complete enough to show decisive results. The other projects require a future remeasurement to enable a comparison to be made. From a report and discussion of these projects (in which we use our common instruments and techniques) I would like to give an idea of the magnitude of movements we believe we can detect and the distance over which the movements could be determined with an estimate of their accuracy. I will also mention what we believe we may be able to do in the near future with new techniques.

2. Horizontal Motion Studies

A. ROBESON CHANNEL NET

In this network the lengths of all the lines were measured as were the angles between the lines.

It was measured following a suggestion of Tuzo Wilson in 1963 that there might be motion between two plates of the earths' crust along that channel. It was a joint

L. Mansinha et al. (eds.), Earthquake Displacement Fields and the Rotation of the Earth, 269-275.

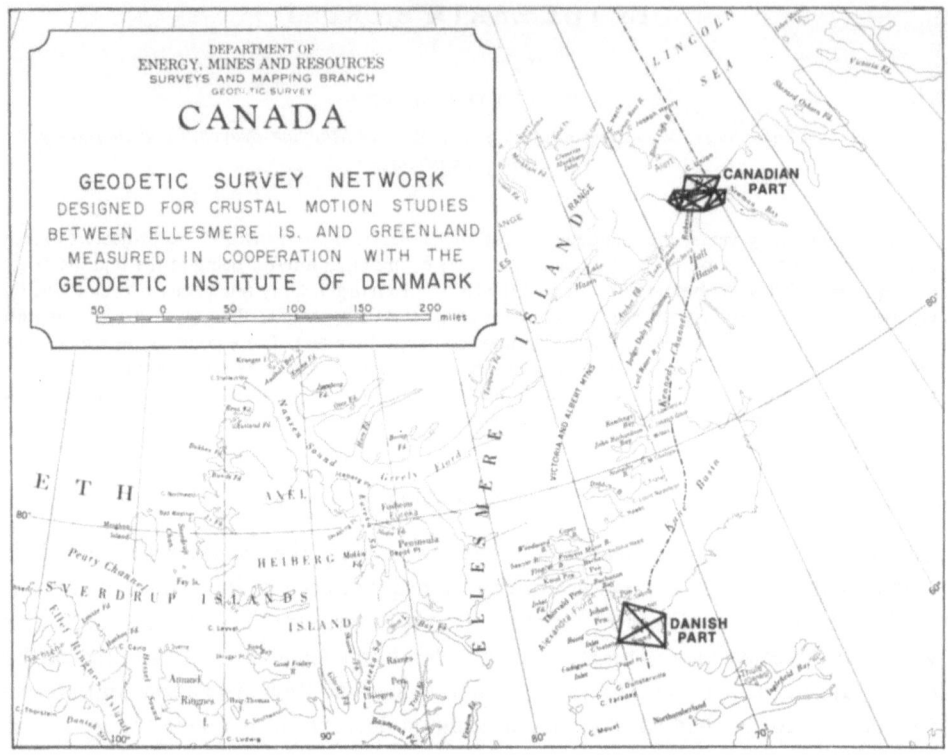

Fig. 1.

project with the Danish Geodetic Institute (Figure 1). The Geodetic Survey of Canada measured the northern part in 1967. One of the reasons for the more compli- cated network in the north, is because it was not possible to find solid bedrock for markers. When the net is remeasured in a few years, we want to have some measure of certainty that any difference is not due to local surface movement. We hope that any surface movement would be indicated by differential movement between the many marks on each site. The southern (Danish) net is monumented by reliable marks in bedrock. We believe that a motion of 4 inch in any direction would be detectable with reasonable certainty. A 1 inch motion would probably show up, but we would certainly not be prepared to rule out measurement errors of this magnitude. The two nets com- bined when remeasured might give some indication of twist if the motions detected were different.

B. STRAIT OF GEORGIA NET

This net was measured in 1966, mainly to strengthen the old geodetic net in the area. (Figure 2). It was made more rigid than normal in hopes that we might detect any differential movement between the two sides of the strait. Various authors had in- dicated a row of earthquake epicentres, on each side of Vancouver Island, which we thought could indicate a possible fault along which motion might occur.

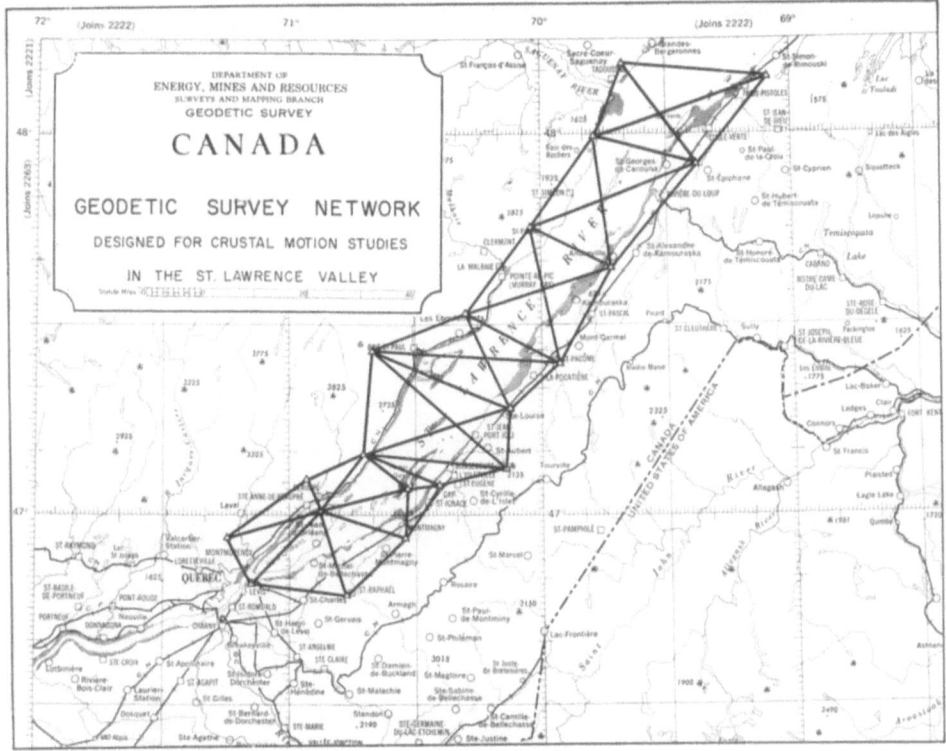

Fig. 2.

The accuracy of this net is not as good as the Robeson Channel Net, but differential movements of one or two inches might be indicated and an 8 inch movement could certainly be detected.

C. ST. LAWRENCE VALLEY NET

This net was measured in 1965, in the process of strengthening old work in the area, but in the designing and measuring of the net, the possibility of detecting crustal movements was considered (Figure 3). We estimate that a 6 inch movement could be detected with reasonable certainty. Althought the Strait of Georgia and the St. Lawrence nets were, in part, remeasurements of older nets, the accuracy of the old work was adequate to detect changes in the distances only if they were several feet, and no clear evidence evolved. We plan to remeasure these three nets in a few years.

In the process of adding to and remeasuring the precise level net of Canada, a few indications of crustal tilting have shown up.

D. QUEBEC – LAKE ST. JOHN NET

In this area, old work had been done as early as 1919 and a fairly strong net established by 1938. It was completely relevelled in the early 1960's. The levelling indicated that the high land between Quebec City and Lake St. John had settled and/or the low land

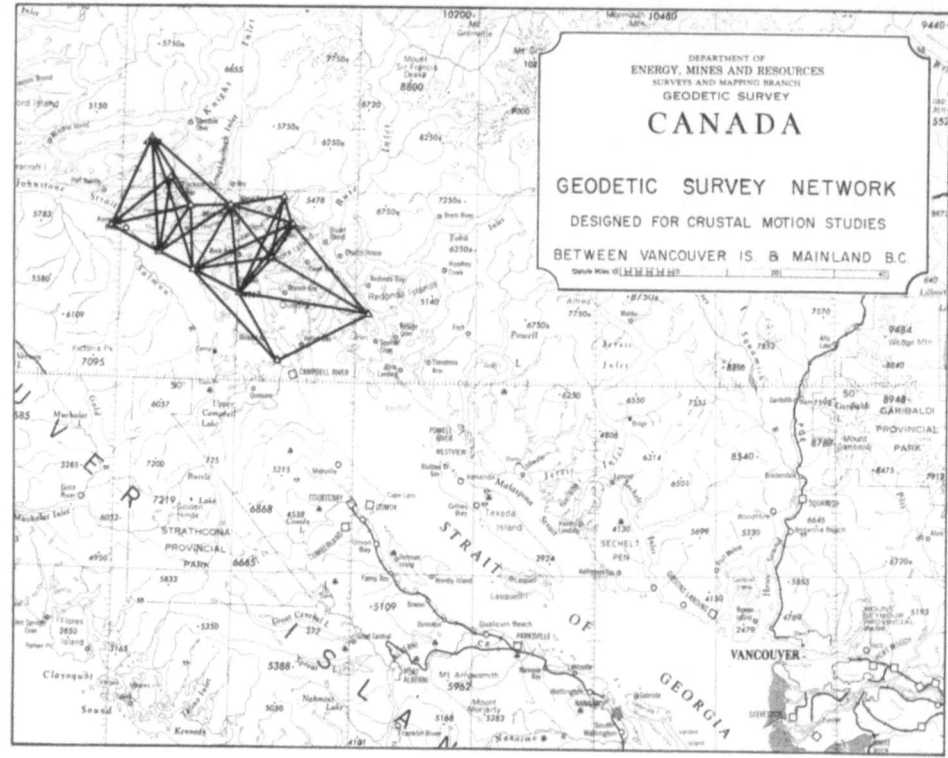

Fig. 3.

surrounding raised, the difference being of the order of 1.3 feet in 40 years (Figure 4). The standard deviation of about 0.2 feet for this difference indicates the sort of accuracy attainable. This was reported by Frost and Lilly in an article in the *Canadian Surveyor* in September 1966, and more recently (with a little more data) by Gale (1970).

E. GREAT LAKES AREA

The Coordinating Committee on Great Lakes Basic Hydraulic and Hydrologic Data for the Reassessment of the International Great Lakes Datum (1955) is active in directing an extra precise level net in the Great Lakes area, which could show crustal movements. A new line of extra precise levels and water transfers, from Father Point on the St. Lawrence to Lake Superior, is to be completed within a few years with many lines between and around the Lakes. This might confirm the hypothesized tilt of the land surface down to the south.

F. OTHER STUDIES

There seems to be some evidence that the land about 30 miles east of Winnipeg has risen a few tenths of an inch with respect to the land in the general area of Winnipeg in the last 20 years.

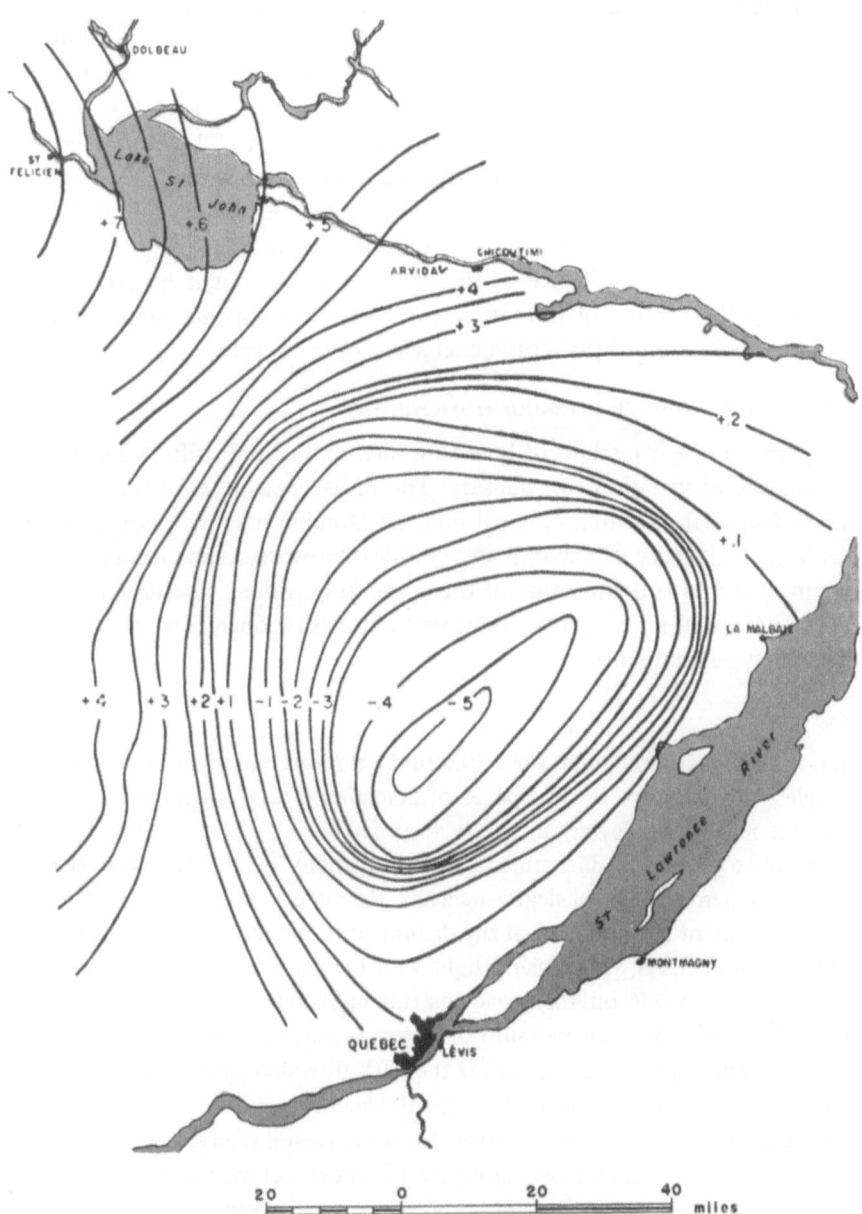

Fig. 4. Crustal Movement in Quebec. – There is a circular area in the centre of Laurentides Park that has sunk to the maximum of 0.56 foot and which seems to be centred in the area of highest elevations. The country rises from Quebec City at approximately sea level to an elevation of 2800 feet. Once outside this area, the elevation of the country gradually slopes down to an average of about 300 feet above sea level. At these lower elevations in the Lake St. John area, the crust has risen to a maximum of 0.76 foot. This means that the elevation of the highest ground, relative to the lower land has changed by 1.32 feet in a period of 40 years.

Two studies, to determine local crustal movements in the vicinity of large water storage reservoirs behind dams, are in progress. These reservoirs are hundreds of feet deep and cover hundreds of square miles. At the Bennet Dam on the Peace River, the Geodetic Survey, in cooperation with the University of British Columbia, has measured some levelling lines before the reservoir started to fill, with the intention of relevelling when the reservoir is full in a few years. The same sort of thing will be done at the Mica Creek Dam on the Columbia River. Studies by consultants, who have used assumptions of the elastic coefficient that they admit are very questionable, have indicated that crustal subsidence of the order of a foot might be expected for the whole area in the vicinity of the reservoirs. Our measurements should give a more accurate determination of the elastic coefficient of the crust.

G. RADIO INTERFEROMETER INTERCONTINENTAL TIES

There appears to be a method of detecting intercontinental drift by means of radio interferometric observations on quasars. The measurements are being made by the Canadian National Research Council and the Dominion Observatory, in cooperation with the University of Toronto and Queens University. The Geodetic Survey has been consulted and is doing some of the computing. Initial measurements have not been very successful but they have indicated that with refinements, accuracies of one foot or so may be attainable.

H. ACCURACY

In making their measurements, for horizontal positions, surveyors normally measure both angles and distances. For purposes of detecting crustal movement, it seems that the distance measuring devices limit the accuracy.

The distance measuring instruments, used normally by the Geodetic Survey, are ranging instruments which basically measure the time it takes for a signal to travel from the instrument to a reflector at the distant station and back. They can be divided into two majors classes, those using light as a carrier and those using microwaves. Their accuracy depends on the basic instrumental accuracy and, in addition, the accuracy with which one can measure the average index of refraction (ratio of actual velocity over velocity in a vacuum) over the path travelled by the signal. For the best microwave instruments, the basic instrumental accuracy is such that 1 inch errors will occur occasionally (despite better claims by some manufacturers) and even with extreme care one cannot confidently compute the mean velocity to better than 3 ppm. This is mainly because of the difficulty in measuring the average water vapour content of the air. Also, microwaves are liable to unknown reflections and for this, or some other, reason large errors occasionally occur which are much greater than could be predicted from fitting a Gaussian curve to the error frequency distribution. For this reason, when microwave instruments are used for high precision work, a network with many redundant measurements is absolutely essential. The maximum range of these instruments is typically 30 miles (radio line of sight). For these longer lines there are appreciable errors introduced by the uncertainty of the radius of curvature of the path of the signal.

For the instruments, using light as a carrier, the basic instrumental error is such that $\frac{1}{2}$ inch errors are rare. With great care one can compute the mean velocity to an accuracy of about 1 ppm. However, up until a couple of years ago the range of these instruments was at best about 10–15 miles. There are now several new instruments on the market, using a laser for light source, with 30 mile range (if the air is clear of haze and fog).

It is expected that in the near future it will be possible to increase, perhaps by a factor of 10, the accuracy of the measurements using light as a carrier by using the dispersion effect to measure the mean velocity at the same time that the distance is being measured. This involves measuring the distance using lights of two (or three) different, accurately known, colours. Since the speed depends on the colour, in a known relationship, it will be possible to deduce the speed from the difference in the two travelling times.

The accuracy of precise levelling has been hard to assess. We believe that with normal first-order techniques, and a network of level lines, a motion of a few tenths of a foot, at a distance of 100 miles from a fixed point could be detected and with extra precautions the accuracy might be improved.

It seems that in the future we may well have to measure and take into consideration the rate of tilt of the Earth's crust for levelling computations. It seems reasonable that these measurements should be of interest to geophysicists.

As far as we know, we have nothing in Canada comparable to the San Andreas Fault with its appreciable motion; but we may know better in a few years when we reobserve the three nets. This meeting has served to direct our attention to some new areas of study where our measurements would be interesting and valuable in the broader fields of geophysics as well as to survey control.

References

Frost, N. H. and Lilly, J. E.: 1966, 'Crustal Movement in the Lake St. John Area, Quebec', *Canad. Surveyor* **20**, 292–299.

Gale, L. A.: 1970, 'Geodetic Observations for the Detection of Crustal Movement', *Can. J. Earth Sci.* **7**, 602–606.

Discussion

Nason: Are there any levelling or relevelling nets crossing the Rocky Mountains in Western Canada or anywhere in British Columbia which might show uplift or compression of the land there.

Jones: We will have some in a year or so. We have put on a crash programme to relevel right across Canada within about three years. There are bits and pieces of relevelling in the mountains but not enough to make a definite statement as to what is happening.

PRECISE MEASUREMENT OF THE EARTH'S ROTATION AND POLAR MOTION BY NEW METHODS

THE ROLE OF LONG BASE-LINE INTERFEROMETRY IN THE MEASUREMENTS OF EARTH'S ROTATION

N. W. BROTEN

Radio and Electrical Engineering Division, National Research Council, Ottawa, Canada

Abstract. The principles of radio interferometry utilizing an independent time and frequency standard at each station is treated. The application of this technique to the measurement of the earth's rotation rate and the advantages and limitations is discussed.

Previous publications (Gold, 1967; MacDonald, 1967) have extolled the advantages of long base-line interferometry for the determination of the rate of rotation of the Earth. It seemed that it might be useful therefore, if some explanation of this technique were given in order that geophysicists might better understand any potential experiments.

Let two antennas be situated a distance D apart and be connected by a suitable transmission line. If a plane wavefront falls upon the antennas the signal recorded will be the sum of what is received by each antenna.

At the time when the path difference between the wavefront and each of the antennas is equal and opposite the output voltage will be

$$E_0 e^{jwt} = k(E_1 e^{j\psi/2} + E_2 e^{-j\psi/2}) e^{jwt}.$$

e^{jwt} is the time variation at signal frequency; ψ is the phase difference of the wavefront at each antenna and if the distance apart is equal to D then $\psi = 2\pi(D/\lambda)\cos\theta$; θ is the angle between the base-line joining the antennas and the normal to the wavefront.

For simplicity let $E_1 = E_2$, then

$$E_0 = k_1 E \frac{(e^{j\psi/2} + e^{-j\psi/2})}{2}$$
$$= k_1 E \cos(\psi/2).$$

We measure output power and

$$P_0 \sim E^2 \cos^2 \psi/2$$
$$\sim E^2 (1 + \cos\psi).$$

The output power fluctuates between zero and the sum of the power received by both antennas as a function of the angle of incidence, θ of the incoming wave. More generally, if the antennas have power gain given by $A_{(\theta)}$ and $B_{(\theta)}$ the output power will be

$$P_0 \sim \sqrt{A_{(\theta)} B_{(\theta)}} (1 + \cos\psi).$$

In long base-line interferometry the change in θ is due to the rotation of the Earth. The variation in output is frequently termed 'fringe rate', and is, of course, due to doppler shift of the incoming radiation.

If the radiating source has finite size it will not radiate a plane wavefront. Wave-

L. Mansinha et al. (eds.), Earthquake Displacement Fields and the Rotation of the Earth, 279–283.

fronts emanating from different parts of the source will arrive at the two antennas at different times, that is with different phase angles ψ. As the dispersion of ψ increases, and ψ is proportional to the distance separating the two antennas, the depth of the fringe diminishes and reaches zero when the average phase dispersion of ψ is $\pi/2$. The source is then said to be completely resolved.

The signal which is being received at high radio frequencies must be translated to low frequencies. A receiver in its simplest form will consist of a radio frequency amplifier, an oscillator to translate the incoming signal to a lower frequency, further amplification, detection, and a suitable recorder.

Separation of the two antennas up to about 100 km can be achieved by techniques such as radio links carrying the common local oscillator signal and the signal information. At distances greater than this, however, radio propagation anomalies limit useful extension of the base-line distance between antennas.

To extend the base-line further two requirements have to be met; a suitable means must be found of obtaining coherent oscillator signals at each site and, a suitable method devised for storing signal information for reduction at a later date.

The first of these can be met with atomic frequency and time standards whose stability is such that independent standards can be used at each site. Small frequency differences between the standards can be taken into consideration during reduction of the data. The second requirement can be met by using magnetic tape recorders. Precise time from the atomic standard can be used for signal storage and retrieval. Two such systems have been in use in North America for the last couple of years (Broten *et al.*, 1967; Bare *et al.*, 1967).

The time difference between the signal arrival at the two antennas, the delay τ, is equal to the path difference $D \cos\theta$ divided by the velocity of propagation:

$$\tau = (D/C)\cos\theta.$$

The doppler frequency is the time rate of change of delay multiplied by the frequency of observation

$$F = (\mathrm{d}\tau/\mathrm{d}t)\cdot f.$$

The general relationship for a two-station interferometer can be shown if D is produced to intercept the celestial sphere at hour angle h and declination d. If the source has hour angle H and declination δ then

$$\cos\theta = \cos d \cos\delta \cos(H - h) + \sin d \sin\delta,$$

so that

$$\tau = (D/C)\{\cos d \cos(H - h)\cos\delta + \sin d \sin\delta\}$$

and

$$F = -(D/\lambda)\cos d \cos\delta \sin(H - h)(\mathrm{d}H/\mathrm{d}t).$$

Allowing for the figure of the Earth, latitude, elevation of each site, longitude, the general equation becomes

$$\tau = (R_0/C)\{(\varrho_1 \cos\theta_1 \cos h_1 - \varrho_2 \cos\theta_2 \cos h_2)\cos\delta$$
$$+ (\varrho_1 \sin\theta_1 - \varrho_2 \sin\theta_2)\sin\delta\}$$

where θ_i is the latitude of station i; R_0 is the mean radius of the Earth; ϱ_i is the radius vector of the Earth at station i; h_i includes the hour angle of observation and the longitude of each station.

The fringe frequency is given by

$$F = -(R_0/\lambda)(\varrho_1 \cos\theta_1 \sin h_1 - \varrho_2 \cos\theta_2 \sin h_2) \cos\delta(dH/dt).$$

Both frequency and delay are continually changing and any successful observation or data reduction must compensate for these changes.

The magnitude of these changes can be shown by

$$\tau = (D/C)\cos\theta.$$

The maximum delay for $D = 3 \times 10^6$ m is

$$\tau_{max} = (3 \times 10^6)/(3 \times 10^8) = 10 \text{ msec}.$$

The rate of change of delay

$$\frac{d\tau}{dt_{max}} = \frac{D}{C}\frac{dH}{dt} = 10^{-2} \times 7.3 \times 10^{-5} \cong \tfrac{3}{4} \, \mu\text{sec/sec}.$$

For a frequency of observation of 4 GHz, the maximum doppler frequency

$$F_{max} = f(d\tau/dt) = \tfrac{3}{4} \times 10^{-6} \times 4 \times 10^9 \simeq 3000 \text{ Hz}.$$

Thus, for a distance of 3000 km – the distance between the telescopes in Algonquin Park, Ontario and Penticton, B. C. – the delay can be any value between ± 10 msec depending on the hour angle of observation of a source. The delay can be changing at a rate up to $\tfrac{3}{4}$ of a microsec per sec. The recording technique must be capable of being adjusted to accommodate the delay, and the rate of change of delay, either during recording or playback. Further, for a frequency of 4 GHz, the maximum fringe frequency would be 3000 Hz. Since sensitivity requirements dictate as long an integration period in the output as is possible, almost all this fringe frequency is accounted for on recording by off-setting one oscillator from the other by almost this amount.

Information about the signal is obtained only after the tapes have been brought together and the data on the two tapes have been cross-correlated. In the analogue method used by the Canadian L.B.I. group, signal from one recorder is passed through a delay line and is brought out to multiple correlators. In this fashion a band of delays can be examined at one time, reducing search time and assisting in the precise determination of the delay τ. If, during recording, the local oscillators were exactly compensated for the doppler frequency F the output would be d.c. in the appropriate delay channel. To assist in the detection of small outputs one local oscillator is offset from the required value by 0.10 Hz. The presence of correlation is indicated by a sinusoidal output that has a period of 10 sec.

This fringe frequency can be determined to about $1/T$, the total period of the observation. Our technique allows us to observe for up to 90 min, thus the fringe frequency can be determined with an accuracy of about ± 0.001 Hz. If the phase of the output is well defined it may be possible to increase the accuracy by an order of magnitude. For a base-line of 3×10^6 m and a frequency of observation of 4×10^9 the fringe frequency f is 3000 Hz. Thus

$$\Delta F/F = (1 \times 10^{-4})/(3 \times 10^3) \simeq 3 \times 10^{-8}.$$

The accuracy of determining fringe rate does not appear to involve the accuracy of the atomic clocks. However, this is illusionary. The fringe rate is obtained by offsetting one oscillator from the other by a slight amount. Hence the stability of the clocks must be

$$(1 \times 10^{-4})/(4 \times 10^9) = 2.5 \times 10^{-14},$$

about 2 orders of magnitude better than Rb or Cs clock stability and taxing the stability of hydrogen masers. 3×10^{-8} is approximately 3 msec per day. Thus, if no other errors were involved, a measurement of fringe frequency could only give answers to the rate of rotation of the Earth to about 3 msec per day.

What about measurement of delay? Assuming that we can put observations on our tapes and take them off with little error, then we have the accuracy to which the alignment of tapes can be made. That is, we can only determine alignment to some fraction of the effective correlation width.

The change in delay for a 3000 km base-line is about a microsec per sec of time. What this means is that, if the Earth had changed its rate by 1 sec in one day, then the delay that is measured would only have changed by 1 microsec. Now if we want to know the rate of rotation to a millisec per day, then we would have to be able to distinguish between correlated events differing by about a nanosec. The correlation half width is equal to $1/2\Delta f$ hence the effective bandwith must be $1/(2 \times 10^{-9}) = 500$ MHz. This effective bandwith is not impossible even though the bandwith of the recorder is very much less (Hinteregger, 1968).

What does a nanosec a day mean with regard to the atomic clock time? It means that the difference in rates of the clocks must be stable and be known to the order of $10^{-9} (8 \times 10^4) \simeq 10^{-14}$. This certainly is 2 orders better stability than can be obtained with Cs or Rb clocks and may be better than can be obtained with hydrogen masers.

The accuracy in determining fringe frequency and delay time is proportional to the distance separating the telescopes. Hence it would appear that the telescopes could be separated by 10 km and achieve 3 times the accuracy. At such a spacing however the telescopes would have zenith angles of about 65° and atmospheric irregularities may limit the accuracy.

It has been shown that, to obtain measurements of Earth's rotation rate to an accuracy of about 1 msec per day, time and frequency stabilities of the order of 10^{-14} are required. Hydrogen masers may have sufficient stability to allow this order of accuracy but it is believed that more must be known about hydrogen maser rates over long periods of time.

References

Bare, C., Clark, B. G., Kellerman, K. I., Cohen, H., and Jauncy, D. L.: 1967, 'An Interferometer Experiment Using Independent Local Oscillators', *Sci.* **157**, 189.

Broten, N. W., Legg, T. H., Locke, J. L., McLeish, C. W., Richards, R. S., Chisholm, R. M., Gush, H. P., Yen, J. L., and Galt, J. A.: 1967, 'Long Base-line Interferometry, a New Technique', *Sci.* **156**, 1592.

Gold, T., 1967, 'Radio Method for the Precise Measurement of the Rotation Period of the Earth', *Sci.* **157**, 302.

Hinteregger, H. F.: 1968, 'A Long Base-line Interferometer System with Extended Bandwidth', NEREM Record 1968, p. 66.

MacDonald, G. J. F.: 1967, 'Implications for Geophysics of the Precise Measurement of the Earth's Rotation', *Sci.* **157**, 304.

GEOPHYSICAL APPLICATIONS OF
LONG-BASELINE RADIO INTERFEROMETRY

IRWIN I. SHAPIRO* and CURTIS A. KNIGHT

*Dept. of Earth and Planetary Sciences, Massachusetts Institute of Technology,
Cambridge, Mass., U.S.A.*

Abstract. Effective wide-bandwidth techniques for making precision phase-delay measurements (errors ≲ 0.1 nsec) over intercontinental baselines are under development by an MIT and Lincoln Laboratory group. The point sources of radio radiation for such interferometric measurements can be either natural (e.g., quasars) or artificial (e.g., beacons placed in synchronous orbit or on the Moon). Possible applications of this technique include: precision determination of global geodetic ties; measurements of tidal oscillations, crustal-block motions (including continental drift), and Earth polar motion and rotation; refinement of values for the precession and nutation constants and for the rate of change of the obliquity of the ecliptic (the latter when combined with 'times-of-arrival' observations of pulsars); measurements of the shape of the sea surface; determination of the geopotential and its time dependence; and global time synchronization at the subnanosecond error level. In this paper we describe the basic technique, the limitations on accuracy, useful antenna systems and sources of radiation, geophysical applications, and briefly, the recent experiments performed by the MIT and Lincoln Laboratory group.

1. Introduction

A marriage of convenience between the disparate fields of geophysics and radio astronomy is now being consummated. The new technique of long-baseline radio interferometry promises to have a profound effect on studies of the Earth. As examples, we could cite direct measurement of intercontinental drift and prediction of earthquakes through ultraprecise measurements of polar motion. Whether such promises will be fulfilled remains to be seen.

In this paper, we will outline the basic principles involved in long-baseline interferometry, discuss the important factors limiting the accuracy of the technique, describe some of the possible applications in more detail and, finally, mention the experiments currently underway. Minor deviations from the truth will occasionally be allowed to accompany explanations so as to emphasize the main points without adding the confusion that usually accompanies too many qualifications.

2. Basic Technique

Interferometry is certainly not new nor is radio interferometry. What is new? The technique of *long-baseline* interferometry, i.e., the use of widely separated radio antennas in an interferometric mode. The previous limit on separation, up until 2 years ago, was set by the imagined need to connect the two antennas electrically in order to properly compare the received signals: it was felt that the same local oscillator, in the language of electrical engineering, had to be used to convert incoming radio signals down to video. However, a direct electrical connection between antennas is

---

* Also Department of Physics and Lincoln Laboratory.

L. Mansinha et al. (eds.), Earthquake Displacement Fields and the Rotation of the Earth, 284–301.

not really necessary. Atomic-frequency standards are sufficiently stable so that a separate one can be used at each site to control both the video conversion and a suitable recording of the signals (e.g., onto magnetic tape). The magnetic tapes from the separate sites can then be brought together and processed to obtain the desired results. What actually do we seek from the data? Primarily, we wish to determine the difference Δt in the times of arrival at each antenna of the same wave front from a celestial (point) source of radio radiation. The source is usually sufficiently far away so that the wave fronts can be considered to be plane at the Earth. (For certain applications where, e.g., the source is an earth satellite, spherical wave fronts must be considered.) The error in determining Δt is essentially independent of the length, d, of the baseline separating the two antennas. This simple fact is the key to the potential usefulness of this technique for geophysical applications: a determination of the distance between two antennas 1 km apart with an error of the order of a few centimeters would not be so terribly impressive. But the fact that this error can be kept so low even though the antennas be separated by intercontinental distances is the impressive feature of the technique.

How is Δt deduced from the recorded signals from a point source of radiation? Bypassing the quantum-mechanical aspects of radiation, we simply remark that the source is composed of atoms each of which, when undergoing a transition to a lower energy state, sheds electromagnetic radiation of a characteristic frequency. (In fact, the radiation will not be monochromatic but will have a spectrum distributed about the characteristic frequency with the spread determined by the duration of the emission.) At a given distance from the atom at a suitable time after emission, i.e. along a particular wave front, the phase of the electric field vector will have a particular value. This particular phase, or wave front, propagates (in vacuum) at the speed of light, c. Hence if we record the signal at two different – but constant – distances from the source, as a function of time, and cross-correlate these recordings we would get a maximum correlation if the recordings were offset before calculation (i.e., shifted with respect to one another) by an amount equal to the time required for the wave to propagate the difference in distance between the source and the two sites. As the offset is changed from this amount, the two recordings would become more and more out of phase (up to a point) and the cross-correlation would decrease. The actual radiation comes not from the transition of a single atom but from many atoms whose transitions are in general quite independent of one another. But, by applying the principle of superposition, we see that the exact same argument as given above applies to the total radiation field and that in this situation, too, the cross-correlation will be maximized when the offset is Δt.

How can we best minimize the error in determining Δt? Clearly, if we have, say, one essentially instantaneous recording at a single frequency then, if the signal-to-noise ratio is high enough, we can obtain a good estimate of Δt by determining the offset time which yields a maximum for the cross-correlation – but only modulo λ/c (where λ is the wavelength of the radiation). That is, the cross-correlation will be a maximum for many values of Δt (depending on the length of the recording) each of which differs from the other by the equivalent of 2π in the phase of the signal. Thus

$$c \, \Delta t = n\lambda + \delta\lambda; \quad 0 \leqslant \delta\lambda < \lambda$$
$$2\pi f \, \Delta t = 2\pi n + \phi; \quad f = c/\lambda; \quad 0 \leqslant \phi < 2\pi,$$

where n (a positive or negative integer) would be essentially indeterminate from this instantaneous single-frequency measurement. Of course, in practice, one does not record only a single frequency. But, if the bandwidth Δf of the recording is only narrow instead of infinitesimal, we are not helped much. We will still be plagued by ambiguities which, however, will not be essentially infinite in number. Rather, the ambiguities will spread only over a time offset interval roughly equal to $(\Delta f)^{-1}$. The cross-correlation in the noise-free case as a function of the difference, $\delta \, \Delta t$, between the actual and true offset is proportional to

$$\frac{\sin(2\pi \, \Delta f \, \delta \, \Delta t)}{\delta \, \Delta t}(\cos 2\pi f \, \delta \, \Delta t),$$

where the first term provides the 'sin x/x' envelope that restricts the ambiguities (maxima of $\cos 2\pi f \, \delta \, \Delta t$) to lie in a region of approximate extent $(\Delta f)^{-1}$.

Suppose, however, that the only restriction on the experiment is that the incoming signal can be sampled at a rate no higher than a prescribed maximum. What then would be the optimum strategy to minimize the error in the determination of Δt? That is, it would not be necessary to sample the signal only over a single band of extent Δf; one could sample, e.g., several distinct parts of the spectrum with the bandwidth recorded at each part being, of course, less than Δf. We have not examined this general mathematical problem in detail. However, a workable scheme, which is probably close to optimal, was suggested by A. E. E. Rogers of the MIT Lincoln Laboratory. In this scheme, samples are recorded from narrow bands distributed throughout an extremely wide band. From the relation

$$2\pi f \, \Delta t = 2\pi n + \phi$$

it follows that

$$\frac{\mathrm{d}\phi}{\mathrm{d}f} = 2\pi \, \Delta t,$$

i.e., the delay sought is simply the slope of the curve of phase vs. frequency: at a given instant, the cross-correlation will have a maximum at a different phase (but the same time offset!) for each different frequency. Therefore, if we sample the ϕ vs. f curve over a very wide band we can determine Δt very accurately. Thus the error in the estimation of Δt is

$$2\pi \, \delta \, \Delta t \approx \frac{\delta\phi}{f_{max} - f_{min}},$$

where $f_{max} - f_{min}$ is the effective wide band and $\delta\phi$ is related to the error in the estimation of ϕ from each of the separate narrow bands. These latter must be distributed so as to insure no ambiguity. (The satisfaction of this criterion will depend on the achievable signal-to-noise ratio.) One useful distribution scheme, easy to describe, is as follows: two narrow bands, centered at f_1 and f_2, are chosen close enough together so that the *a priori* difference in the phases which maximize the cross correlation for each is much less than 2π. The spacing between f_3 and f_2 can be increased over the $f_2 - f_1$ difference since the error in extrapolating the phase to $\phi(f_3)$ can be kept tolerable with f_3 satisfying

$$f_3 - f_2 \leqslant \frac{f_2 - f_1}{\delta\phi}.$$

The centers for succeeding bands are chosen in an analogous manner such that

$$(f_m - f_{m-1}) \approx \frac{S}{N}(f_{m-1} - f_{m-2}) \approx \cdots \approx \left(\frac{S}{N}\right)^{m-1}(f_2 - f_1),$$

since $\delta\phi \sim (S/N)^{-1}$, where S/N represents the relevant signal-to-noise ratio for the phase determination in a single band. For this scheme, then, the narrow bands, or 'windows', are spaced in a geometric series with the ratio of successive window separations being given approximately by the signal-to-noise ratio obtainable in the determination of ϕ from a single window. For typical values of S/N, effective bandwidths of 100 MHz could be utilized with six windows. The m separate windows can be sampled sequentially, or simultaneously; in the latter case the instantaneous recorded bandwidth would be divided equally among the m windows. For this type of extended-bandwidth to work it is of course essential that strict account be kept of the instrumental phase relations for the different frequencies. Several suitable phase-calibration schemes have already been tested successfully by the MIT-Lincoln Laboratory group. If the overall system is carefully calibrated, it would be possible to eliminate entirely the 'fringe ambiguity' so that the error in Δt would be reduced to $\delta\phi/f$, i.e. to the time-equivalent of a fraction of a wavelength. No technological breakthrough is required to reduce $\delta\,\Delta t$ below 0.1 nsec (≈ 3 cm), where the most interesting geophysical applications are anticipated. The useful interpretation of this Δt observable is however limited by the Earth's atmosphere, as discussed in the next section.

What in fact can we deduce from measurements of Δt? A single (instantaneous) measurement, involving two antennas, suffices in essence to locate the radiation source on the surface of a cone whose axis is the intersite vector, **d**, and whose half-angle, θ, is determined by $c\,\Delta t$ and the intersite distance:

$$\cos\theta = \frac{c\,\Delta t}{d}.$$

This result is interesting but does not help much by itself. However, with a series of measurements, we can infer not only the vector **d** but also the direction to the source, any proper motion of the source, variations in the motion of the Earth's crust with respect to its axis of rotation, the motion of this axis with respect to a stellar reference frame, etc. The procedure is based largely on the 'bootstrap principle' and its success is based on the following fact: the theoretical expression for Δt is a function of the source direction, the antenna locations, the time of day, the model atmosphere employed, etc. By parameterizing this theoretical expression in terms of the (*a priori* partially unknown) source direction and antenna locations and by making a sufficiently large number of observations of each of a number of celestial radio sources, we can estimate 'best-fit' values for these parameters from, say, a weighted-least-squares analysis. The effects of tidal motions, continental drift, motion of the Earth's rotation axis, etc., are incorporated since these all affect the time-dependence of the locations of the antenna sites in inertial space. The extraction from phase-delay observables of estimates of the relevant geophysical quantities is thus a complicated, but perfectly feasible, exercise in parameter estimation. Each geophysical effect under consideration introduces a characteristic time variation in the Δt observable which allows the corresponding parameters to be estimated unambiguously. Of course, nonzero correlations between the parameters causes the standard deviations of the parameter estimates to be increased. However, our experience with multi-parameter estimates of solar-system constants (up to 100) from radar and optical observations of the Moon and planets has convinced us that no insurmountable problems will arise in the actual reduction of interferometer data.

As an illustration of the extraction method, let us consider a very simple example in which the Earth is assumed rigid with a known and constant rotation vector. Then the unknowns comprise three for the intersite vector $\mathbf{d}$, two for the direction $\mathbf{e}_0$ in space of each observed radio source, and one for the error in initial clock synchronization. (Because of the stability limit of the hydrogen-maser frequency standard, discussed below, only data taken within about 1 day can be processed under the assumption that a single parameter suffices for the description of the relative error in clock settings with no degradation of results.) By assuming that the axis direction and rate of the Earth's rotation are known, the coordinate system in terms of which $\mathbf{d}$ and $\mathbf{e}_0$ are to be expressed is only partially defined. The origin of longitude in the Earth's equatorial plane remains arbitrary but can be specified, e.g., by choosing the longitude of one of the sources to be zero. Hence, for observations of n sources, there will be a total of $2n+3$ unknowns to be determined. For each source, the time dependence of Δt will then be (if we ignore aberration and other corrections):

$$\Delta t(t) = \delta t_c + \frac{1}{c}\mathbf{d}\cdot\mathbf{e}_0$$

$$= \alpha + \beta \sin \Omega t + \gamma \cos \Omega t,$$

where δt_c is the clock synchronization error; α, β, and γ are constants; and Ω is the Earth's spin velocity. The dot product $\mathbf{d}\cdot\mathbf{e}_0$ represents, of course, the component of the intersite vector along the direction to the source. The constant α contains δt_c and the (constant) component of $\mathbf{d}\cdot\mathbf{e}_0$ along the Earth's axis of rotation. The constants β and γ are the coefficients of the (sinusoidally changing) components of $\mathbf{d}\cdot\mathbf{e}_0$ along two orthogonal directions in the Earth's equatorial plane. Hence arbitrarily large numbers of measurements of Δt for a single source can determine no more than three constants (α, β, and γ); the remainder of the measurements will be redundant. To determine **d** at least three observations must be made of at least n^* sources where n^* is the smallest integer that satisfies the equation

$$3n^* \geqslant 2n^* + 3,$$

the left side giving the number of 'knowns' and the right the number of 'unknowns'. Clearly, $n^* = 3$. Of course, once the source directions e_0 have been determined to the required accuracy or if three or more antennas are involved, a few observations of each of two sources will provide the necessary data. (The 'proper motions' of the sources, if discernible, could easily be taken into account.)

In the above discussion we assumed, for simplicity, that the difference in the distance of a given source from each antenna site remained constant during a single observation. In fact, the sites move differentially with respect to the source causing Δt to vary and thereby giving rise to the 'fringe rate'. The *a priori* fringe rate is, of course, taken into account in the data processing; the residual rate is estimated along with Δt and is useful in the deduction of geophysical and astronomical information. Accurate estimates of fringe rate (error $\leqslant 0.001$ Hz) may even be useful in relating the fringe phases from separate tape recordings, taken sequentially, of signals from a given source. An extensive mathematical discussion of the detailed processing procedures required for the analysis of the interferometry data is being prepared for separate publication.

3. Limitations

There are many influences on the Δt measurements that affect the interpretation of these data. Such influences can be divided into two classes: those that are of geophysical interest and those that can be viewed as unwanted 'noise'. Here we consider only the latter which in effect govern the measurement 'errors' and establish the limitations of the technique. We discuss, in turn, the effects of the neutral atmosphere, charged particles, frequency standards, and antenna flexure.

The total delay introduced by the neutral atmosphere in the zenith direction is equivalent to the time taken by a radio wave to traverse about 2.5 m in vacuum. The main difficulty in accounting for the atmospheric delay – the most important error source – is caused by the variability of the water-vapor content. With the use of model atmospheres and surface measurements of temperature, pressure, and humidity at each site, the uncertainty in the electrical path length through the atmosphere can be reduced below 1 m for the zenith direction. A reduction of the uncertainty to the order of 10–20 cm can be accomplished for many localities with the use of standard weather bureau data of temperature, pressure, and humidity as a function of altitude. (These soundings are performed daily at a large number of sites.) The use of special techniques will most likely allow the uncertainty to be reduced much further. A particularly promising method employs radiometers to monitor the sky brightness temperature along the line-of-sight to the object being observed. A linear relation between sky brightness temperatures near the 22.235 GHz resonance line of water vapor and the electrical path length through the atmosphere has been determined semi-empirically by Staelin and Waters at MIT (see, e.g., Staelin, 1969). Their studies indicate that the zenith-direction uncertainty in the atmospheric delay can be reduced to 1–2 cm by use of the radiometric and surface meteorological data.

Simultaneous observations of the thermal emission near the 60 GHz resonance band of oxygen are also helpful for calibration of the electrical path length. Feeds for such observations can be 'nested' with the main feeds because of the large separation in frequency. (For water-vapor interferometry, discussed in Section 4, there is also no problem with feeds.) Redundancy of measurements, permitting averaging, may enable the error caused by atmospheric variations to be reduced further. Still greater reductions are achievable, at least in principle, with more sophisticated – and expensive – systems such as measurements at three-frequencies (two optical, one radio) of the round-trip delays of signals sent from the site to retro-reflectors or transponders on a powered balloon moving near the line of sight.

The ionosphere, in particular, and the plasma between source and observing site, in general, cause an important change in phase delay. Fortunately, the index of refraction n of a plasma depends strongly on frequency $(n^2 - 1 \approx \text{const}/f^2)$ so that by making simultaneous measurements at two widely separated frequencies, the plasma effects can be deduced and subtracted. The residual error, due in part to the slightly different paths followed by waves of different frequency, can be reduced to well under 1 cm.

The frequency standards serve two functions: (i) to provide sufficient phase stability to insure that the signals recorded separately at each site can be cross-correlated without significant loss of coherence. For the present limit of 3 min on each digital tape recording, the newest rubidium standards are adequate since their long-term stability is about 5×10^{-13}. A hydrogen maser, operating within presently advertised specifications (long-term stability of 7×10^{-15}), will serve this function admirably. (ii) to keep time with sufficient accuracy to ensure that the 'clock offset error' remains constant – within the accuracy otherwise achievable – over the interval required for the determination of the 'instantaneous' distance vector between interferometer sites. This interval, to start, would be about 8 h leading to a corresponding change in clock offset error of ± 0.1 nanosec, or equivalently, to about a ± 3 cm error in distance. As the estimates of source directions are refined, the time required for the determination of the intersite vector will be decreased substantially. The effect of this change in clock offset error could be reduced further by taking enough measurements, by no means an inordinately large number, to solve for the clock error more frequently. Thus the newest model hydrogen masers appear capable of satisfying this second requirement for frequency standards. It may, however, turn out that the drift of rubidium standards is sufficiently smooth that a suitable parameterization can be employed. Several more measurements (i.e., sets of tape recordings) could then be made on each day of observations to allow these parameters to be estimated along with the others of geophysical interest. Measurements are needed on the characteristics of the standards to assess the feasibility of this approach. It is clear, of course, that improvements in the basic standards will be of great help for geophysical applications.

Changes with antenna orientation and temperature in the effective distance to the feed could introduce systematic errors in the geophysical interpretations of the Δt data. Fortunately, these effects are all negligible. Detailed measurements carried out, e.g., on the 140-foot diameter antenna of the National Radio Astronomy Observatory, show that these effects are each very small and, furthermore, can be calibrated with a residual error less than 1 mm. For a smaller, portable antenna, the effects will be even less, provided that the base is secured properly to the Earth's crust.

Thus, if hydrogen-maser frequency standards are used and measurements made simultaneously in two widely separated frequency bands, the main source of error will be caused by atmospheric variations which may produce an uncertainty of a few centimeters in distance determinations. (Note that any uncertainty in expressing the speed of light, say, in kilometers per second, is irrelevant for the purposes of interferometry since all measurements are of time.)

4. Radiation Sources and Antenna Systems

Before discussing the potential of long-baseline interferometry for geophysical studies, we shall first describe the various radiation sources and antenna systems that could be used for these applications. Imbedded in the description, however, will be some brief comments about applications, not necessarily geophysical, that follow specifically from the system being described.

The radiation sources for the interferometry measurements can be either natural or artificial. The most desirable natural sources are strong, broad-band ('continuum') emitters of negligible angular size ('point' sources) and sufficiently distant to have negligible proper motions. Certain quasars, such as 3C273, 3C279, 3C454.3, and 3C345, seem to satisfy all of these requirements admirably. A list containing some of the presently known, suitable natural sources of continuum radiation and the fluxes of the unresolved ('point' source) part of each at L-band and X-band is given in Table I. Most of these are in the northern celestial sphere since almost all of the pertinent observations have been made in this hemisphere. There is no reason to believe that there are not at least as many suitable sources in the southern hemisphere.

How easy to detect are these quasar noise signals? The background celestial noise level is not a serious problem between 1 and 10 GHz, nor, except at the higher frequencies in inclement weather, is atmospheric noise. The total background plus the receiver noise constitutes, by definition, the system noise temperature. Thus, for a 5 flux unit source, a signal-to-noise ratio of 20 can be achieved for an interferometer with the characteristics shown in Table II. These numbers can be scaled by using the simple formula:

$$\frac{S}{N} \sim P \left[\frac{(A_1\eta_1)(A_2\eta_2)(B\tau)}{T_{s1}T_{s2}} \right]^{1/2},$$

TABLE I

Celestial sources of 'point' radiation

Name	Right ascension			Declination			Unresolved flux density	
	Hr.	Min.	Sec.	Deg.	Min.	Sec.	L-Band	X-Band
CTA 21	3	16	09	16	17	40	5	
3C 84	3	16	30	41	19	52	3	16
3C 120	4	30	32	5	15	0	4	5
4C 39.25	9	23	55	39	15	24	2.5	8
1127-14	11	27	36	−14	−32	−55	6	3
3C 273	12	26	33	2	19	34	14	20
3C 279	12	53	36	− 5	−31	− 6	8	10
3C 345	16	41	18	39	54	11	3	6
CTA 102	22	30	8	11	28	23	6	
3C 454.3	22	51	30	15	52	54	10	25

Note: 1 flux unit $= 10^{-26}$ w/m²-Hz.

TABLE II

Parameters for a typical radio interferometer

Antenna diameter (each)	30 m
Antenna efficiency (each)	50 %
System noise temperature (each)	200 K
Recording bandwidth	0.4 MHz
Integration time	100 sec
Signal-to-noise ratio for 5 flux unit source	20

where P is the source strength (in flux units), A_1 the antenna area, η_1 the efficiency, T_{si} the system noise temperature $(i=1, 2)$, B the instantaneous recorded bandwidth, and τ the integration time. Since two antennas are involved in an interferometric meas-urement, the signal-to-noise ratio depends only on the square root of the relevant an-tenna parameters. The presently used NRAO Mark-I digital recording system is limited to an instantaneous bandwidth of only 0.36 MHz. However, a new analog-digital system currently under development for the National Radio Astronomy Ob-servatory will have about a 2 MHz recordable bandwidth. In the near future further improvements in recording systems such as the EVR technique developed at CBS Laboratories, will probably enable bandwidths of the order of 10 MHz to be recorded for periods of time up to several hours on a single 'tape'. Although at present very expensive ($\approx$ \$ 100000 each), cryogenic maser receivers are available which have reasonable gain with noise temperatures as low as 10 K at X-band and bandwidths of the order of 30–40 MHz. A useful system, then, could be composed of one, or several, fixed antennas in the 30-m diameter size range and a number of transportable antennas only 3–5 m in diameter. Alternatively, the system could consist exclusively of 10–12 m diameter antennas each of which could be made transportable. In both cases we assume the use of low-noise receivers ($< 50 \, \text{K}$) and wideband recorders ($\simeq 2 \, \text{MHz}$) so that a good delay measurement could be obtained after 1 min of integration when the source has a strength of 3 or more flux units.

Besides continuum sources, a number of very strong, point sources of celestial radiation have been discovered at *discrete* frequencies associated with hydroxyl and water molecule spectral lines. Such emissions appear to be the result of cosmic maser action; the frequencies are in the 1.6–1.7 GHz range for OH and near 23 GHz for the very recently discovered H_2O. The latter, for several of the sources so far discovered, have enormous fluxes ($\simeq 10^3$ flux units) and can be detected easily with very small, portable antennas – diameters of, say, 1–2 m. The H_2O rotational 'line' is often a series of separated lines, presumably caused by emissions from neighboring regions with differing velocity components along our line of sight. A series of such lines, which extend over about 10 MHz, could be used under appropriate circumstances for effective wide-bandwidth phase-delay measurements. The OH masers are probably of less interest for geodetic applications for several reasons: the frequencies are low enough for there to be ionospheric distortions, the emissions are generally weaker, and the separation of lines are mostly inappropriate for effective wide-bandwidth measurements. Defects common to both are their tendency to cluster near the galactic plane and possibly to have significant proper motions.

Far more powerful sources can be provided from artificial broad-band noise sources placed either on Earth satellites or on the lunar surface. Noise beacons on the moon can be used, in addition, for the precise study of the lunar orbit and librational motions. Because lunar dynamics are involved, the antenna sites can be located with respect to the Earth's center of mass as well as with respect to each other. Tidal effects which have been estimated to cause an increase of several centimeters per year in the mean Earth–Moon distance should be easily discernible from several years of interferometric beacon observations through the corresponding variations of the orbital longitude with time. After perhaps 5 years, the so-called geodesic precession of the lunar orbit predicted by general relativity might be detected reliably. This precession manifests itself as a 0.02 arcsecond per year rotation of both the perigee position and the ascending node. Although the direction of the beacon, with respect to an inertial frame defined by quasars, could be followed in time from the Earth with a precision of about $0\rlap{.}''001$, the geodesic precession is still hard to discern because of (1) the small eccentricity ($\simeq 0.05$) and low inclination ($\simeq 5°$) to the ecliptic of the lunar orbit; and (2) the correlations between the many parameters which must be estimated to determine the lunar orbit from the beacon position measurements. Because of the strongly interacting three-body system formed by the Earth, Moon, and Sun, such beacon observations might also enable a test to be made of the principle of equivalence by the placement of a limit on the relative contributions of gravitational binding energy to inertial and gravitational mass. A careful analysis must be carried out before a reliable quantitative estimate can be given of the usefulness of this test. The precise measurements of the lunar librations would provide improved estimates of the Moon's principal moments of inertia and might even disclose a free precession, analogous to the Chandler wobble, or place a stringent limit on its possible amplitude. Greatly increased accuracy in measurements of lunar motions and surface distortions would be possible if two or more beacons were placed at different locations on the moon and observed simultaneously with an interferometer (H. Hinteregger, private communication). The lines of sight from the beacons to each antenna would pass through the earth's atmosphere along almost identical paths. The accuracy in the determination of the *relative* locations of the beacons would therefore not be affected appreciably by the atmosphere and ionosphere. Preliminary calculations indicate that the relative changes in beacon location normal to the line of sight could be monitored in this manner with present ground-based equipment with errors $\lesssim 10^{-6}$ arcsec, a distance of approximately 1 mm on the lunar surface. It is to be hoped that such beacons can be placed on the Moon during the later Apollo flights.

Radio beacons placed aboard artificial Earth satellites offer other very attractive features. Because even at synchronous altitude the beacons will be more than 10 times closer to the Earth-based antennas, the flux at reception will be over 100 times as high. Therefore the same signal-to-noise ratios can be achieved with antennas 10 times smaller in diameter. Truly transportable systems then become feasible. In fact, there is at present a synchronous satellite – 'TACSAT' – that can transmit a noise signal at X-band over a 10 MHz band which yields a flux at the Earth of about 5×10^6 flux units! With such sources, even small portable horn antennas with apertures of only a few inches could yield useful signal-to-noise ratios after less than a minute of integration. A source in Earth orbit, however, has certain aspects of the traditional 'double-edged sword'. While providing more powerful signals and the opportunity to determine accurately the geopotential, as well as the station locations with respect to the Earth's center of mass, the satellite will perforce suffer partly random orbital perturbations, due to variations in radiation pressure, gas leakage, and atmospheric drag (for satellites in lower altitude orbits), which make more difficult a proper modeling of its motion. One can nevertheless create in principle at least a system for geophysical research which would combine the best features of the natural and artificial sources: small, portable units can be distributed globally – some aboard ship – to monitor almost continuously a set of suitably distributed satellites. (The thermal, power-source, and lifetime problems are less severe for these than for the lunar beacons.) The satellite positions can then be related to the quasar-defined inertial reference frame through periodic observations by a few relatively large antenna systems that are sufficiently sensitive to obtain accurate quasar position data. The satellites could also act as conduits (on a different frequency band) for relaying in real time the video signals and associated timing information to one or more data-processing centers, thus eliminating the tape-recording and tape-transportation systems at the cost of adding a transmitter to each unit. Much more prohibitive economically, but certainly technically feasible, would be a system for tracking satellites in which in addition to possible satellite-to-satellite measurements, satellite pairs are used to observe interferometrically strong signals from lunar beacons and/or quasar signals if the latter remain unresolved for the larger baselines involved. This system, of course, would be free from the nuisance effects of the Earth's atmosphere. From here, it is but a small step to contemplate interferometers formed by antennas with one on the Moon and one on the Earth. Scintillations caused by the interstellar medium may then set the limit on the attainable resolution at a given radio frequency.

5. Geophysical Applications

Having completed our cursory examination of the possible radiation sources and antenna systems, we turn to a discussion of some of the important geophysical applications which we first list:

(1) Global Geodesy.
(2) Tidal Oscillations.
(3) Crustal-Block Motions (including continental drift).
(4) Polar Motion.
(5) Earth Rotation.
(6) Precession and Nutation (including a test of general relativity).
(7) Obliquity of Ecliptic.
(8) Shape of Sea Surface.
(9) Geopotential.
(10) Global Time and Frequency Synchronization.

Interferometric phase-delay measurements can yield improvements in geodetic 'ties', especially over long baselines, by several orders of magnitude. For example, the relative positions of the stations of the world-wide net of Baker-Nunn satellite-tracking cameras have uncertainties of the order of 10–20 m. Interferometry offers the possibility of reducing such uncertainties to a few centimeters. Moreover, the present camera-location uncertainties were reduced to the 10–20 m level only after extensive analysis based on long series of observations of a large number of satellites. With the interferometry technique, the intersite vector should be determinable more accurately with respect to an inertial frame from a relatively simple analysis of data accumulated in less than 1 day.

Due to the differential attraction of both the Sun and Moon, the Earth undergoes periodic distensions. The maximum amplitude of such tides in the solid parts of the Earth reach about 20 cm. The vertical tidal motions have never been measured directly but are inferred from observations of variations of g. Diurnal and semi-diurnal tides can be determined in this manner with errors of only a few percent. But, because of the lack of stability in the equipment, fortnightly tidal effects in the solid Earth have not been observed. Local tilts in the Earth's crust have been monitored using tilt meters; also lasers have been used to measure local displacements in the horizontal directions. The results appear to depend importantly on location. With interferometry, direct global measurements of the solid Earth's tidal responses – both 'vertical' and 'horizontal' – could be made with errors in the 1-cm range. Variations in response due to changes in snow loading, proximity to oceans, etc., will probably be too small to observe in the near future.

Results from sea-floor spreading analyses have spawned a revolution in geophysical thinking. It is now widely believed that the crust is composed of a number of 'plates' that are moving with respect to one another. (See, e.g., Le Pichon, 1968 and Isacks *et al.*, 1968.) Average rates for these movements over the past tens of millions of years have been estimated from magnetic and other geophysical evidence and seem to vary from 1 to 2 cm/yr up to 15 cm/yr for some island arc regions. Interferometry offers the possibility of actually measuring directly the present rates of such drifts (Shapiro, 1967). A long series of observations will be required (at least several years duration) to establish reliably any secular character of the motions. Nevertheless measurements of variations in intercontinental distances and directions might be discerned in some cases from less than 1 year of observations. Such measurements may also prove of value for earthquake prediction. If this method proves successful, it will probably lead to the establishment of an international system under the aegis of the IUGG for the long-term monitoring of crustal-block movements – at least until a better method is developed.

Polar motion – the movement of the Earth's crust as a whole with respect to the axis of rotation – can be monitored using the interferometry technique with spatial and temporal resolutions both being between one and two orders of magnitude better than is currently achieved by the IPMS. Determination of polar position with an error of several centimeters from less than 1 day's data seems feasible. By contrast, results from a comparison of IPMS and BIH data show (Gaposchkin, private communication) an rms deviation of about 1.5 m for the pole position determinations between January and November 1968. Precise estimates of polar motion may prove useful for earthquake prediction (Mansinha and Smylie, 1968).

Monitoring UT.1, i.e. the Earth's orientation about the rotation axis (Gold, 1967), is somewhat more difficult than, say, determining polar motion. The diurnal signature of the time variation of the phase-delay measurements, because of its involvement in the determination of other unknowns (see Section 2), is not so accurately translatable into UT.1 values. Nonetheless, although careful error analyses have not yet been carried out, a 'bootstrap' approach seems capable of yielding errors perhaps a factor of 10 or more less than the 5 m (equivalent distance on the surface) uncertainty that results from one night's observations using current photographic zenith tube techniques. Random, or quasi-random, fluctuations in the Earth's rotation with periods less than 1 day could interfere with some of the geophysical inferences from the interferometric measurements. If the rms amplitudes of these fluctuations, when converted to distances on the Earth's surface, were to be of the order of several centimeters or larger, the basic vector distance determinations could be limited in accuracy by this effect. However, there is at present no indication of the existence of such large-amplitude, short-period fluctuations. If there are, these will be observed and would be of great geophysical interest in their own right. Effects of longer period on the angular momentum of the solid Earth, due to large-scale motions in the atmosphere or oceans, might also be discernible.

The torques exerted on the 'flattened' Earth, primarily by the Moon and Sun, cause the Earth's angular momentum vector and, hence, the rotation axis, to move in inertial space. This motion is conventionally labelled 'precession and nutation' and described by matrices given, e.g. in the *Explanatory Supplement to the American Ephemeris and Nautical Almanac* (1961). The precession, which is the approximately 50″/yr conical motion of the rotation axis about the normal to the ecliptic, contains, according to general relativity, a 0″.02/yr or 0.04% contribution due to the Earth's orbital motion. This contribution has not been isolated experimentally since the Newtonian part is not known with nearly the necessary accuracy. The difficulty lies in the determination of the Earth's characteristics. However, almost the same characteristics enter in almost the same manner into the Newtonian expressions for both precession and nutation. (The general relativistic contribution is affected negligibly by the nutation.) Thus, by measuring both the precession and nutation with the high precision made possible by interferometry, one could determine both the Earth's contribution to the Newtonian expressions – essentially the fractional difference between the polar and equatorial moments of inertia – and the relativistic contribution to the precession. The nutation constant, which approximates the amplitude of the 18.6-year period nutational motion, is about 9″.2. Taking into account only a modest error reduction through averaging, we can conclude that the nutation constant and the yearly precession can both be estimated with errors under 0″.001 from observations extending over a substantial fraction of the nutation period. If the difference in the theoretical Newtonian contributions to the nutation and precession can be determined with comparable accuracy, the relativistic precession could be separated with an error less than 25%. A theoretical study to ascertain the extent to which the Newtonian contributions to both the precession and nutation can be reduced to only one unknown constant is still in progress. Clearly a formidable effort is required to detect the relativistic contribution to precession – even with such a formidable weapon as atomic-clock interferometry. Another approach to this 'gyroscope' test of general relativity has been proposed by Schiff and involves an artificial Earth-orbiting gyroscope instead of the Sun-orbiting Earth as in our proposal. The artificial gyroscope approach holds more promise but is fraught with technical difficulties that have not yet been surmounted fully despite almost a decade of clever and sophisticated development efforts at Stanford, primarily by Fairbank and Everitt.

The obliquity of the ecliptic – the angle between the Earth's equator and orbital plane – is one of the fundamental astronomical constants. Its rate of change has been monitored for several centuries through continual stellar and planetary observations. This change appears to differ from theoretical predictions by about 0″.35/century and represents one of the outstanding problems of modern astrometry. Recent attempts to explain the discrepancy have involved geophysics through postulates concerning core-mantle interactions (Aoki, 1969). It is clearly desirable to check the measurements. How can interferometry contribute? Interferometry data are exquisitely sensitive to motions of the Earth's equator but tell us almost nothing of value about the ecliptic. Pulsar data, on the other hand, are exceedingly sensitive to ecliptic motion but insensitive to equator motions. By observing in two modes – collecting pulse arrival time data and monitoring the emissions interferometrically – the obliquity of the ecliptic could be determined to within $0\rlap{.}{''}01$ from 1 year's data. A change of the order of $0\rlap{.}{''}35$/century might be verifiable reliably after about 5 years. To maximize the signal-to-noise ratio for a given amount of data, the interferometry recording systems should be activated only during pulse reception. Such cyclical control is easily accomplished. Through the use of template-matching techniques (Counselman and Shapiro, 1968), the pulse arrival times for some pulsars can be measured with errors of the order of only several tens of microseconds. For the crab pulsar – NP0532 – the uncertainties can be reduced below 10 μsec (Counselman, Rankin, and Richards, private communication). Since the radius of the Earth's orbit is 500 light-sec, the orientation of the ecliptic can be determined with respect to a pulsar reference frame from about 1 year's data with an uncertainty of $0\rlap{.}{''}01$ if we allow some error reduction through averaging to compensate for the inaccuracy introduced by correlations with the other parameters that must be estimated simultaneously. Interplanetary radar echo time delay data are useful here since the errors are often as small as a few microseconds and hence these data determine all parameters of the Earth's orbit with extremely high precision, except of course for the orientation of the ecliptic with respect to a stellar reference system. Proper motions, if not compensated for accurately, could render the pulsar reference frame unsuitable. Other interferometry observations can, however, be used to relate the pulsar directions to the basic quasar reference frame. (If the quasars exhibit relative proper motion, that by itself would constitute a major scientific discovery! Relatedly, we might mention that relativistic deflection effects will be important in parallax determinations based on the interferometry measurements.)

The shape of the sea surface is of great interest to both geophysicists and oceanographers. The determination of the height of the sea surface with respect to say, the Earth's center of mass with an uncertainty reliably below 10 m is an important present goal. It might well prove feasible to install interferometry equipment aboard oceanographic vessels to make such determinations. If we assume that the interferometry array consists of at least one land-based antenna and that the only unknowns are those associated with the shipboard antenna, then four quantities need be estimated: three scalars associated with the ship's (vector) position and one scalar which provides the ship's clock offset, or *a priori* synchronization error. A phase-delay measurement on each of four well-distributed sources will suffice to determine these quantities. With only one antenna available at each terminal, three-axis shipboard accelerometers will be required to track the antenna's motion over the sea during the course of the observations. If two-fold redundancy of observations is adequate and if the antenna slewing rates are reasonable a single sea-height determination would require about 1 h. Presently available accelerometers could be used without degrading appreciably the accuracy of the results. If suitably strong and well-distributed sources – either natural or artificial – are available, a set of four independently steerable small ($\approx$ 1 m-diameter) antennas could be used on the ship (Rosenberg, private communication) and a complementary set at the land-based site. For such a system the time required to make a single sea-height measurement could be reduced to several minutes. The economic feasibility of this arrangement depends crucially on the antenna sizes required. For TACSAT-strength sources (see Section 4), multiple antennas per site would probably be practical. The recording system can be multi-plexed between antennas; the time saving arises because it would no longer be necessary to steer the antenna from one source to another while making a height determination. The uncertainty in each could probably be reduced to about 1 m. (This technique can, of course, also be used for navigation.)

The measurement of the geopotential and its time variation can be accomplished not only through sea-height determinations but also through the tracking of artificial satellites that possess wide-band noise beacons (see Section 4). The direction of such satellites in the quasar inertial reference frame can be determined from interferometry measurements with errors of only about 0".001. Independent measurements could, under reasonable conditions, be made as often as every few seconds. A powerful combination for tracking would involve both the interferometric measurements to provide direction and laser, or radio transponder, data to provide high-precision range (more properly, echo time delay). A careful choice of satellite and orbit configuration can be made to maximize sensitivity to geopotential variations and to minimize the adverse effects of non-gravitational forces such as sunlight pressure and air drag.

Finally we note that a byproduct of the interferometer data reduction is the correction for the clock offset, or original clock synchronization, error. In a well-developed system the uncertainty in synchronization should be reducible to the 0.1 nanosecond level – almost four orders of magnitude better than is achievable over continental distances with current techniques. (Analysis of clock comparisons over such distances with such accuracies discloses a number of pseudo relativistic paradoxes which will be discussed in a separate publication.) Unless a real-time inter-site communications link via, say, satellite is available, the clock offset determination would be available only long after the epoch to which it applies and might thereby be less useful. In either event, interferometry clearly offers the possibility for the routine global synchronization of clocks and frequency standards with unprecedented precision. Such precision may actually be needed for certain air traffic control systems.

6. Results and Conclusions

Long-baseline interferometry has so far been applied mainly to the precision determination of the angular size of distant celestial radio sources. The first successful experiments were conducted in 1967 by groups in Canada (see, e.g., the preceding paper by N. Broten) using an analog recording system and by groups in the U.S.A. (Bare *et al.*, 1967; Moran *et al.*, 1967) using a digital recording scheme. Preliminary precision experiments directed partly towards geophysical applications were conducted by the MIT-Lincoln Laboratory group in October 1968 and January 1969. The antennas used were the Lincoln Laboratory 120-ft diameter Haystack dish in Tyngsboro, Mass., and the National Radio Astronomy Observatory 140-ft diameter dish in Green Bank, W.Va. The two are separated by about 845 km. The October experiment involved measurements at both L-band (1.6 GHz) and X-band (8 GHz). The narrow (0.36 KHz) windows that were sampled, for example, at X-band extended over a band of about 40 MHz. Fringes were obtained in this experiment at both L-band and X-band; it marked the first time in fact that fringes had been obtained at as high a frequency as 8 GHz. At both frequency bands, the measurements were mainly of the radio sources 3C279 and 3C273 since the prime purpose of that experiment was to detect the gravitational bending of light (Shapiro, 1967) by monitoring the angular separation – about 8° – between these two sources as the emissions detected from one (3C279) passed close to the limb of the Sun. Observations were also made of other sources so that a precise determination of the intersite vector **d** could be obtained from an analysis similar to that described in Section 2. Unfortunately, no results are yet available from the reduction of these data. The January experiment, by contrast, was undertaken primarily to determine **d**. The measurements, which included phase calibrations, were at L-band and extended over about 24 h. The windows sampled spanned from 1.60 to 1.71 GHz, i.e. the effective bandwidth was 110 MHz. The sources observed included 3C273, 3C279, 3C345, and 3C454.3. About 30 different measurements of Δt from these sources were used to solve for 12 parameters: 3 baseline parameters, 7 source-position parameters, and 2 parameters describing the initial clock synchronization error and the difference in frequency between the Haystack and NRAO hydrogen masers which were used for the experiment. The estimated error in the determination of each value of Δt was 1 nsec. Due to the correlations between the various parameter estimates, the resultant formal standard error in the determination of d is 5 nsec (≈ 5 ft). The 'true' error may be somewhat greater. A complete description of this experiment and the analysis procedures used are now being prepared for publication. A more ambitious experiment involving, in addition, one of the Owens Valley, Calif., 90-ft diameter antennas is planned for October 1969. Both L-band and X-band measurements will again be made.

Clearly but a small start has been made in the application of long-baseline radio interferometry for geophysical studies. Nevertheless, these first results are very promising and further work will probably keep a number of radio astronomers busy for several years with intercontinental negotiations for antenna usage. The field is also particularly attractive for graduate students who would not have to join the Navy to see the world.

Acknowledgments

The development of long-baseline radio interferometry techniques at MIT and Lincoln Laboratory has involved the active participation of a large number of people including, in addition to the authors, B. F. Burke, J. C. Carter, H. F. Hinteregger, J. M. Moran, D. S. Robertson, A. E. E. Rogers, and A. R. Whitney. This work was supported in part by NSF Grant GA 4494. Lincoln Laboratory is operated with support from the U.S. Air Force.

References

Aoki, S.: 1969, 'Friction between Mantle and Core of the Earth as a cause of the Secular Change in Obliquity', *Astron. J.* **74**, 284.

Bare, C. C., Clark, B. G., Kellerman, K. I., Cohen, M. H., and Jauncey, D. L.: 1967, 'Interferometer Experiment with Independent Local Oscillators', *Science* **157**, 189.

Counselman, C. C. and Shapiro, I. I.: 1968, 'Scientific Uses of Pulsars', *Science* **162**, 352.

Explanatory Supplement to the American Ephemeris and Nautical Almanac 1961, Her Majesty's Stationery Office, London.

Gold, T.: 1967, 'Radio Method for the Precise Measurement of the Rotation Period of the Earth', *Science* **157**, 302.

Isacks, B., Oliver, J., and Sykes, L. R.: 1968, 'Seismology and the New Global Tectonics', *J. Geophys. Res.* **73**, 5855.

Le Pichon, X. J.: 1968, 'Sea Floor Spreading and Continental Drift', *J. Geophys. Res.* **73**, 3661.

Mansinha, L. and Smylie, D. E.: 1968, 'Earthquakes and the Earth's Wobble', *Science* **161**, 1127.

Moran, J. M., Crowther, P. P., Burke, B. F., Barrett, A. H., Rogers, A. E. E., Ball, J. A., Carter, J. C., and Bare, C. C.: 1967, 'Spectral Line Interferometry with Independent Time Standards at stations Separated by 845 kilometers', *Science* **157**, 676.

Shapiro, I. I.: 1967, 'New Method for the Detection of Light Deflection by Solar Gravity', *Science* **157**, 806.

Staelin, D. H.: 1969, 'Passive Remote Sensing at Microwave Lengths', *Proc. IEEE* **57**, 427.

INDEX OF NAMES

INDEX OF SUBJECTS

ASTROPHYSICS AND SPACE SCIENCE LIBRARY

Edited by

J. E. Blamont, R. L. F. Boyd, L. Goldberg, C. de Jager, Z. Kopal, G. H. Ludwig, R. Lüst,
B. M. McCormac, H. E. Newell, L. I. Sedov, Z. Švestka, and W. de Graaff

1. C. de Jager (ed.), *The Solar Spectrum. Proceedings of the Symposium held at the University of Utrecht, 26–31 August, 1963.* 1965, XIV + 417 pp.

2. J. Ortner and H. Maseland (eds.), *Introduction to Solar Terrestrial Relations. Proceedings of the Summer School in Space Physics held in Alpbach, Austria, July 15–August 10, 1963 and Organized by the European Preparatory Commission for Space Research.* 1965, IX + 506 pp.

3. C. C. Chang and S. S. Huang (eds.), *Proceedings of the Plasma Space Science Symposium, Held at the Catholic University of America, Washington, D.C., June 11–14, 1963.* 1965, IX + 377 pp.

4. Zdeněk Kopal, *An Introduction to the Study of the Moon.* 1966, XII + 464 pp.

5. Billy M. McCormac (ed.), *Radiation Trapped in the Earth's Magnetic Field. Proceedings of the Advanced Study Institute, Held at the Chr. Michelsen Institute, Bergen, Norway, August 16– September 3, 1965.* 1966, XII + 901 pp.

6. A. B. Underhill, *The Early Type Stars.* 1966, XIII + 282 pp.

7. Jean Kovalevsky, *Introduction to Celestial Mechanics.* 1967, VIII + 427 pp.

8. Zdeněk Kopal and Constantine L. Goudas (eds.), *Measure of the Moon. Proceedings of the Second International Conference on Selenodesy and Lunar Topography held in the University of Manchester, England, May 30–June 4, 1966.* 1967, XVIII + 479 pp.

9. J. G. Emming (ed.), *Electromagnetic Radiation in Space. Proceedings of the Third ESRO Summer School in Space Physics, held in Alpbach, Austria, from 19 July to 13 August, 1965.* 1968, VIII + 307 pp.

10. R. L. Carovillano, John F. McClay, and Henry R. Radoski (eds.), *Physics of the Magnetosphere. Based upon the Proceedings of the Conference held at Boston College, June 19–28, 1967.* 1968, X + 686 pp.

11. Syun-Ichi Akasofu, *Polar and Magnetospheric Substorms.* 1968, XVIII + 280 pp.

12. Peter M. Millman (ed.), *Meteorite Research. Proceedings of a Symposium on Meteorite Research held in Vienna, Austria, 7–13 August, 1968.* 1969, XV + 941 pp.

13. Margherita Hack (ed.), *Mass Loss from Stars. Proceedings of the Second Trieste Colloquium on Astrophysics, 12–17 September, 1968.* 1969, XII + 345 pp.

14. N. D'Angelo (ed.), *Low-Frequency Waves and Irregularities in the Ionosphere. Proceedings of the 2nd ESRIN-ESLAB Symposium, held in Frascati, Italy, 23–27 September, 1968.* 1969, VII + 218 pp.

15. G. A. Partel (ed.), *Space Engineering. Proceedings of the Second International Conference on Space Engineering, held at the Fondazione Giorgio Cini, Isola di San Giorgio, Venice, Italy, May 7–10, 1969.* 1970, XI + 728 pp.

p.t.o.

16. S. Fred Singer (ed.), *Manned Laboratories in Space. Second International Orbital Laboratory Symposium.* 1969, XIII + 133 pp.

17. B. M. McCormac (ed.), *Particles and Fields in the Magnetosphere. Symposium Organized by the Summer Advanced Study Institute, held at the University of California, Santa Barbara, Calif., August 4–15, 1969.* 1970, XI + 450 pp.

18. Jean-Claude Pecker, *Experimental Astronomy.* 1970, X + 105 pp.

19. V. Manno and D. E. Page (eds.), *Intercorrelated Satellite Observations related to Solar Events. Proceedings of the Third ESLAB/ESRIN Symposium held in Noordwijk, The Netherlands, September 16–19, 1969.* 1970, XVI + 627 pp.

SOLE DISTRIBUTORS FOR U.S.A. AND CANADA:

SPRINGER-VERLAG NEW YORK, INC., 175 Fifth Ave., New York, N.Y. 10011